T0093760

Intelligent Systems Reference Library

Volume 237

The aim of this series is to publish a Reference Library, including novel advances and developments in all aspects of Intelligent Systems in an easily accessible and well structured form. The series includes reference works, handbooks, compendia, textbooks, well-structured monographs, dictionaries, and encyclopedias. It contains well integrated knowledge and current information in the field of Intelligent Systems. The series covers the theory, applications, and design methods of Intelligent Systems. Virtually all disciplines such as engineering, computer science, avionics, business, e-commerce, environment, healthcare, physics and life science are included. The list of topics spans all the areas of modern intelligent systems such as: Ambient intelligence, Computational intelligence, Social intelligence, Computational neuroscience, Artificial life, Virtual society, Cognitive systems, DNA and immunity-based systems, e-Learning and teaching, Human-centred computing and Machine ethics, Intelligent control, Intelligent data analysis, Knowledge-based paradigms, Knowledge management, Intelligent agents, Intelligent decision making, Intelligent network security, Interactive entertainment, Learning paradigms, Recommender systems, Robotics and Mechatronics including human-machine teaming, Self-organizing and adaptive systems, Soft computing including Neural systems, Fuzzy systems, Evolutionary computing and the Fusion of these paradigms, Perception and Vision, Web intelligence and Multimedia.

Indexed by SCOPUS, DBLP, zbMATH, SCImago.

All books published in the series are submitted for consideration in Web of Science.

Sandeep Kumar Panda · Vaibhav Mishra ·
Sujata Priyambada Dash · Ashis Kumar Pani
Editors

Recent Advances in Blockchain Technology

Real-World Applications

Editors
Sandeep Kumar Panda
Department of Data Science and Artificial Intelligence
Faculty of Science and Technology (IcfaiTech)
ICFAI Foundation for Higher Education
Hyderabad, Telangana, India

Sujata Priyambada Dash
Department of Management
Birla Institute of Technology, Mesra
Ranchi, Jharkhand, India

Vaibhav Mishra
ICFAI Business School
ICFAI Foundation for Higher Education Hyderabad
Hyderabad, Telangana, India

Ashis Kumar Pani
Information Systems and Production
XLRI—Xavier School of Management
Jamshedpur, Jharkhand, India

ISSN 1868-4394 ISSN 1868-4408 (electronic)
Intelligent Systems Reference Library
ISBN 978-3-031-22834-6 ISBN 978-3-031-22835-3 (eBook)
https://doi.org/10.1007/978-3-031-22835-3

This Springer imprint is published by the registered company Springer Nature Switzerland AG
The registered company address is: Gewerbestrasse 11, 6330 Cham, Switzerland

Dedicated to my sisters Susmita, Sujata, Bhaina Sukanta, nephew Surya Datta, wife Itishree (Leena), my son Jay Jagdish (Omm), and Late father Jaya Gopal Panda, and Late mother Pranati Panda.

Sandeep Kumar Panda

Preface

Introduction

In 2008, when financial crisis on the verge and when economic trust was at its low, Satoshi Nakamoto formulated the concept of decentralized chaining of transaction blocks. Using the same concept, the first cryptocurrency Bitcoin was created. Since then, the idea of implementing every possible use case is being evolved. Blockchain technology is identified as one of the disrupting technologies that will transform the business ecosystem. The blockchain is an ever-growing distributed ledger that protects data against tampering and modification. Blockchain works on the Merkle Tree technique with consensus included. Consensus provides the guidelines based on which the data can be included and verified. Secure and efficient verification of huge data can be achieved by Merkle Tree hashing technique. Based on verification, the transactions are added to the blocks and are chained together and thus the name Blockchain. It is an emerging technique that provides individuals and businesses to collaborate with trust and transparency. Blockchain technology and its application domains are continually being developed and enhanced despite the way Bitcoin booms. Smart Contracts, the distinct feature of Blockchain, is no different from an agreement and are identified to execute partially or completely without any human intervention. Blockchain and smart contracts allow transactions to be recorded and prevent the data from any change. This becomes the most significant principle to build a stable and reliable infrastructure. Thus, all the possible blockchain-based applications are sought and built on the basis of privacy and credit risk.

Objective of the Book

This book aims to provide a depth knowledge of how the technology behind Bitcoin and other cryptocurrencies is changing the world. The primary objective of this Edited book is to uncover the revolutionary innovation of the twenty-first century,

Blockchain Technology, and exhibit its wide range of applications in different domains. Although Bitcoin and Cryptocurrencies are the first popularly known applications of Blockchain, at present it has several other applications as well. While some still believe it is overhyped but in real, it is transforming almost every industry. For example, healthcare, retail, supply chain, business management, cloud management, tracking systems, and IoT are to name a few. This technological breakthrough not only paved the way for cryptocurrencies but provides useful solutions for other areas of technologies. The unique nature of the technology like peer to peer, distributed, decentralized, accountable, anonymous, transparent, and secure makes it suitable for all real-world applications. The technology has lot of potential to transform the business and companies are already in race for different product offerings. This book provides the working principles of blockchain technology and recent advancement in its applications.

Organization of the Book

It includes the fundamentals of Blockchain Technology, features, working principles, and its application in different sectors. To meet the objectives, this book contains 15 chapters contributed by accomplished writers as follows:

Chapter 1
The chapter begins with defining blockchain, including its characteristics, benefits, types, recent, and future trends, in order to improve grasp of this new technology. The purpose of this chapter is to illustrate the key characteristics and benefits of blockchain technology when used in conjunction with supply chain, healthcare, tourism, education, human resource, space, and finance industry. The study attempts to make propositions about how technology will evolve and impact the industry.

Chapter 2
This chapter elaborates the largest cryptocurrency created ever in the trade market, Bitcoin: Beginning of the Cryptocurrency Era, A Case study. This chapters consists of case study as well as teaching notes. This case illuminates the distinction between blockchain and bitcoin, working model of Bitcoin, business benefits, and challenges of Bitcoin, how Bitcoin taking over the mainstream process and reformation of ecommerce using Bitcoin.

Chapter 3
This chapter presents the study on different characteristics and the importance of blockchain technology in varied HR practices. This chapter has attempted to discuss certain research questions such as (a) what is human resource and why human resource? (b) What is digital transformation? (c) What is blockchain? (d) Why blockchain? (e) What are the various types of blockchain? (f) Why blockchain in HR? (g) How blockchain has transformed HR? (h) How far blockchain is relevant for HR? and (i) HR digital transformation vis-à-vis blockchain is examined in this study along with the challenges and criticisms.

Chapter 4
This chapter introduced blockchain technology in cloud environments to securely transfer EHR data between authorized entities. Blockchain technology is important in such cases because it provides data-ledger-based features that can be distributed to all companies on the network. This approach is a tamper-proof mechanism because every health transaction information in the blockchain is stored as hash values. This approach provides tamper-proof, trust, confidentiality, integrity, authentication, availability, and easy access control to the cloud user data at the time of data transfer and rest modes.

Chapter 5
This chapter intends to provide a preliminary information of metaverse and blockchain and how these two technologies can be related to each other. The impact of blockchain technologies on metaverse is also discussed in brief.

Chapter 6
The chapter has aimed to design the literature survey studies incorporating online education and blockchain security. In addition, the performance of the blockchain has been validated with different online education features like user verification, certificate sharing, and online teaching facilities which are also discussed. Finally, the performance has been measured based on the execution time, and the merits and limitations have been described separately.

Chapter 7
The authors of this chapter discuss the use of blockchains in public procurement, land registration, and electronic voting.

Chapter 8
This chapter provides a thorough review of studies carried out to demonstrate the benefits of blockchain technology that have been utilized in the domain of healthcare, in addition to the pandemic, COVID-19, which led to a massive and pervasive repercussion on healthcare and has significantly accelerated the implementation of digital technology. This chapter also depicts how researchers have presented the use cases for adopting Blockchain technology in the healthcare sector. The state-of-the-art blockchain application development for healthcare has also been described in this chapter, along with any inadequacies and potential future study topics.

Chapter 9
The chapter describes how businesses may use blockchain to fully capitalize on the fourth industrial revolution. This work attempts to close this gap by outlining a collection of pertinent research difficulties that result from the implementation of blockchain technologies in business process monitoring solutions, describing the current methodologies to address these challenges, and presenting a reference architecture for doing so.

Chapter 10
This chapter presented recent development in transportation industry considering the blockchain technology. Next, the author has further discussed the scope of blockchain

technology adoption to enhance the security, trust, and operational performance of the transportation business. In addition to this, potential benefits of blockchain technologies in payment, immutability, flexibility, and security have been highlighted. Next, the author has presented the role of Blockchain Technology (BT) considering the Indian trucking industry. This chapter helps in strategic planning and decision-making to the transport managers in effective implementation of the BT. At the end of this chapter, the future scope of BT adoption in trucking industry has been discussed.

Chapter 11
The study proposes a security level in a high value creation process of personalized healthcare through remote patient monitoring. This chapter also proposes the framework on how patient layer, physician layer, and technological layer interacts to provide better value of healthcare with the maximum privacy and security.

Chapter 12
This chapter is to examine the pertinent part of blockchain technology in different areas of human resource management, ranging from the recruitment and selection process for certificate verification and skill mapping to payroll processing and employee data protection. Finally, the chapter discusses the challenges behind the implementation of blockchain technology in the area of human resource management.

Chapter 13
This chapter studies the role of P2P in De Fi Financial Transactions (DFFT).

Chapter 14
In this chapter, the authors have emphasis on each and every significant attribute and weakness of blockchain technology. The authors have also conducted a literature analysis on blockchain-related topics and highlighted some of the most current sectors in which blockchain technology has found the greatest utility in Industry.

Chapter 15
In this chapter, the authors suggest an Ethereum-based retail commodity management and tracking. It identifies user and supplier and track node to node delivery status of product without the help of any third-party service. Moreover, the impact of smart contract in transaction in blockchain will bring revolution in retail sector.

Topics presented in each chapter of this book are unique to this book and are based on unpublished work of contributed authors. In editing this book, we attempted to bring into the discussion all the new trends and experiments that have made on blockchain technology. We believe this book is ready to serve as a reference for larger audiences such as system architects, practitioners, developers, and researchers.

Hyderabad, India Sandeep Kumar Panda

Acknowledgments

The preparation of this edited book was like a journey that we had undertaken for several months. We wish to express our heartfelt gratitude to our families, friends, colleagues, and well-wishers for their constant support throughout this journey. We express our gratitude to all the chapter contributors, who allowed us to quote their remarks and work in this book. In particular, we would like to acknowledge the hard work of authors and their cooperation during the revisions of their chapters. We would also like to acknowledge the valuable comments of the reviewers which have enabled us to select these chapters out of the so many chapters we received and also improve the quality of the chapters. We wish to acknowledge and appreciate the Springer team for their continuous support throughout the entire process of publication. Our gratitude is extended to the readers, who gave us their trust, and we hope this work guides and inspires them.

About This Book

Recent Advances in Blockchain Technology—Real-World Applications presents the latest information on adapting and integrating Blockchain technology in real-world business, supply chain, healthcare, education, HRM, retail, logistics, and transport industries. It will elucidate the significance of adapting a blockchain-based solution to the existing real-world problems. The book's editors address the latest advancements in the current business model by adapting blockchain techniques. Several use cases with healthcare, supply chain, metaverse, retail management, business model, and HRM is presented in this book. Since Blockchain is a fairly new concept that exploits decentralized networks and is used for secure, cost-effective, and fast business activities in many sectors, this book is a welcome addition to current knowledge.

Contents

About the Editors

Dr. Sandeep Kumar Panda is currently working as an Associate Professor and Head of Department in the Department of Data Science and Artificial Intelligence, Faculty of Science and Technology (IcfaiTech) at ICFAI Foundation for Higher Education (Deemed to be University), Hyderabad, Telangana, India. His research interests include Blockchain Technology, the Internet of Things, AI, and Cloud Computing. He has published 50 papers in international journals and international conferences and book chapters in repute. He received the "Research and Innovation of the Year Award 2020" from MSME, Government of India and DST, Government of India at New Delhi in 2020 also. He received "Research Excellence Award" from Brand Honchos, in the year 2022. He has 17 Indian Patents on his credit. He has four Edited books named *Bitcoin and Blockchain: History and Current Applications*, CRC Press USA, *Blockchain Technology: Applications and Challenges* Springer ISRL, *AI and ML in Business Management: Concepts, Challenges, and Case Studies* CRC Press, USA, *The New Advanced Society: Artificial Intelligence and Industrial Internet of Things Paradigm*, Wiley Press USA, in his credit. He has 10 lakh Seed money projects from IFHE. His professional affiliations are MIEEE, MACM, and LMIAENG.

Dr. Vaibhav Mishra is currently working as an Assistant Professor at the IBS Hyderabad, IFHE, India. He has completed his engineering in computer science from Uttar Pradesh Technical University, India. Subsequently, he received his MBA and Ph.D. from the Indian Institute of Information Technology (IIIT), Allahabad, India. During his Ph.D., he received the MHRD scholarship. He has also published research articles in international journals of repute (indexed in SCOPUS and ABDC), such as *International Journal of Bank Marketing*, *American Business Review*, *Pacific Asia Journal of AIS*, *International Journal of Electronic Business*, etc. He has also reviewed journal articles for various journals listed in ABDC, ABS, and Scopus. His areas of interest are Management Information System, Database Management System, Artificial Intelligence and Machine learning, cloud computing, e-Commerce, ERP Systems, and Data Mining.

Dr. Sujata Priyambada Dash is currently working as an Assistant Professor in the Department of Management, Birla Institute of Technology, Mesra, Ranchi, Jharkhand, India. Her research interests include Human Capital Management, Emotional Intelligence, Blockchain and Artificial Intelligence in Human Resource Management. She has published her research work in many National and International Journals (Scopus and ABDC indexed) and attended various National and International Conferences, Workshops, and Faculty Development Programmes. She has published book chapters in Scopus indexed Edited Book Volume in International Publication houses such as Springer, CRC Press, Taylor and Francis, and Wiley. Her professional affiliation is AIMS—Associations of Management Scholar.

Dr. Ashis Kumar Pani is the Dean of XLRI—Xavier School of Management, Jamshedpur, Jharkhand, the oldest B-school in India. He is a gold medalist in Computer Science from IIT Madras and Ph.D. in Computer Science from IIT Kharagpur. Published and presented several papers in leading journals and international conferences. Current areas of interest are Social Mobile Analytics and Cloud (SMAC), Artificial Intelligence, and Digital Transformation. Presently a life member of Indian Association of Research in Computing (IARCS), Computer Society of India (CSI), and IEEE Computer Society. He has received IBM best faculty award 2008; best paper award in International Academy of e-Business in Waikiki, Hawaii, 2009; EURECA project award by European Commission 2009; and J. L. Batra best research paper award in AIMS 2010. Fulbright fellow in 2018. Participated in the international workshop by invitation from UNO, ICTP, Microprocessor Laboratory, Trieste, Italy. Participated in the course curriculum design of e-Business program for International University of Japan, Negata, Japan.

Chapter 1
An Introduction to Blockchain Technology: Recent Trends

Sujata Priyambada Dash

Abstract Blockchain is a buzzword in this knowledge era. It is a burgeoning concept emerged as a cutting-edge technology with lots of features being applied in varied domains. Features are inevitable in that it has evolved as a major player of the future digital economy. The Corporates are trying to understand the concept and exploring to their optimum level in formulating policies for the implementation and execution of the blockchain technology in their work system. This makes ease not only at the work system but also makes hassle free from many unwanted tasks. The purpose of this article is to illustrate the key characteristics and benefits of blockchain technology when used in conjunction with supply chain, healthcare, tourism, education, human resource, space, and finance industry. The study attempts to make propositions about how technology will evolve and impact the industry. The goal of this article is to identify and discuss blockchain technology as it applies to the many domains stated. To address the aforementioned, the author begins with defining blockchain, including its characteristics, benefits, types, recent and future trends, in order to improve grasp of this new technology.

Keywords Blockchain technology · Human resource · Supply chain · Space · Education · Tourism · Healthcare · Finance

1.1 Introduction

Our country India is making its all possible way to reach out the young minds for the awareness and inculcating the benefits and applications of blockchain technology through various workshops, events, and development programmes, and events being conducted by AICTE and MHRD, Government of India. Blockchain technology is a burgeoning field of study in this information era. Blockchain has developed as a trend in itself as it is utilized as a decentralized network, thereby, eliminating the intermediaries. Internet fraud, malware, and hacking across the globe having

S. P. Dash (✉)
Department of Management, Birla Institute of Technology, Mesra, Ranchi 835215, Jharkhand, India
e-mail: spdash@bitmesra.ac.in

S. K. Panda et al. (eds.), *Recent Advances in Blockchain Technology*,
Intelligent Systems Reference Library 237, https://doi.org/10.1007/978-3-031-22835-3_1

appropriate information and verification of transaction of value is of significant. To make this happen, we'll need dependable systems that are temper-proof, temper-evident, and trustworthy. A technology known as blockchain can be used to verify the accuracy and quality of the data we're processing. It's also known as the 'Internet of Value,' because it operates as a single, immutable, unchangeable, and verifiable source of truth. It is a digital ledger that preserves all permanent recorded transactions in a more secure and chronological manner. Transactions are assembled into blocks and added to the end of the current chain, which is one of the distinctive aspects of blockchain. Cryptographic hashing (SHA 256) is used in blockchain for security purpose which creates a hash equivalent for every stored transaction [1]. Hence, each block is cryptographically linked to the one before it and cannot be changed [2]. Even if hackers try to add, delete, or change any transaction in the midst of the ledger, they must first construct hash equivalents for all following transactions in the block. The ability to create a decentralized network means of storing data in 'blocks' is one of the benefits of blockchain. A block is added to the blockchain as a permanent record once it has been agreed upon [3]. Blockchain is based on the principles of providing value, establishing trust, and ensuring reliability through consensus. The genesis block, which was produced on January 3rd, 2009, is the first block in a blockchain that is a transaction. Every 10 min, new blocks with the most recent transactions are added. The blockchain protocol verifies the validity of each new block added to the chain. This also conforms to the blockchain rule [4]. In addition, each new block contains the cryptographic hash number of the previous block, producing a blockchain. This assures that the newly inserted blocks are identical to the ones that came before them. If the hash value of one block is altered, then the other blocks are crushed.

The blockchain, often known as the trust-machine, is a distributed database of records. Public, private, and hybrid blockchains are the three forms of blockchain. Permissionless blockchains, such as Ethereum, may be able to read, write, and update data without the need for permission on public blockchains, which can be useful for developing applications. This is a slow permanent ledger, more secure due to the feature like decentralization. The data in this blockchain can be verified and added by anyone. However, with a private blockchain, the consensus process creates a permission blockchain that can only be read, written, and changed with the permission of an organisation. The development of applications is faster and more scalable, and they are utilized to reach an agreement between two parties [5]. The hybrid blockchain is a blend of both public and private blockchains. The most accurate element of both public and private blockchain solutions is used in hybrid blockchain. It operates by using a private network to generate hashed data blocks, which are subsequently stored in a public blockchain. As a result, transactions are completed swiftly and at a substantially reduced cost. Furthermore, they safeguard the network from approximately 50% of threats by preventing hackers from breaking in.

Blockchain technology is used by businesses and organisations for a variety of reasons, including data history, data validity and security, cost reduction and smart contract which is without third parties.

This study aims to illustrate the fundamental features of blockchain technology in relation to supply chain, healthcare, tourism, education, human resource, finance, space [6, 7] while expressing on how technology might evolve and impact the sectors.

The goal of this article is to identify and explore blockchain technology in relation to the various domains stated above. To address the aforementioned, the author begins with discussing the concept of blockchain, as well as its characteristics, benefits, types, and recent as well as future trends, in order to better grasp this new technology.

1.2 Blockchain Technology Background

1.2.1 Blockchain Defined

Blockchains are distributed ledgers of transactions that are not under the jurisdiction of a single central authority, according to the OECD and Tapscott & Tapscott, 2016 [8, 9].

Satoshi Nakamoto argued in 2008 [10] that 'the network's unstructured character makes it robust.' Little coordination exists between the nodes as they all work simultaneously.

In this particular study, the significance of blockchain technology is examined. Blockchain is a shared, immutable ledger that makes it simpler to record transactions and track assets in a business network. Intangible assets include intellectual property, patents, copyrights, and brands. Tangible assets include things like houses, cars, cash, and land. Virtually everything of value may be tracked and traded on a blockchain network, reducing risk and expenses for all parties. The business world runs on information. The better it is accepted, the quicker and more precisely. Blockchain is a great way to convey such information since it provides instant, shareable, and fully transparent data stored on an immutable ledger that can only be read by individuals with authorization to access the network. A blockchain network can be used to track orders, payments, accounts, production, and much more. Since all participants have access to the same information, it is possible to examine all the details of a transaction from beginning to end, increasing confidence as well as opportunities and efficiencies.

1.2.2 Blockchain Characteristics

The five major features of blockchain that need identification are encryption, immutability, tokenization, decentralization, and distribution [11] are described below.

1. Encryption: Technology safely and semi-anonymously records data, with participants using pseudonyms. Participants have complete control over their personal

identification and other information, sharing only what is necessary for a transaction [12].
2. Immutability: Cryptographically signed are the completed transactions, time-stamped and added to the ledger sequentially [13]. It is not possible to change or modify records unless all participants agree that it is necessary.
3. Tokenization: Value is traded in the form of tokens, which can represent a wide range of asset types such as 'money,' data units, or a user's identity. Tokenization, or the creation of tokens, is how a blockchain expresses and allows for a 'native value' that can be traded in the 'currency' [14].
4. Decentralization: No single entity controls or dictates the rules to a majority of nodes. A consensus method verifies and authorises transactions, obviating the requirement for a central network administrator. For instance, member pseudonymity [15], the automation usability [16, 17], data redundancy [18], and peers' participation in 'versatility' development [19] are the few built-in attributes into blockchain technology.
5. Distributed network: Blockchain members are connected via a distributed network and run 'nodes,' which are computers that run a programme to enforce the blockchain's business rules. Nodes also store a full copy of the ledger, which updates on its own as new transactions occur [20, 21].

There are some existing and emerging technologies which enables blockchains essential capabilities includes are decentralized app (DApps), where data, instructions, and records of operations are kept cryptographically in a distributed ledger technology (DLT). Similarly, automated trust mechanisms for external contacts and transactions between companies or individuals are used in distributed business terms and conditions (T&C). Furthermore, smart contracts are programmes or protocols that make it easier to automate, validate, and execute business activities. Finally, smart assets are digital representations of actual assets that exhibit programmable activity. Organizations will need to combine existing technical capabilities in new ways and learn new know-how in order for blockchains to dramatically modify present operating models or corporate processes and build new commercial, social, and other governing paradigms.

1.2.3 Blockchain Benefits

As depicted by [22], there exists varied blockchain benefits categorized as: (i) economic [23–35], (ii) informational [30, 36–46], (iii) organizational [17, 28, 47–58], (iv) strategic [59–61], and (v) technological [22, 23, 26, 27, 30, 37, 38, 41, 55]. Table 1.1 depicts the benefits of blockchain technology in various category.

Table 1.1 Blockchain benefits

Sl. No.	References	Type	Category
1	[23–32]	Cost reduction	Economic
	[9, 23, 32, 33]	Energy management	
	[26, 32, 34, 35]	Cheaper and quicker shares acquisitions	
2	[37–40]	Data quality and integrity	Informational
	[9, 30, 41, 42]	Reduction of human errors	
	[23, 37–39]	Privacy	
	[23, 37–39]	Data sharing	
	[9, 36, 42, 43]	Reliability	
	[23, 41]	Information anonymity	
	[31, 44– 46]	Information access	
3	[31, 37, 39, 47]	Trust	Organizational
	[9, 23, 38, 47]	Transparency and auditability	
	[9, 48, 52]	Prediction capability	
	[17, 39, 49, 50]	Control of transactions	
	[26, 30, 37, 51]	Clear ownership	
	[26, 28, 30, 37, 52]	Voting accuracy	
	[37, 52, 53, 55]	Secure storage and transmission	
	[56, 57]	New organizational model	
	[25, 39, 50, 58]	Decentralization	
4	[22, 29, 31, 38, 54]	Record keeping and ensured trusted and immutable transaction	Strategic
	[23, 37–39]	Identity management	
	[23, 38, 45, 47]	Avoid fraud	
	[28, 35, 37, 38]	Corruption reduction	
	[22, 23]	Improve justice services	
	[22, 26, 29, 38]	Holding self-executing smart contracts	
	[23, 59–61]	Regulation measurements	
5	[9, 37]	Flexibility	Technological
	[23, 37, 38, 55]	Security	
	[22, 26, 38]	Smart contract applicability	
	[27, 30, 41]	Increased resilience	
	[23, 26, 30]	High speed	
	[37, 55]	High degree of authenticity	
	[22, 31]	Complexity reduction	
	[37, 38]	Proof of identity	
	[30, 37]	Automatization process	

Source Author's representation

1.3 Propositions: Recent Progresses

1.3.1 Proposition 1

Blockchain Technology and Human Resource: A glance of blockchain technology in the world of human resource management.

Blockchain technology is a strong form of *disruptive innovation* in the world of *human resource management*. According to Rubio [62], disruptive innovation in human resources may be defined as developing and delivering a new value proposition to change operations and increase agility while using technology as a tool. Blockchain has the potential to completely alter and interfere with every activity a human resources manager undertakes. Affordably and effectively growing businesses will be made possible by the blockchain. The results of the disruption caused by blockchain in this industry will benefit both employers and employees. Human resource is expected to use the *public, permissionless blockchain network* to disrupt functions such as recruitment, payroll, and training.

In *recruitment process,* credential verification is very important. During the process of recruitment, falsification of information is vulnerable. The introduction of blockchain eliminates all doubt regarding credentials and employment history, saving both time and money. Traditional CVs could be eliminated by blockchain. Blockchain will hold the most recent information on things like a candidate's educational background, skill sets, and rewards [63]. Companies use the CV process to verify the accuracy of the information on the blockchain. In order to make sure that candidates are providing accurate background information, verification is a development of blockchain technology that interacts with decentralized verification programmes and platforms.

Additionally, blockchain enables hiring managers to view a candidate's whole job history. It is no surprise that people can offer employment histories to make their resumes more attractive. Millennial's tendency to job-hop, reference checks are very time-consuming process. Therefore, blockchain can fully revitalize this process. Blockchain will verify the information inputted, and can also be verified by the employers to keep it from false information [64]. Once this information is verified by users, it is unable to be altered.

Blockchain will fully revolutionize the act of *procurement* both in a positive and negative share of information such as title changes, reasons for dismissal and negative performance reviews. This will allow a complete understanding of a candidate's strengths and weaknesses, allowing the recruiters to make a better decision.

Blockchain can also help revolutionize the world of *payroll* reduces mistakes and discrepancies that can be made with human error. Companies employ blockchain technology to track time and attendance, payment choices, fraud protection, and benefit administration. Blockchain could disrupt the payroll function is the use of cryptocurrency. Users can send and receive money through the blockchain with cryptocurrency, creating an untamperable ledger along with it. This is a safe and effective way to send money across individuals. Blockchain has the ability to eliminate all the constraints that the original way of conducting payroll was associated with.

With blockchain technology, when an employee learns a new subject, it will be made visible in the form of block. As a result, a verifiable and accurate history of *training* initiatives is created, which is essential for the company's success. Skill scores are given to the employees in this system. These scores will allow employees with expertise, to be noticed by hiring managers and other stakeholders. Not only with tracking employee training will blockchain increase the opportunities for employees, but it will also create a type of competition and motivation.

There will be an increase in employee motivation and competition because participants may earn tokens in the form of cryptocurrency by providing value. This value can be in the form of actions such as providing consultancy services, answering questions, or becoming an 'expert'. Tokens can create an immense source of motivation for employees to become a subject matter expert. When tokens are earned, they can be used in online market places. 'People can make better decisions based on expert recommendations and reviews, all while earning tokens for their participation, and even more token for overachievement well described by Marcia Hales [65]. This benefit both the employee and the company. The employee has the ability to earn tokens by doing exercises, and the organizations from having a higher quality employee.

Smart contracts include blockchain technology agreements. They define the terms for transactions, and the penalties that would be apparent if the terms were violated. These contracts outline and enforce immutable obligations that are automatically followed [66]. They can streamline functions such as payroll and onboarding by using the idea of 'If This, Then, That' [67]. Once a contract and code is set, it is designed to work entirely dependent on that code.

There are few examples of smart contracts in actions, disrupting and changing the original ways of operation. First, when an employee completes the hours that he/she is contracted to work recorded in the blockchain, the employee is paid automatically. A second implementation could include a new hire. When the new hire successfully meets the documentary requirements, such as background checks or terms and conditions agreement, they can be given access to the company's system. Smart contract payments are irreversible, so if a mistake were to be made there would be no way to fix it.

Blockchain in human resource maintains *transparency* which is the largest strength relating to disruption. In the world of human resource, it is extremely important to have facts, rather than facetious statements employees can create to give themselves a step ahead, especially in the hiring process. Between employers and employees, it raises the level of *trust*. Blockchain also offers higher efficiency, lower cost, and lower risk of hacking [68]. The potential of blockchain technology is endless. Human resource can be revolutionized with the implementation of this decentralized technology, this possibility spreads across the business world as a whole.

1.3.2 Proposition 2

Blockchain Technology and Education: A glance of blockchain technology in the world of education sector.

Blockchain is a *distributed data storage technology* which is the root of Bitcoin cryptocurrency. Bitcoin alters the financial landscape that was previously governed by financial institutions as a third party. To be open and transparent, allowing all peers inside a valid network to conduct transactions directly with one another without the use of intermediaries [69]. This lowers and accelerates the cost of user transactions. In this section, the author goes through how blockchain technology was used to construct an information system for college students, covering design, consensus algorithm, and smart contract deployment. The technology offers a novel concept for information sharing among college students and has strong *security* and *traceability*. Analysis of college students' physical characteristics, academic records, reward and punishment data, and other recruitment-related factors is crucial for management and organisations. In many schools and universities today, the aforementioned information is not shared, the student's data can only be examined by the institution to which it belongs, and the value of the data cannot be completely utilized. The fundamental cause of the aforementioned situation is the requirement for data sharing to address issues with data security and traceability. If a college student's academic record has been altered, the person who altered it needs to be identified. It is challenging to accomplish the goals with the conventional college student information management system. The realization of information sharing among college students is appropriate given that blockchain technology is *tamper-proof* and *decentralized.*

Setting up a secure, trustworthy, and safe system for retrieving and storing information is crucial. Data access considerations and secure data storage are essential. *Smart consensus* methods are employed to the fullest extent possible for the information data of college students in order to address the issue of *data security* while offering good services to the outside world. Data chaining, data maintenance, and information query authorization are included in the primary function. Security and traceability become necessary when multiple layers are taken into account. The layers are *application layers* where student grades are uploaded. The verification process is carried out in the *contract layer* after which the student's information is uploaded and the new data is stored in the blockchain. Similar to the previous example, the *consensus layer* guarantees that the system can quickly reach a consensus on the blockchain network and guarantee the data consistency of each network node. Additionally, the *network layer* uploads the student grades, and the uploaded grades are then broadcast to the network for peer-to-peer verification. The *data layer*, which is the final component, also comprises hash functions, digital signatures, asymmetric encryption, block data, Merkel trees, chain data, etc.

Consensus algorithms like PoW, PoS, DPoS, and PBFT are related versions that are always being improved. These modifications improve the level of efficiency, lessen significant centralization issues, guarantee security, and raise energy consumption. After the contract code is completed in the smart contract deployment, the entire contract deployment procedure occurs. As a result, this guarantees the veracity, security, and traceability of student data and offers a standard against which businesses can choose employees, thereby easing the workload on the management department.

1.3.3 Proposition 3

Blockchain Technology and Healthcare: A glance of blockchain technology in the world of healthcare management.

Blockchain technology is very much useful in medical records management system. Due to the finite capacity and fixed number of credentials offered by this centralized authentication in traditional systems, it is difficult for several users to use the database concurrently. Medical test, X-ray, and other data and information are kept in physical and online records that patients cannot view without authorization. Even doctors cannot access the information at critical times. By providing patients' medical information to doctors and patients directly, this issue can be prevented. The current manner of storing is also vulnerable to security threats from tampering and illegal access. In order for a patient or doctor to make better and more efficient judgments while assuring immutability, the blockchain network resolves concerns with the medical record system. If the patients have given their consent, doctors may view previous visits for their medical records. Hyperledger Compose, Composer REST server, Hyperledger fabric, and Angular 4 are used to tackle the aforementioned issue. The integrity of the patient records is ensured through blockchain use. When necessary, the patient will provide access to a certain doctor and will revoke it when the doctor has finished their visit. Similar to this, a doctor can only examine a patient's data with their consent and change the patient's history after that.

Electronic health records have shown to work well on the blockchain with smart contracts. It serves as an interface to the patient's record and establishes the framework for a decentralized medical service stage that is shared by patients and suppliers. Blockchain is divided into patient-centric and research-centric blockchain network categories, according to "Blockchain Technology in Healthcare" [70]. The privacy concerns with regard to medical records are addressed by a patient-centric blockchain network that provides the patient authority over sharing their medical information with different users. Blockchain technology has the potential to revolutionize healthcare administration by enabling transparent and unambiguous data access across all stakeholders involved, including therapists, hospitals, general practitioners, and medical experts. By doing this, many medical stakeholders would no longer need to go through time- and resource-consuming verification and information processes. Due to coordination problems, this would also lessen medical malpractice and aid in the early detection of health problems. It gives people trust in their overall care because the network cannot tamper with their medical history from prior visits to other doctors. Furthermore, blockchain is seen as a tool for creating patient-centric systems in a variety of ways, such as boosting data availability, establishing distinct patient identifiers, setting up digital access controls, and enhancing data liquidity

[71]. A system called Hyperledger Fabric is used to create permissioned blockchain networks [72]. There are two different kinds of peers: a validating peer is a node in charge of managing the ledger, operating the consensus process, and validating transactions. A node known as a non-validating peer may validate transactions but does not actually execute them. This is better suited to creating a network with permissions that is appropriate for a patient-centric healthcare system. The assumption is that data is kept in a distributed management system and can be used to upload the patient's medical records, check their medical history, or grant or request access to data.

It is not advisable to disclose patient medical records on a public network because they are private information. In addition, it should be up to the patient to decide who has access to their medical records. As a result, a permissioned network built on the Hyperledger fabric can be used to analyze, but not disclose, medical health records and show the patient how well the doctor is performing so that they can make an informed choice regarding the doctor's prescribed medical treatments.

1.3.4 Proposition 4

Blockchain Technology and Tourism: A glance of blockchain technology in the world of tourism industry.

One of the most recent network-based technologies, blockchain technology, had a big impact on the travel and tourism sector [73, 74]. Although blockchain technology is still in its infancy, tools like cryptocurrencies, smart contracts, and decentralized applications have already started to affect travel-related transactions. Blockchain has increasingly gained relevance over the past ten years since the introduction of Bitcoin [25]. According to trends, efforts to build more integrated and collaborative cities are being made [75]. These cities have the potential to offer more effective and efficient ways to live, work, and network through integrative platforms that connect different stakeholders and entities, including government [76]. In order to provide travelers with better integrated services/solutions and comprehensive experiences, the tourism sector has reportedly been investigating the development of smart destinations [77].

Three generations of blockchain technology could be distinguished in its evolution. The creation of Bitcoin, the first digital currency, came first. The second generation—smart contracts—was created as a result of the quick realization that the blockchain technology that powers Bitcoin may be utilized for other things. These smart contracts (blockchain-based code that enables self-executing, self-enforcing contracts) are developed and run-on platforms like Ethereum [78–81]. DApps are what define the third generation, or the most recent technology. DApps give users a more frequent and comfortable way to connect with blockchain technology, such as through cellphones or web browsers. Disintermediation, security, automation, immutability, trust, costs, and traceability are the main characteristics of blockchain that are applied in the tourism business [74, 78, 82].

Tourism destinations have benefited greatly from the creation of smart cities, which are cooperative communities that offer integrated systems that facilitate efficiency, citizen participation, and improved quality of life [75]. According to [76], and [77], a number of examples of smart cities include Amsterdam, Barcelona, Brisbane, Dubai, and Seoul. These cities demonstrate traits like cutting-edge information and communication technology (ICT), co-creative platforms that involve key stakeholders, and efficient data use [83]. The smart contract has the potential to simplify the advantages that accrue to both citizens and travelers. The connection between smart tourism and smart cities has long been acknowledged. The six layers of the smart city and smart tourist architecture are the IT infrastructure, data, platform, policy, citizen, and traveler, according to [83].

As per the nature of blockchain, cryptocurrency is a digital currency that operates without a central bank or middleman. This allows for improved accuracy, traceability, and security [78] at low transaction costs [84].

It is conceivable for citizens of a smart city who have accepted cryptocurrencies to make payments in a secure and effective manner while also having access to all the advantages of blockchain. Several tiny island locations have been attempting to implement current currencies or create their own cryptocurrencies in the context of tourism [73]. In this context, cryptocurrencies are employed to broaden a number of payment options for locals and visitors [85]. Tourists will be able to book reservations and pay for their travel using blockchain and cryptocurrencies without exchanging money when this is coupled with an increase in the number of travel companies adopting blockchain [86–88].

Due to the fact that these actions are already included in smart contracts as an autonomous escrow function, businesses might use them to autonomously place orders, complete activities, and send payments without the need for authorizations or instructions. The escrow feature of smart contracts enables two unidentified parties to conduct transactions using trust as the foundation. Applications are capable of automating relatively simple and time-consuming operations, such as programming a hotel room key [89], which might be expanded to handle more complicated and large-scale transactions. They might include giving out incentives and points [89], as well as making payments, purchases, and salary payments [78].

Blockchain technology may potentially be used to streamline the flow of documents like passports and travel visas. Excellent examples of blockchain applications that eliminate manual passport control for tourists include those in Dubai and Switzerland [90, 91]. Both smart cities and smart travel destinations have the potential to be utilized. Medical tourism is affected by the development and application of *smart contracts*, which are directly related to the management of data and rights [78]. Access to a patient's complete medical history is helpful in medical tourism transactions where travel is done for medical reasons. This is especially true when the location or setting is unknown or unfamiliar [89, 92, 93].

Additionally, domestic and international travel loyalty programmes can benefit from the advantages that blockchain brings to records management [89, 94]. Travelers can use token-based loyalty advantages anywhere in the world without incurring exchange loss and have easy access to their loyalty information.

Globally, smart cities and smart travel destinations are emerging, and they are starting to adopt blockchain technology in a variety of ways, from cryptocurrency payments to business automation using smart contracts to sustainable and ethical sourcing. These changes are being welcomed by the tourism industry as well and may have a substantial impact on current business procedures. Reduced costs, cryptocurrency adoption, and the creation of new, all-encompassing eco-systems were found to be three traits of existing DApps through examination.

1.3.5 Proposition 5

Blockchain Technology and Supply Chain: A glance of blockchain technology in the world of supply chain management.

Blockchain technology is a new paradigm for computing and information flow that will have significant effects on how supply chain management and logistics will grow in the future [95–97]. The design, organisation, operation, and overall administration of supply chains could be significantly disrupted by blockchain technology. Supply chain management will likely undergo a significant rethinking as a result of blockchain technology's ability to guarantee the reliability, traceability, and authenticity of information as well as smart contractual agreements for a trustless environment.

Blockchain-based supply chains involve four primary actors, some of which are absent from conventional supply chains. Registrars, who give distinct identities to network actors. Organizations that develop standards programmes like Fairtrade for ecologically friendly supply chains or technological requirements on the blockchain. Those who certify individuals for involvement in supply chain networks. A registered auditor or certifier must certify participants, such as producers, sellers, and buyers, in order to retain the system's credibility [98]. There are additional influences on the product and material movements in the supply chain. To allow all relevant stakeholders to have rapid access to the product profile, every product may have a presence on the digital blockchain. Access can be restricted by security measures so that only those having the proper digital keys can use a certain product. There are many other types of information that can be gathered, such as the product's state, type, and the standards that need to be applied to it [99]. A product's information tag serves as an identification that connects tangible goods to their virtual identities in the blockchain [95]. How a product is "owned" or transferred by a specific actor is an intriguing structural and flow management feature. A key regulation will probably be the requirement of smart contract agreements and consensus for actors to obtain authorization before adding new data to a product's profile or starting a trade with a third party. Both parties may sign a digital contract or fulfil a smart contract condition before a product is transferred (or sold) to another actor in order to validate the exchange. Transaction information updates the blockchain ledger once all parties have complied with their contractual duties and procedures. When a change is made, the system will automatically update the data transaction records

[95]. At least five essential aspects of a product can be highlighted and described using blockchain technology: its nature, its quality, its quantity, its location, and its ownership. The blockchain accomplishes this by removing the need for a dependable central organization to operate and administer the system while allowing customers to observe the continuous chain of custody and transactions from the raw materials to the final sale. When transactions occur on these many blockchain information dimensions, this data is logged in ledgers with verifiable changes. With automated governance standards, blockchain dependability and transparency are designed to more efficiently allow the flow of materials and information along the supply chain. The economy may change more broadly as a result of this revolution, moving from one based on industrial durable, commodity, and product to one based on information and personalization. Knowledge, communication, and information—rather than the properties of the materials themselves—will be more important in the production process [100]. Customers can track the specific information about products, for instance, increasing their confidence in the product's features [96]. The interaction of network actors inside the system can be defined via smart contracts, which are written rules maintained in the blockchain. Continuous process improvement and network data exchange between supply chain actors are influenced by smart contracts. Examples include the digital verification of actor profiles and products by certifiers and standards organisations. On the network, actors and items each have a digital profile that includes details like a description, location, credentials, and affiliation with particular goods. Each participant in the supply chain can log crucial data about a specific product and its status on the blockchain network [99]. An actor's certification and approval as well as the processes they can access and use for execution can be managed by smart contract governance and process rules in a blockchain-based supply chain. Depending on the supply chain type, position, and trigger specified by a smart contract, changes to actor data may take place. Without a consensus-building mechanism, the actors are unable to change the rules [97]. The use of smart contracts in procurement is an additional example. A smart contract between two trade partners can update the automated record of the goods that were bought, sold, and delivered in real-time by customers across all lines of business.

1.3.6 Proposition 6

Blockchain Technology and Space industry: A glance of blockchain technology in the world of space industry.

The use of small satellite architectures like constellations and swarms is the latest trend in Earth Observation (EO) applications. These architectures are affordable, straightforward, and took a minimal amount of time to develop. Due to its dependability, confidentiality, and decreased latency, the Blockchain has demonstrated significant efficiency in terrestrial systems. Blockchain represents a revolution in the way *data integrity* is verified and secure prioritized data exchange is created in mobile satellite communication systems.

The importance of Earth Observation (EO) satellite and space systems has increased recently. Images and other types of remote sensing data are made available by EO for a growing number of applications that have a significant impact on the global economy and business. According to the United Nations, EO is essential to accomplishing the 17 sustainable development objectives [101]. Small satellite classes are getting more popular than the larger ones. Only six enterprises presently fall under the "big satellite" category, while sixteen fall under the "small satellite" category [102]. In satellite communication systems for EO applications, the constellation of many satellites plays a crucial role. This is as a result of its advantageous characteristics, including high throughput, low cost, low latency, global connectivity, security, and resilience. The usage of satellite swarm is necessary as the necessity for environmental pollution management, border surveillance, and disaster monitoring increases. As well as the data needed for "supporting decision making" in close to real-time, satellite swarms frequently produce images with the requisite geographic and temporal resolution for analysis [103].

Decentralization, persistence, anonymity, integrity, and command tracking are the key characteristics of Blockchain [104]. The existence of the Blockchain in space supports open economic models for the development of new initiatives, particularly space-based ones. Through openness, immutability, and consistency, blockchain improves the quality of the data. Blockchain is being used in the space sector, especially for small satellite systems' intersatellite communications. A number of the difficulties that face the space industry can be solved in part with blockchain technology. Here are explanations of a few of these solutions:

Self-reconfiguration: To strengthen the adaptability of the network space information, blockchain may employ smart contracts, a piece of software capable of self-execution. The orbits of satellites are established prior to launch, allowing for the prediction of their positions at any given time. Approaching satellites can automatically cluster at a set moment with smart contracts. In addition to safeguarding satellite organisation schemes from interference attempts, they also establish communication linkages on their own in the case that the ground station is taken down by a cyberattack. The space information network may quickly self-reconfigure if adjacent satellites step in to replace a failing satellite after an attack [105].

Identity identification and the integrity of transmitted data: These are two areas where the Low Earth Orbit (LEO) satellite network varies from traditional satellite networks. These areas include the dynamic topology and frequent connection switching. Therefore, they should utilize the lightest and most efficient authentication technique available to assure security [106]. A quick and effective authentication methodology was presented and is based on the Identity-based Encryption (IBE) mechanism and Blockchain. Blockchain and smart contracts served as the inspiration [107]. It is a virtual trust zone construction approach that makes use of capability-based access restriction and decentralized authentication. The distributed components in these zones should find and update one another in a trustless network environment. Data collection and data recording are made easier by the inherent safety features of blockchain technology. It is also possible to verify and record each change made to the initial data set [108]. To create traceability for the satellite data, the system does such. Orbital debris poses a serious threat to satellites. To anticipate and avoid collisions, blockchain technology could offer a high level of verified positional data. These result in enormous financial and time savings [109].

1.3.7 Proposition 7

Blockchain Technology and Finance industry: A glance of blockchain technology in the world of finance industry.

Increased transparency and trust among players are made possible by blockchain technology and smart contracts, which also result in significant infrastructural, transaction, and administrative cost savings [110–112]. The following are a few typical uses for smart contracts in the financial sector.

Complex methods used in the *security* sector are time-consuming, expensive, burdensome, and risky. By removing middlemen from the chain of custody for securities, smart contracts can minimize operational risks and enable automatic liability management, dividend payments, and stock splits. The clearing and settlement of securities can also be made easier via smart contracts. Complex internal and external reconciliations as well as labor-intensive activities are a part of centralized clearing. Bilateral peer-to-peer clearing business logic execution is made possible by blockchain technology. The Australian Securities Exchange is creating a DLT-based post-trade platform to replace its equities settlement system [113–116].

The processing of claims costs the insurance business tens of millions of dollars annually, and false claims cost it millions more dollars. In order to speed up claim processing, eliminate fraud, and avoid potential dangers, smart contracts can be used to automate claim processing, verification, and payment [117–120]. Flight insurance is being added to smart contracts, for instance, by the French airline AXA14. Passengers will immediately be informed of their choices for compensation if their flight is more than two hours late. Due to their ability to store insurance terms, driving histories, and accident reports, smart contracts may also be employed in the auto insurance industry. This will enable IoT-enabled automobiles to process claims quickly following an accident.

The current state of trade finance is one of severe inefficiency and fraud vulnerability. Additionally, it is urgently necessary to modernize or replace trade finance's paper-based processes with digital ones. Smart contracts give businesses the ability to autonomously initiate business operations based on established criteria. By streamlining procedures, this improves efficiency and lowers costs associated with fraud and compliance. Australia and Japan concluded a trade agreement in July 2017. This trade transaction used the Hyperledger Fabric platform exclusively for all trade-related operations, including the issuance of a letter of credit and the delivery of trade documents, which decreased the labour and other costs as well as the time needed to send documents [121, 122].

Apart from the above-mentioned recent advances in blockchain technology, there are some more advancements as well relating to this field of study. The most prominent study these days are going on fusing *blockchain and AI with metaverse.* Neal Stephenson introduced the idea of the *metaverse* in his science fiction novel Snow Crash approximately thirty years ago [123–126]. Due to the quick development of Blockchain, Internet of Things (IoT), VR/AR, Artificial Intelligence (AI), and cloud/edge computing. Many IT industries are interested in metaverse. In order to assist in bringing the metaverse to life and enable people to interact, learn, work together, and have fun in ways they could never have imagined, the social media business Facebook is renamed as *Meta* [127]. The metaverse smoothly merges the real world and the virtual world, enabling avatars to engage in a wide range of activities such as creation, display, entertainment, social interaction, and trading. In the ever-expanding metaverse, blockchain and AI technology are expected to be crucial. For instance, the metaverse employs AI and blockchain to build a virtual environment where anyone can securely and freely engage in social and commercial activities that go beyond the bounds of the real world. The use of these most recent blockchain and AI technologies will also be accelerated by utilizing the metaverse.

1.4 Future Trends

The development of standards and interoperability options is another trend that is accelerating. These should make it possible for various blockchains to communicate. Blockchain is a developing technology, just like any other, and standards are crucial to its success. When the correct standards are established at the right point in the development of a technology, interoperability, trust, and ease of use are all enhanced. By doing this, they encourage its growth and pave the way for widespread adoption. Chains of all kinds are expected to emerge as a result of the blockchain's rapid development. Cross chain technology, a new technology that aims to enable the movement of value and information between several blockchain networks, is one such technology that is coming into greater and greater focus. As the best way to improve interoperability between blockchains, this technology is fast becoming a major topic of discussion.

Due to numerous improvements in this blockchain-based environment, blockchain-as-a-service (*BaaS*) has become a driver of adoption across commercial firms. Companies have a strong demand for BaaS, which allows third parties to build and administer cloud-based networks for businesses that develop blockchain applications. Microsoft, Amazon, and R3 are the major participants in this market. BaaS makes it easier for its customers to use the tools to create hosts. Based on the cloud and give them the ability to run relevant operations on the blockchain and their applications without having to deal with operational overhead, technological challenges, or the need to invest in additional infrastructure upgrades or lack of skills. BaaS providers aid clients in concentrating just on blockchain functionalities and their primary tasks.

When the potential for growth in the blockchain industry and the growing dominance of blockchain across multiple industries serve as a key factor for the rising demand for these talents, a stronger demand for blockchain and crypto skills is noticed. Blockchain professionals are in greater demand due to the potential benefits of blockchain technology for businesses in terms of cost reduction and performance enhancement, as well as the explosive growth of the cryptocurrency markets. Blockchain has been ranked as one of the most in-demand skills for the upcoming years in research by LinkedIn. As a result, businesses require blockchain experts that have the knowledge and expertise to assist them make the most of blockchain technology to further their corporate goals.

Integration of blockchain with other technologies, like *Big Data and Artificial Intelligence*, is on the rise. Corporate entities are becoming more interested in using blockchain for *Internet of Things (IoT)* applications. The recent adoption of the 5G network has significantly boosted the growth of the IoT market. The current immensely fragmented IoT ecosystem, however, limits the predicted potential of the 5G IoT market. The many 5G IoT challenges appear to be best and most effectively solved by blockchain technology. Due to the automated encryption and immutability of blockchain, it may be able to assist in solving a number of scalability and security-related issues.

Decentralized finance, or DeFi, is swiftly establishing itself as a transparent and permissionless method for users to communicate directly with one another. DeFi's asset worth this year exceeded $180 billion, and in the years to come, it is anticipated that this value would increase much more. We will witness a further expansion of the DeFi market as well as the introduction of more specialised DeFi apps as there is an increasing need to reproduce tangible things' attributes like uniqueness and ownership evidence. More convergence between traditional or centralised finance (CeFi) and decentralised finance may eventually result from impending regulation as well as the rising acceptance that cryptocurrencies are here to stay (DeFi).

The NFT market is anticipated to maintain its impressive growth in the years to come. There is an increasing desire to recreate physical things' characteristics like greater uniqueness, ownership evidence, and scarcity as nearly everything becomes digital. The previously discussed Metaverse notion will open up a lot of new possibilities for creative NFT application cases. The NFT sector is seeing an influx of new use cases due to the numerous advantages and potential earnings, including gaming, music, ticketing, posting on social media, etc.

Future trends have been carefully cited by De Meijer [128] and are predicated on a variety of assumptions. Older patterns so play a significant role in these forecasts.

1.5 Conclusions

In this study, an introduction of blockchain fundamentals is provided before diving into the details of the blockchain technology. The author also lists blockchain's benefits and the development of blockchain applications. The author also provides a summary in the form of propositions on the use of blockchain technology in several fields related to recent advancements. This study also discusses potential future trends. It will be a fantastic year for blockchain technology and the cryptocurrency market if all of these predictions come true. The study findings are intended to assist experts, practitioners, and stakeholders who want to manage blockchain-related transformation projects in their industries. Additionally, the organisation's decision-making processes would benefit from a deeper knowledge of the associated factors. In recent years, blockchain technology has grown in popularity in both business and academia. Due to the fact that this technology is still in its infancy, there is the potential for future study to demonstrate usability and user perceptions for the application of blockchain technology. Additional empirically tested studies in various fields should be the focus of high calibre academic investigations.

References

1. Naik, R.P.: Optimising the SHA256 Hashing Algorithm for Faster and More Efficient Bitcoin Mining. UCL, London (2013)
2. Murck, P.: Who controls the blockchain? Harv. Bus. Rev. (2017)
3. Iansiti, M., Lakhani, K.A.: The truth about blockchain. Harv. Bus. Rev. (2017). https://hbr.org/2017/01/the-truth-about-blockchain
4. Aishwarya, N.: Potential impact of blockchain on HR and people management. JETIR **5**(9), 127–130 (2018)
5. Sakho, S., Zhang, J., Mbyamm, M.J.K., Kouassi, A.B., Essaf, F.: Privacy protection issues in blockchain technology. Int. J. Comput. Sci. Inf. Secur. (IJCSIS) **17**(2), 124 (2019)
6. Vijai, C., Elayaraja, M., Suriyalakshmi, S.M., Joyce, D.: The blockchain technology and modern ledgers through blockchain accounting. Adalya J. **8**(12), 545–557 (2019)
7. Lin, I.C., Liao, T.C.: A survey of blockchain security issues and challenges. Int. J. Netw. Secur. **19**(5), 653–659 (2017)
8. OECD Blockchain Primer. http://www.oecd.org/finance/blockchain
9. Tapscott, Tapscott: The impact of the blockchain goes beyond financial services. Harv. Bus. Rev. (2016). https://hbr.org/2016/05/the-impact-of-the-blockchain-goes-beyond-financial-services
10. Nakamoto, S.: Bitcoin: A Peer-to-Peer Electronic Cash System, pp.1–9. https://bitcoin.org/bitcoin.pdf
11. Gartner Blockchain Spectrum. http://www.gartner.com/smarterwithgartner/the-4-phases-of-the-gartner-blockchain-spectrum
12. Wu, B., Li, Y.: Design of evaluation system for digital education operational skill competition based on blockchain. In: IEEE 15th International Conference on E-Business Engineering, pp. 102–109 (2018)
13. Yi, C.S.S., Yung, E., Fong, C., Tripathi, S.: Benefits and use of blockchain technology to human resource management: a critical review. Int. J. Hum. Resour. Stud. (Macrothink Institute) **10**(2), 131–140 (2020)
14. Wang, G., Nixon, M.: SoK: tokenization on blockchain. In: IEEE/ACM 14th Internal Conference on utility and cloud computing (UCC'21) Companion (UCC'21 Companion), 6–9, Leicester, United Kingdom. ACM, New York, NY, USA, 9. https://doi.org/10.1145/3492323.3495577)
15. Zyskind, G., Nathan, O., Pentland, A.S.: Decentralizing privacy: using blockchain to protect personal data. In: Proceedings of the IEEE Security and Privacy Workshops, pp. 180–184, May 2015

16. Xu, J.J.: Are blockchains immune to all malicious attacks? Financ. Innov. **2**(1), 1–9 (2016)
17. Guo, Y., Liang, C.: Blockchain application and outlook in the banking industry. Financ. Innov. **2**, 24 (2016)
18. Hull, R., Batra, V.S., Chen, Y.M., Deutsch, A., Heath, F.F.T., Vianu, V.: Towards a shared ledger business collaboration language based on data-aware processes. In: Sheng, Q.Z., Stroulia, E., Tata, S., Bhiri, S. (eds.) Service-Oriented Computing. Lecture Notes in Computer Science, vol. 9936, pp. 18–36. Springer, Cham (2016)
19. Zhao, J.L., Fan, S., Yan, J.: Overview of business innovations and research opportunities in blockchain and introduction to the special issue. Financ. Innov. **2**(1), 1–7 (2016)
20. Glaser, M.: Pervasive decentralization of digital infrastructures: a framework for blockchain enabled system and use case analysis. Presented at the 50th Hawaii International Conference on System Sciences, Waikoloa, HI, USA (2017)
21. Beck, R., Müller-Bloch, C., King, J.L.: Governance in the blockchain economy: a framework and research agenda. J. Assoc. Inf. Syst. **19**(10), 1020–1034 (2018)
22. Olnes, S., Ubacht, J., Janssen, M.: Blockchain in government: benefits and implications of distributed ledger technology for information sharing. Gov. Inf. Q. **34**(4), 355–364 (2017)
23. Jaoude, J.A., Saade, R.G.: Blockchain applications–usage in different domains. IEEE Access **7**, 45360–45381 (2019)
24. Alketbi, Q.N., Talib, M.A.: Blockchain for government services—use cases, security benefits and challenges. In: 5th Learning and Technology Conference (L and T), pp. 112–119 (2018)
25. Bohme, R., Christin, N., Edelman, B., Moore, T.: Bitcoin: economics, technology, and governance. J. Econ. Perspect. **29**(2), 213–238 (2015)
26. Yermack, D.: Corporate governance and blockchains. J. Rev. Finance **21**(1), 7–31 (2017)
27. Kshetri: Blockchain's roles in strengthening cybersecurity and protecting privacy. Telecommun. Policy **41**(10), 1027–1038 (2017)
28. Shen, C., Pena-Mora, F.: Blockchain for cities—a systematic literature review. IEEE Access **6**, 76787–76819 (2018)
29. Sun, M., Zhang, J.: Research on the application of block chain big data platform in the construction of new smart city for low carbon emission and green environment. Comput. Commun. **149**, 332–342 (2020)
30. Agustin, R.F., Susilowati, D.: Preventing corruption with blockchain technology (case study of Indonesian public procurement). Int. J. Sci. Technol. Res. **8**(9), 2377–3238 (2019)
31. Palfreyman, J.: Blockchain for Government? IBM Government Industry (2015). Blog. [Online]. https://www.ibm.com/blogs/insights-on-business/government/blockchain-forgovernment/. Accessed 7 Jan 2020
32. Zhu, H., Zhou, Z.Z.: Analysis and outlook of applications of blockchain technology to equity crowdfunding in China. Financ. Innov. **2**(1), 1–11 (2016)
33. Kouhizadeh, M., Sarkis, J.: Blockchain practices, potentials, and perspectives in greening supply chains. Sustainability **10**(10), 1–16 (2018)
34. Zohar, A.: Bitcoin: under the hood. Commun. ACM **58**(9), 104–113 (2015)
35. Möser, M., Böhme, R.: Trends, tips, tolls: a longitudinal study of bitcoin transaction fees. In: Brenner, M., Christin, N., Johnson, B., Rohloff, K. (eds.) Financial Cryptography and Data Security. Lecture Notes in Computer Science, vol. 8976, pp. 19–33. Springer, Heidelberg (2015)
36. Saberi, S., Kouhizadeh, M., Sarkis, J., Shen, L.: Blockchain technology and its relationships to sustainable supply chain management. Int. J. Prod. Res. **57**(7), 2117–2135 (2019)
37. Casino, F., Dasaklis, T.K., Patsakis, C.: A systematic literature review of blockchain-based applications: current status, classification and open issues. Telemat. Informat. **36**, 55–81 (2019)
38. Alketbi, A., Nasir, Q., Talib, M.A.: Blockchain for government services—use cases, security benefits and challenges. In: Proceedings of the 15th Learning and Technology Conference (L and T), pp. 112–119 (2018)
39. Kosba, A., Miller, A., Shi, E., Wen, Z., Papamanthou, C.: Hawk: the blockchain model of cryptography and privacy-preserving smart contracts. In: Proceedings of the IEEE Symposium on Security and Privacy (SP), pp. 839–858 (2016)

40. Ateniese, G., Goodrich, M.T., Lekakis, V., Papamanthou, C., Paraskevas, E., Tamassia, R.: Accountable storage. IACR, This report also discusses potential future trends. It will be a fantastic year for blockchain technology and the cryptocurrency market if all of these predictions come true. Cryptol. ePrint Arch. **2014**, 886 (2014)
41. Monrat, A.A., Schelen, O., Andersson, K.: A survey of blockchain from the perspectives of applications, challenges, and opportunities. IEEE Access **7**, 117134–117151 (2019)
42. Cai, Y., Zhu, D.: Fraud detections for online businesses: a perspective from blockchain technology. Financ. Innov. **2**(1), 20 (2016)
43. Biswas, K., Muthukkumarasamy, V.: Securing smart cities using blockchain technology. In: Proceedings of the IEEE18th International Conference on High Performance Computing and Communications; IEEE 14th International Conference on Smart City; IEEE 2nd International Conference on Data Science and System (HPCC/SmartCity/DSS), Sydney, NSW, Australia, pp. 1392–1393 (2016)
44. Chen, H.S., Jarrell, J.T., Carpenter, K.A., Cohen, D.S., Huang, X.: Blockchain in healthcare: a patient-centered model. Biomed. J. Sci. Tech. Res. **20**(3), Art. no. e15017 (2019)
45. Swan, M.: Blockchain: Blueprint for a New Economy. O'Reilly Media, Newton, MA, USA (2015)
46. Hölbl, M., Kompara, M., Kamisalić, A., Nemec, L., Zlatolas: A systematic review of the use of blockchain in healthcare. Symmetry **10**(10), 470 (2018)
47. Batubara, F.R., Ubacht, J., Janssen, M.: Challenges of blockchain technology adoption for e-government: a systematic literature review. In: 19th Annual International Conference on Digital Government Research, Governance Data Age, pp 1–9 (2018)
48. Agbo, C., Mahmoud, Q., Eklund, J.: Blockchain technology in healthcare: a systematic review. Healthcare **7**(2), 56 (2019)
49. Zhang, P., Schmidt, D.C., White, J., Lenz, G.: Blockchain technology use cases in healthcare. Adv. Comput. **111**, 1–41 (2018)
50. Lindman, J., Tuunainen, V.K., Rossi, M.: Opportunities and risks of blockchain technologies: a research agenda. In: Proceedings of the 50th Hawaii International Conference on System Sciences, Honolulu, HI, USA, pp. 1533–1542 (2017)
51. Toyoda, K., Mathiopoulos, P.T., Sasase, I., Ohtsuki, T.: A novel blockchain-based product ownership management system (POMS) for anti-counterfeits in the post supply chain. IEEE Access **5**, 17465–17477 (2017)
52. Wright, A., DeFilippi, P.: Decentralized Blockchain Technology and the Rise of Lex Cryptographia. Cardozo School L Yeshiva Univ., Paris, France, Tech. Rep. (2015)
53. Boucher, P.: What if Blockchain Technology Revolutionized Voting? European Parliament (2016). [Online]. https://www.europarl.europa.eu/thinktank/en/document.html?reference=EPRS_ATA(2016)581918)
54. Zheng, Z., Xie, S., Dai, H., Chen, X., Wang, H.: Blockchain challenges and opportunities: a survey. Int. J. Web Grid Serv. **14**(4), 352–375 (2018)
55. Siyal, A.A., Junejo, A.Z., Zawish, M., Ahmed, K., Khalil, A., Soursou, G.: Applications of blockchain technology in medicine and healthcare: challenges and future perspectives. Cryptography **3**(1), 3 (2019)
56. Perboli, G., Musso, S., Rosano, M.: Blockchain in logistics and supply chain: a lean approach for designing real-world use cases. IEEE Access **6**, 62018–62028 (2018)
57. Treiblmaier, H.: The impact of the blockchain on the supply chain: a theory-based research framework and a call for action. Supply Chain Manag. Int. J. **23**(6), 545–559 (2018)
58. Seebacher, S., Schüritz, R.: Blockchain technology as an enabler of service systems: a structured literature review. In: Proceedings of the 8th International Conference on Exploring Service Science, pp. 12–23 (2017)
59. Noyes, C.: BitAV: Fast Anti-malware by Distributed Blockchain Consensus and Feedforward Scanning (2016). arXiv:1601.01405. [Online]. http://arxiv.org/abs/1601.01405
60. Tseng, J.H., Liao, Y.C., Chong, B., Liao, S.W.: Governance on the drug supply chain via Gcoin blockchain. Int. Environ. Res. Public Health **15**(6), 1–8 (2016)

61. Allen, D.W.E., Berg, C., Davidson, S., Novak, M., Potts, J.: International policy coordination for blockchain supply chains. Asia Pacific Policy Stud. **6**(3), 367–380 (2019)
62. Rubio, E.: The Case for Disruptive Innovation in Human Resources (2017). https://www.linkedin.com/pulse/case-disruptive-innovation-human-resources-enrique-rubio-csm-cspo/
63. Biswas, S.: 3 Ways Blockchain Will Transform Recruitment (2018). https://www.hrtechnologist.com/articles/recruitment-onboarding/3-ways-blockchain-will-impact-recruitment/
64. Francis, F.: Blockchain: What Is Next for HR and Payroll? (2018). https://www.hrtechnologist.com/articles/digital-transformation/blockchain-what-is-next-for-hr-and-payroll/
65. Carmody, B.: This ICO Is Counting on Blockchain to Disrupt Employee Training (2017). https://www.inc.com/bill-carmody/this-ico-is-counting-on-blockchain-to-disrupt-employee-training.html
66. Ranosa, R.: 'Smart' Contracts: Are You Using Them? (2018). https://www.hrtechnologynews.com/news/hris/smart-contracts-are-you-using-them/117291
67. Day, N.: What are smart contracts and how will they affect payroll and HR? Blockchain Article 4/10 (2018). https://jgarecruitment.com/what-are-smart-contracts-and-how-will-they-affect-payroll-and-hr-blockchain-article-410/
68. Niranjanamurthy, M., Nithya, B.N., Jagannatha, S.: Analysis of Blockchain Technology: Pros, Cons, and SWOT (2019). https://www.researchgate.net/publication/323865742_Analysis_of_Blockchain_technology_pros_cons_and_SWOT
69. Juricic, V., Radosevic, M., Fuzul, E.: Creating students profile using blockchain technology. In: 42nd International Convention on Information and Communication Technology, Electronics and Microelectronics, pp. 521–525 (2019)
70. Mettler, M.: Blockchain technology in healthcare: the revolution starts here. In: 18th International Conference IEEE on e-Health Networking, Applications and Services (Healthcom), pp. 1–3 (2016)
71. Gordon, W.J., Catalini, C.: Blockchain technology for healthcare facilitating the transition to patient-driven interoperability. Comput Struct Biotechnol J. **16**, 224–230 (2018)
72. Cachin, C.: Architecture of the hyperledger blockchain fabric. In: Workshop on Distributed Cryptocurrencies and Consensus Ledgers, vol. 3, no. 10 (2016)
73. Kwok, A.O.J., Koh, S.G.M.: Is blockchain technology a watershed for tourism development? Curr. Iss. Tour. 1–6 (2018)
74. Önder, I., Treiblmaier, H.: Blockchain and tourism: three research propositions. Ann. Tour. Res. **72**, 180–182 (2018)
75. Snow, C.C., Håkonsson, D.D., Obel, B.: A smart city is a collaborative community. Calif. Manag. Rev. **59**(1), 92–108 (2017)
76. Khan, M.S., Woo, M., Nam, K., Chathoth, P.K.: Smart city and smart tourism: a case of Dubai. Sustainability **9**(2), 2279 (2017)
77. Gretzel, U., Werthner, H., Koo, C., Lamfus, C.: Conceptual foundations for understanding smart tourism ecosystems. Asia Pacific J. Tour. Res. Comput. Human Behav. **50**, 558–563 (2015). https://doi.org/10.1916/jchb/2015.03.043
78. Boucher, P., Nascimento, S., Kritikos, M.: How blockchain technology could change our lives. In: European Parliamentary Research Service, Scientific Foresight Unit, pp. 1–28. European Parliament, Brussels (2017)
79. Ethereum (2018). https://www.ethereum.org/
80. Fedrer, P.: The three generations of the Blockchain technology. BTCWires [News] (2018). https://www.tcwires.com/block-o-pedia/the-three-generations-of-the-Blockchain-technology/
81. Smith, B.: What are the three generations of blockchain, and how are they similar to the web? Coin Insider [News] (2018). https://www.coininsider.com/threegenerations-of-Blockchain/
82. Seffinga, J., Lyons, L., Bachman, A.: The Blockchain (R)evolution—The Swiss Perspective. Deloitte (2017)
83. Koo, C., Park, J., Lee, J.N.: Smart tourism: traveler, business and organizational perspectives. Inf. Manag. **54**, 683–686 (2017)

84. Schlegel, M., Zavolokina, L., Schwabe, G.: Blockchain technologies from the consumers' perspectives: what is there and why should who care? In: Proceedings of the 51st Hawaii International Conference on System Sciences, Hawaii (2018)
85. Sov (2018). https://www.sov.global/
86. Travala: Decentralized Travel Booking Marketplace. Travala (2018). https://www.travala.com/
87. Travelblock (2018). Whitepaper. https://www.travelblock.io/
88. Travelchain: Decentralized Data Exchange for the Travel Industry (2018). https://travelchain.io/#/
89. Pilkington, M.: Can blockchain technology help promote new tourism destinations? The example of medical tourism in Moldova. SSRN Electron. J. 1–8 (2017)
90. D'Cunha, S.D.: Dubai sets its sight on becoming the world's first blockchain-powered government. Forbes [News] (2017). https://www.forbes.com/sites/suparnadutt/2017/12/18/dubai-sets-sights-on-becoming-theworlds-first-Blockchain-powered-government/#374ea161454b
91. FINews: Zug to Use Blockchain for Citizen ID. FINews [News] (2017). https://www.finews.com/news/english-news/28087-zug-to-use-Blockchain-for-citizen-id
92. Varaprasada Rao, K., Panda, S.K.: A design model of copyright protection system based on distributed ledger technology. In: Satapathy, S.C., Lin, J.C.W., Wee, L.K., Bhateja, V., Rajesh, T.M. (eds.) Computer Communication, Networking and IoT. Lecture Notes in Networks and Systems, vol. 459. Springer, Singapore (2023). https://doi.org/10.1007/978-981-19-1976-3_17
93. Varaprasada Rao, K., Panda, S.K.: Secure electronic voting (E-voting) system based on blockchain on various platforms. In: Satapathy, S.C., Lin, J.C.W., Wee, L.K., Bhateja, V., Rajesh, T.M. (eds.) Computer Communication, Networking and IoT. Lecture Notes in Networks and Systems, vol. 459. Springer, Singapore (2023). https://doi.org/10.1007/978-981-19-1976-3_18
94. Ereiqat, S.: Blockchain in Dubai: Smart Cities from Concept to Reality. IBM (2017). https://www.ibm.com/blogs/Blockchain/2017/04/Blockchain-in-dubai-smart-cities-fromconcept-to-reality/
95. Abeyratne, S.A., Monfared, R.P.: Blockchain ready manufacturing supply chain using distributed ledger. Int. J. Res. Eng. Technol. **5**(9), 1–10 (2016)
96. Tian, F.: An agri-food supply chain traceability system for China based on RFID & blockchain technology. In: 13th International Conference on Service Systems and Service Management (ICSSSM) (2016)
97. Maurer, B.: Blockchains are a diamond's best friend: Zelizer for the bitcoinmoment. In: Nina Bandelj, F.F.W., Zelizer, V.A. (eds.) Money Talks: Explaining How Money Really Works, pp. 215–230. Princeton University Press, Princeton (2017)
98. Steiner, J., Baker, J.: Blockchain: The Solution for Transparency in Product Supply Chains (2015). https://www.provenance.org/whitepaper
99. Tian, F.: A Supply chain traceability system for food safety based on HACCP, blockchain & internet of things. In: International Conference on Service Systems and Service Management (ICSSSM) (2017)
100. Pazaitis, A., De Filippi, P., Kostakis, V.: Blockchain and value systems in the sharing economy: the illustrative case of backfeed. Technol. Forecast. Soc. Change **125**, 105–115 (2017)
101. Crisp, N.H., Roberts, P.C., Livadiotti, Oiko, S.V.T.A., Edmondson, S., Haigh, S.J., Schwalber, A.: The benefits of very low earth orbit for earth observation missions. Prog. Aerosp. Sci. **117**, 100619 (2020)
102. https://spacechain.com/wp-content/uploads/2020/03/whitepaper-200320.pdf. Space chain, March 2020 [Online]. https://spacechain.com/wp-content/uploads/2020/03/whitepaper-200320.pdf. Accessed 12 Feb 2021
103. Farrag, A., Othman, S., Mahmoud, T., ELRaffiei, A.Y.: Satellite swarm survey and new conceptual design for Earth observation applications. Egypt. J. Remote Sens. Space Sci. (2019)

104. Tama, B.A., Kweka, B.J., Park, Y., Rhee, K.H.: A critical review of blockchain and its current applications. In: International Conference on Electrical Engineering and Computer Science (ICECOS), pp. 109–113. IEEE (2017)
105. Cheng, S., Gao, Y., Li, X., Du, Y., Du, Y., Hu, S.: Blockchain application in space information network security. In: International Conference on Space Information Network, pp. 3–9. Springer, Singapore (2018)
106. Liu, S., Li, M., Wei, S.: A distributed authentication protocol using identity-based encryption and blockchain for LEO network. In: International Conference on Security, Privacy and Anonymity in Computation, Communication and Storage, pp. 446–460. Springer, Cham (2017)
107. Xu, R., Chen, Y., Blasch, E., Chen, G.: Exploration of blockchain enabled decentralized capability-based access control strategy for space situation awareness. Opt. Eng. **58**(4) (2019)
108. Moeniralam, S.: Blockchain based Data Production Traceability System (2018)
109. Molesky, M.J., Cameron, E.A., Jones, J., Esposito, M., Cohen, L., Beauregard, C.: Blockchain network for space object location gathering. In: 2018 IEEE 9th Annual Information Technology, Electronics and Mobile Communication Conference (2018)
110. Ojetunde, B., Shibata, N., Gao, J.: Secure payment system utilizing MANET for disaster Areas. IEEE Trans. Syst. Man Cybern. Syst. https://doi.org/10.1109/TSMC.2017.2752203
111. Panda, S.K., Satapathy, S.C.: An investigation into smart contract deployment on Ethereum platform using Web3.js and solidity using blockchain. In: Bhateja, V., Satapathy, S.C., Travieso-González, C.M., Aradhya, V.N.M. (eds.) Data Engineering and Intelligent Computing. Advances in Intelligent Systems and Computing, vol. 1. Springer, Singapore (2021). https://doi.org/10.1007/978-981-16-0171-2_52
112. Panda, S.K., Rao, D.C., Satapathy, S.C.: An investigation into the usability of blockchain technology in internet of things. In: Bhateja, V., Satapathy, S.C., Travieso-González, C.M., Aradhya, V.N.M. (eds.) Data Engineering and Intelligent Computing. Advances in Intelligent Systems and Computing, vol. 1. Springer, Singapore (2021). https://doi.org/10.1007/978-981-16-0171-2_53
113. Australian Securities Exchange. CHESS Replacement [Online]. https://www.asx.com.au/services/chess-replacement.htm. Accessed 15 Oct 2018
114. Panda, S.K., Dash, S.P., Jena, A.K.: Optimization of block query response using evolutionary algorithm. In: Bhateja, V., Satapathy, S.C., Travieso-González, C.M., Aradhya, V.N.M. (eds.) Data Engineering and Intelligent Computing. Advances in Intelligent Systems and Computing, vol. 1. Springer, Singapore (2021). https://doi.org/10.1007/978-981-16-0171-2_54
115. Nanda, S.K., Panda, S.K., Das, M., Satapathy, S.C.: Automating vehicle insurance process using smart contract and Ethereum. In: Chakravarthy, V.V.S.S.S., Flores-Fuentes, W., Bhateja, V., Biswal, B. (eds.) Advances in Micro-Electronics, Embedded Systems and IoT. Lecture Notes in Electrical Engineering, vol. 838. Springer, Singapore (2022). https://doi.org/10.1007/978-981-16-8550-7_23
116. Panda, S.K., Elngar, A.A., Balas, V.E., Kayed, M. (eds.): Bitcoin and Blockchain: History and Current Applications, 1st ed. CRC Press (2020). https://doi.org/10.1201/9781003032588
117. Gatteschi, V., Lamberti, F., Demartini, C., Pranteda, C., Santamaria, V.: Blockchain and smart contracts for insurance: is the technology mature enough? Future Internet **10**(2), 20 (2018)
118. Panda, S.K., Mohammad, G.B., Nandan Mohanty, S., Sahoo, S.: Smart contract-based land registry system to reduce frauds and time delay. Secur. Priv. e172 (2021). https://doi.org/10.1002/spy2.172
119. Niveditha, V.R., Sekaran, K., Amandeep Singh, K., Panda, S.K.: Effective prediction of bitcoin price using wolf search algorithm and bidirectional LSTM on internet of things data. Int. J. Syst. Syst. Eng. **11**(3–4), 224–236
120. Panda, S.K., Jena, A.K., Swain, S.K., Satapathy, S.C. (eds.): Blockchain Technology: Applications and Challenges. Springer, Intelligent Systems Reference Library. https://doi.org/10.1007/978-3-030-69395-4
121. Peyton, A.: Mizuho Trials Australia–Japan Trade Transaction on Blockchain (2017) [Online]. https://www.bankingtech.com/2017/07/mizuho-trials-australia-japan-trade-transaction-on-blockchain/

122. Panda, S.K., Satapathy, S.C.: Drug traceability and transparency in medical supply chain using blockchain for easing the process and creating trust between stakeholders and consumers. Pers. Ubiquit. Comput. (2021). https://doi.org/10.1007/s00779-021-01588-3
123. Joshua, J.: Information bodies: computational anxiety in Neal Stephenson's snow crash. Interdiscip. Literary Stud. **19**(1), 17–47 (2017)
124. Sathya, A.R., Panda, S.K., Hanumanthakari, S.: Enabling smart education system using blockchain technology. In: Panda, S.K., Jena, A.K., Swain, S.K., Satapathy, S.C. (eds.) Blockchain Technology: Applications and Challenges. Intelligent Systems Reference Library, vol. 203. Springer, Cham (2021). https://doi.org/10.1007/978-3-030-69395-4_10
125. Lokre, S.S., Naman, V., Priya, S., Panda, S.K.: Gun tracking system using blockchain technology. In: Panda, S.K., Jena, A.K., Swain, S.K., Satapathy, S.C. (eds.) Blockchain Technology: Applications and Challenges. Intelligent Systems Reference Library, vol. 203. Springer, Cham (2021). https://doi.org/10.1007/978-3-030-69395-4_16
126. Panda, S.K., Daliyet, S.P., Lokre, S.S., Naman, V.: Distributed Ledger Technology in the Construction Industry Using Corda, The New Advanced Society: Artificial Intelligence and Industrial Internet of Things Paradigm. https://doi.org/10.1002/9781119884392.ch2
127. Meta. Introducing Meta: A Social Technology Company. https://about.fb.com/news/2021/10/facebook-company-is-now-meta/. Accessed 11 Nov 2021
128. De Meijer, Carlo, R.W.: (2016). https://www.finextra.com/blogposting/21453/main-blockchain-and-crypto-trends-in-2022-unexpected-expectations

Chapter 2
Bitcoin: Beginning of the Cryptocurrency Era

Sandeep Kumar Panda, A. R. Sathya, and Sukanta Das

Abstract This case discusses how the evolution of Cryptocurrency and related technology changing the Business Models in different sectors. In spite of receiving backslash in its earlier days, it is observed that the impacts and consequences of cryptocurrencies were having a growing effect on more on industries and business models. And, there has been a huge funding on digital wallets and payment systems in the past couple of years. Facebook launching its own cryptocurrency is the best example to witness this fact. The technology behind Cryptocurrencies is Blockchain, a distributed network connected through peer-to-peer (P2P) nodes. The distinct features of the cryptocurrency like improved security, transparency and liability makes them more popular and a major reason for replacing the traditional methods. Cryptocurrencies, otherwise termed as digital currencies are cryptographically protected and are extremely difficult to forge. Being digital in nature cryptocurrencies are well suited for online transactions and the business process can be automated through smart contracts, a significant feature of cryptocurrencies. This virtual currency basically is used either as a payment token or as an investment (similar to shares). FinTech and Banking are the pioneers in adapting cryptocurrencies in their business models. However, those are not the only one domain where cryptocurrencies are revolutionizing the business models. This case illuminates the distinction between blockchain and bitcoin, working model of Bitcoin, business benefits, and challenges of Bitcoin, how Bitcoin taking over the mainstream process and reformation of ecommerce using Bitcoin. Most of the time it is interpreted that Bitcoin and Blockchain are the same. Actually, it is NOT. Bitcoin is the most popular digital currency and Blockchain was bundled up in the same solution when Bitcoin was released and Bitcoin was the very first application of blockchain, it is often misinterpreted as the same. It was introduced with the aim of evading government currency controls by eliminating third-party payment processing mediators and streamlining

---

S. K. Panda (✉) · A. R. Sathya
Department of Data Science & Artificial Intelligence, Faculty of Science & Technology (IcfaiTech), ICFAI Foundation for Higher Education, Hyderabad, Telangana, India
e-mail: sandeeppanda@ifheindia.org

S. Das
Department of Computer Science and Engineering, Faculty of Science and Technology (IcfaiTech), ICFAI Foundation for Higher Education, Hyderabad, Telangana, India

S. K. Panda et al. (eds.), *Recent Advances in Blockchain Technology*,
Intelligent Systems Reference Library 237, https://doi.org/10.1007/978-3-031-22835-3_2

online transactions. Paypal, the pioneer in online transactions is efficient in international fund transfers but they have a very high transaction fee. Apart from this P2P transaction has certain constraints like caps on transfer amount, restrictions on locations etc.

Keywords Cryptocurrency · Bitcoin · Peer-to-peer · Business model

2.1 Introduction

In the year 2008, a group of individuals or an individual, pseudonymously known as Satoshi Nakamoto, introduced the first cryptocurrency—'Bitcoin' to the world. Though Bitcoin was founded in the year 2008, the transactions began in the year 2009, after Satoshi Nakamoto released the genesis block. The real identity of Satoshi Nakamoto remains a secret till date, despite the multiple claims made by many people. The unit of this currency is a bitcoin (BTC) and the smallest unit of a bitcoin is termed as a satoshi (named after its creator) and it is equivalent to 0.00000001 BTC. The price of bitcoin in equivalence to the US dollar had many fluctuations, beginning from the initial value of 1 BTC = 0.00003 USD, in the year 2010 to reach a record of 1 BTC ~ 20,000 USD, in the year 2017 (Exhibit I). Contrary to fiat currency, Bitcoin is a virtual currency that cannot be touched or printed but has to be mined instead, to use them for various payments.

The main motive behind Bitcoin is to introduce the concept of a decentralized flow of money. According to Satoshi Nakamoto, "A purely peer-to-peer version of electronic cash would allow online payments to be sent directly from one party to another without going through a financial institution" [1]. This cryptocurrency works on the blockchain technology also invented by Satoshi Nakamoto in 2008. The application of blockchain in this scenario made bitcoin the first solution to the double-spending problem that doesn't require a third-party, thus introducing decentralization.

In this scenario, blockchain represents a public ledger that records all the transactions made in the bitcoin network. In layman terms, a blockchain is a chain of blocks where each block contains a list of successful transactions and all the blocks are connected in a specific order with the help of cryptographic hashes. It is said to be a public ledger as any participant can enter and leave the bitcoin network according to their will as there is no central point of authority here.

2.2 Background Note

In 1983, David Chaum's research paper introduced the concept of digital cash. In 1989, DigiCash, a cash company for electronics, was established in Amsterdam, to promote the concept of digital cash. In 1998 it was brought to bankruptcy. Introduced in 1996, e-gold has expanded to several million users until the US government shut it down in 2008. In 1997, Coca-Cola offered to purchase from the sales system by using mobile payments. In 1998, PayPal launched a US Dollar-denominated service. The Coca-Cola mailing list users used the word “digital money” to describe peers’ to peer payments in many instruments.

In 2009, Bitcoin was introduced to launch a decentralized digital currency [2] based on blockchain, without a central server and with no tangible reserved properties. Known as cryptocurrencies, digital currencies based on blockchain proved resistant to government attempts to regulate them, provided that there was no central agency or person with the ability to turn them out.

On the third of January 2009, the first block or the genesis block of the bitcoin network was mined by Satoshi Nakamoto. The mining reward for each block was 50 bitcoins when bitcoin originated. In the genesis block, a text was embedded that read as “The Times 03/Jan/2009 Chancellor on brink of second bailout for banks” [3]. Satoshi Nakamoto continued to develop bitcoin with other developers until 2010, after which he handed over the source code repository to Gavin Anderson, a Silicon Valley engineer. He was the lead developer for the bitcoin digital currency project. This project aimed to create stable and secure cash for the internet. Bitcoin’s domain was registered on 18th August 2008 and the link to the whitepaper titled Bitcoin: A Peer-to-Peer Electronic Cash System, authored by Satoshi Nakamoto was made public on 31st October 2008. Andersen went onto develop the client network for bitcoin called bitcoin core. He also started the bitcoin foundation which was established in 2012.

2.3 Leveraging Blockchain Technology

Blockchain is an all-embracing technology which creates a digital ledger through built-in blockchain on platforms and hardware and allows independent parties to exchange data across the network around the globe. That is, it is a virtual network made up of data and transaction blocks, which is distributed to users worldwide. Any type of information/data can be stored in data blocks based on blockchain implementation. This digital ledger is a peer-to-peer network that uses consensus algorithms to eliminate dependency on external or internal sources to confirm the validity of records/data/information. Simply put, when it comes to data verification or validity the middlemen are excluded from the equation and the end-users can communicate directly with each other. This leads to the reduction of processing time and money costs for both concerned.

Blockchain decentralizes the database, leaving no central/center points to get hacked or broken in. A reviewed and completed entry by the majority of users is permanently registered in the database and it is difficult to make adjustments. Additionally, making changes to one element of a block would modify the specific hash of the block and inform users that the block has been modified. It is known as a fingerprint. The entire block altered will lead to the blockchain being rejected [4].

For cryptocurrencies like Bitcoin, Blockchain shapes the cornerstone. As we already know, a central authority, normally a bank or government, controls and verifies currencies such as the US dollar. Data and currency of a consumer under the central authority scheme are legally at the discretion of their bank or government. If a user's bank fails or they live in a country with weak governance, their currency's worth can be in danger. These are the fears Bitcoin was born out of. Blockchain enables Bitcoin and other cryptocurrencies to function without a Central Authority by spreading transactions through a network of computers. Not only does this minimize costs, but certain processing and care fees are also reduced. It also provides countries with weak currencies with a more secure currency and a larger network of citizens and organizations with which they can negotiate both domestically and internationally.

Blockchain is a popular public directory on which the whole Bitcoin network is based. The blockchain contains all verified transactions. It enables Bitcoin wallets to measure their expendable balance to check those new transactions guarantee that they are still the property of the spender. Cryptography is used to reinforce the credibility and temporal order of the blockchain.

2.4 Key Blockchain Technology Powered Products

One of the blockchain technology-powered products that have played a major role in the world economy is Bitcoin. Bitcoin is a digital asset and what is most manifesting about it is the proximity of the fact of supply and uncertainty of demand [5]. The pace at which bitcoin is mined has been devilishly predictable and unlike any other currency or asset its end-to-end supply is a known quantity and fixed in advance, not more than 21 million coins. This advantage makes bitcoin supply perfectly inelastic. Bitcoin's limited supply of coins and highly inelastic supply proves to be a major factor when it comes to handling its cost appreciation. Statistics show that the average supply of bitcoins first four years grew by around 2.5 million coins per year. Since then, the supply had kept on growing but the momentum at which it was growing decreased gradually as the demand has sporadically fallen away, even on a year-on-year basis. Bitcoins are mined by computers by solving a cryptographic math problem. Miners inside the network try solving the problem in the exchange of bitcoins [6]. The difficulty of these problems increases over time. With an increase in the difficulty, it requires an immense computational power to solve the problems. If we take a close look at the graph below, we can see that there has been a rise in "difficulty" which means that the supply of bitcoin has become expensive.

Well, we cannot conclude that due to the exponential rise in the difficulty led to the exponential rise in price. Have a close look at what happened to the "difficulty" after the first bitcoin price drop (nearly 93%). Its inevitable rise came to a stop only after a gap of 2 long years until the prices recovered. A similar situation occurred in the fallout of 2013–15 when the market dropped to 84%. Even there the "difficulty"

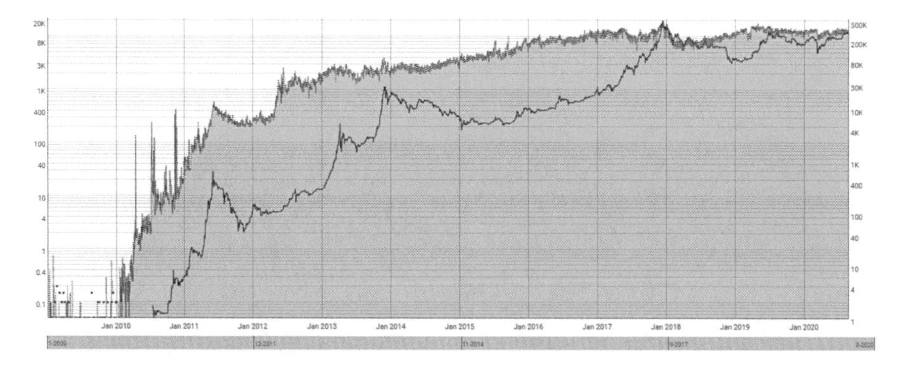

Fig. 2.1 Bitcoin transaction versus price

stopped until the market slowly started to gain its momentum. Intriguingly the "difficulty" never staunched when the bitcoin prices dropped down to 50% in recent times. The relationship between the difficulty and price concludes that bitcoin continues to remain in a heaved manner and that we can never predict how the market is going to behave. When it comes to transactions, it appears to be a vague relationship between the growth of transactions and the price of the bitcoin. Notice that the transactions stopped increasing in the year 2012. It again started gaining momentum in 2014, before bitcoin prices began to recover in steady but it has been declining since the end of 2016. What is particularly noticeable is that though the price declined as they did during Dec 2013–Jan 2015, the number of transactions did not rise (Fig. 2.1).

One thing is for sure that if bitcoin fails to replace the paper money, it will not necessarily be lacking long-term economic impact. Blockchain technology has the potential to allow policymakers to furnish their virtual currencies that help give us real-time information about inflation and real GDP. Moving on to blockchain enhanced cryptocurrencies could furthermore reduce the volatility of the economy, which in turn enables further advantage of the highly bounded global economy as people search for new ways to use capital efficiently (Fig. 2.2).

The volume of trading plays a major role when it comes to deciding the market value of bitcoin. One might raise a question saying how can we be so sure that the high demand that bitcoin has will persist in the future? Because one should not forget that if the bitcoin price remains high then only it can be seriously viewed as a potential currency, product, or asset. To understand this sudden spike in the price of bitcoin we need to take the help of Metcalfe's Law.

Conceived by George Gilder and then attributed to Robert Metcalfe, the law states that the value of the network is proportional to the square of the number of its users, which perfectly applies to bitcoin. Well according to the study published at the Journal of Electronic Commerce Research and Applications, the price of the digital currency as compared to the unique addresses that take part in transactions on the network each day and measured the value of the network. The results were positive and have shown that the networks were fairly well modeled by Metcalfe's Law. The findings were very much useful to spot potential bubbles. As proposed in the

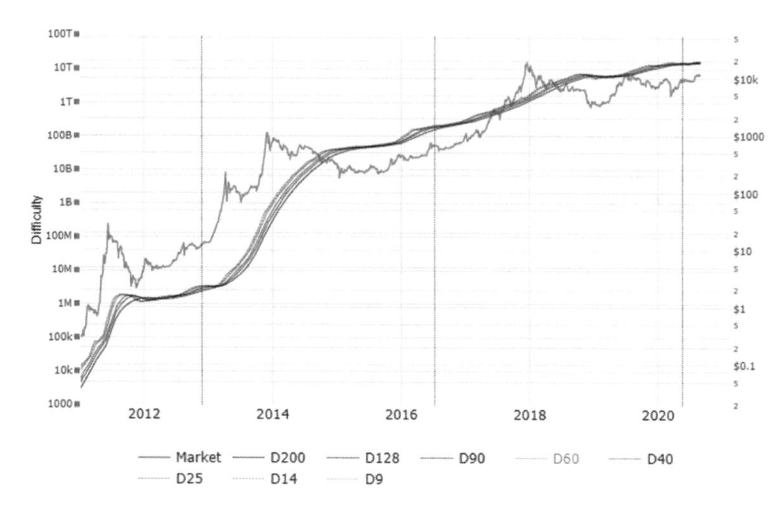

Fig. 2.2 Comparison between difficulty and price of bitcoin

study "*a financial bubble shows up where repeated extremely high-value increases are not accompanied by any corresponding increase in the number of participating users*" [7]. This means that once the price value chart is outraged beyond a certain peak point, eventually causing the graph to move in the opposite direction until the value catches up again. Therefore, Metcalfe's law towards the transaction numbers explicitly has long been suggested. This law quantified 94% movement of the bitcoin prices over the last 4 years and proved that Bitcoin still has a lot to grow (Fig. 2.3).

Bitcoin's network architecture is much more compared to that of a topology choice. Bitcoin is a peer-to-peer digital cash system in terms of design and the network architecture is both an indication and a foundation of the core characteristic (Fig. 2.4).

Bitcoin's peer-to-peer network architecture works above the IP protocol that connects the internet. In general, peer-to-peer (P2P) means a flat network with no single authority controller. The bitcoin P2P ensures that the endow of the currency

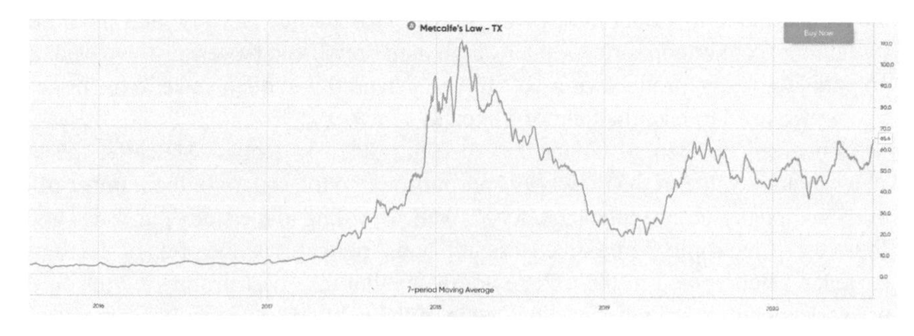

Fig. 2.3 Bitcoin network

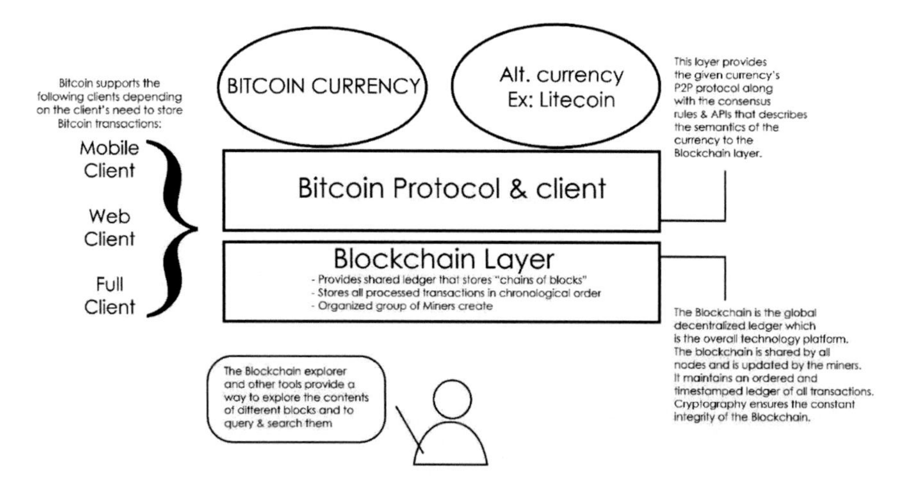

Fig. 2.4 Bitcoin architecture

is managed by no single authority and the economic process of the bitcoin system's currency supply is dispensed uniformly by specialized nodes known as miners who not only produce currency but also makes sure that the network is secured.

The technical architecture of bitcoin is described here which consists of 3 important layers.

1. Blockchain Platform
2. Bitcoin protocol layer
3. Bitcoin client application layer

Let's look at each of the layers a little bit in detail.

2.5 Blockchain Platform

Blockchain, which is a decentralized database that performs as a highly transparent ledger maintains all the records of the transactions carried out by bitcoins. Blockchain consists of millions of networking nodes that participate in the bitcoin network. Each node in the network runs a section of blockchain and makes sure that the transactions carried out by the digital currency are unique across the system (*see* ***Exhibit*** III). The process by which the transactions are confirmed and proceeded for their inclusion in the blockchain is known as bitcoin mining. Mining creates an ideology such that the systems that process the transaction are not only faster but more efficient. The "miners" known as the grids of the servers, create bitcoins by continuously solving the mathematical problems while parallelly processing the transactions. The miners are designed in such a way that for every 10 min a miner is available to validate the transactions and is rewarded with bitcoins.

2.6 Bitcoin Protocol Layer

The two main functions provided by the Bitcoin protocol layer are:

- API's: That acts as a mediator to the underlying blockchain platform.
- Algorithms: That drives and controls the miners across the network.

The Bitcoin API layer uncovers programmatic interfaces and APIs to all the numerous places of interactions with the bitcoin system especially with the blockchain network and the translation services provided by bitcoin.

2.7 Bitcoin Client Application Layer

The topmost layer of the bitcoin consists of end-user software applications that can be installed on a full server or a smartphone or can simply be used over an internet browser. The client–server helps in generating security features like private keys and also provides wallet functionality to send bitcoins using the private key provided to the client. With the help of APIs provided by the bitcoin client, wallets can calculate the balance and the blockchain network can verify the transactions for account limits.

The second most popular cryptocurrency in the market after bitcoin (BTC) is none other than Ether (ETH), the virtual currency of the Ethereum [8, 9]. Launched in July 2015, Ethereum is an open-source blockchain-based platform for its cryptocurrency, ether. It facilitates the trading of cryptocurrencies securely without any involvement of a third party. Even though bitcoin continues to be the most popular cryptocurrency, but its Ethereum's dynamic growth has many guessing that it might overtake Bitcoin in usage very soon.

Both Ether and Bitcoin are similar in many ways like, both of them are virtual currencies that can be traded over the online exchanges and are stored in various cryptocurrency wallets. Both of them are decentralized i.e. they are not issued or owned by central banks or any other authority and both accept the distributed ledger technology, known as Blockchain. Although, there are also many key differences between these two popular cryptocurrencies in the market field [10–13].

Bitcoin has developed itself as a comparatively stable and the most successful cryptocurrency to date, while Ethereum's main objective was to build a multipurpose platform with its cryptocurrency Ether being a part of its smart contract applications. Even when differentiating them in terms of cryptocurrency accept, Ethereum and bitcoin projects appear to be enormously different. For example, the supply of bitcoins is fixed to 21 million meaning that at max only 21 million bitcoins can be supplied. Whereas the supply of Ether can be practically endless. Besides the average block mining time of bitcoin is 10 min' while Ethereum aims to be no more than 12 s which is much quicker compared to bitcoin. Both ETH and BTC are cryptocurrencies but the main purpose of ether was to smoothen and construct the operation of smart contracts and DApps platform of the Ethereum rather than to exhibit itself as an alternative monetary system [14–17].

Moving on to Litecoin [18] (LTC), created by an MIT graduate and former Google engineer Charlie Lee, is an alternative peer-to-peer cryptocurrency based on the model of bitcoin. LTC is fully decentralized and established on an open-source global payment network that does not require any central authority to govern it. The purpose of developing Litecoin is to improve bitcoin's limitations which have gained industry support along with high trade volume and liquidity in recent years.

Both Bitcoin and Litecoin have a lot in common [19–21]. They both are decentralized cryptocurrencies and unlike fiat currencies that depend on the central bank for value, cryptocurrencies depend only on their cryptographic integrity of the network itself. Both currencies can be bought through exchange or can be mined. Further, both require cold storage or something like a digital wallet where the transactions between them can be safely stored.

When it comes to market capitalization, Bitcoin, and Litecoin differ significantly. As per the reports, the total value of all bitcoins in the distribution in May 2020 was near to $128 billion, which was 45 times larger than Litecoin whose value stood close to around $3 billion. Given the fact that bitcoin is much larger than all the other cryptocurrencies, it possesses a higher value than Litecoin is not a surprise. Another major difference is the total number of coins each cryptocurrency produces. Bitcoin cannot exceed more than 21 million coins which is 4 times less compared to Litecoin which can produce up to 84 million coins. In theory, this might give a slight advantage to Litecoin over Bitcoin but in the real world, this turns out to be negligible because both Bitcoin and Litecoin are divided into very small amounts. In mid-November 2013, IBM executive Richard Brown stated that some users chose to transact the coins at once rather than infractions unit, a possible advantage for Litecoin. Still, assuming this to be true, we can solve the problem through some simple software changes which are introduced in the digital wallets through which bitcoin transactions are carried out. Popular bitcoin wallets such as Coinbase and Trezor provide an option to view the Bitcoin values in terms of fiat currencies which help in overcoming the psychological avoidance of dealing with fractions.

Bitcoin and Litecoin maybe the gold and silver of the cryptocurrency space today, but history says it all that the status quo of these currencies can change anytime. All we can do is just wait and watch whether the cryptocurrencies which we are familiar with will retain their satire or not (see Exhibit II).

2.8 Results

"Kaspersky Cryptocurrency", a Moscow-based cryptocurrency firm revealed that 19% of the total world's population have bought cryptocurrencies in the year 2019 which is interesting but at the same time a study reveals that more than 81% of the world's population never brought cryptocurrencies. Only 10% of people assert they "completely understand what cryptocurrency is."

In 2017, at the start of the year, bitcoin was priced somewhere around $1000. It was the highest price of Bitcoin measured till then. That's not all, it continued to rise and at one stage it took everyone by surprise by reaching a value of $20,000 in the same year. Well, due to its unnatural behavior, it suffered a huge drop of 30% decline at $11,000. By the year-end, it marked $15,000.

In 2019, there were $153 million bitcoin active addresses which were open and out of which 147 million of them contained less than 1000 coins. Current statistics revealed that this number might have increased to about 30–40%.

According to cryptocurrency statistics, Coinbase, a digital currency exchange platform has about 12 million users. In 2019, one of the top global market companies released new features related to bitcoin in Q4 (October–December). With this launch, Coinbase added 1,00,000 users in just 24 h.

Bitcoin distribution is not limitless. Its distribution stops once every 21 million units are mined. Having said that, in August 2010, a major event occurred that shook the whole world. An unknown hacker almost destroyed bitcoin which "accidentally" led to the creation of exactly 184,467,440,737.09551616 Bitcoins. Before the situation could slip out of hand, the error was addressed and the limit was fixed and brought back to its original 21 million.

According to Coinmarketcap, the current market cap of bitcoin sits at $210.89 billion. Though there are many restrictions concerning the usage of Bitcoin, many large companies are accepting Bitcoin as a legitimate source of funds (see **Exhibit IV**) (Fig. 2.5).

Throughout the Bitcoin reign, there were many instances when Bitcoin crashed. Some important instances were:

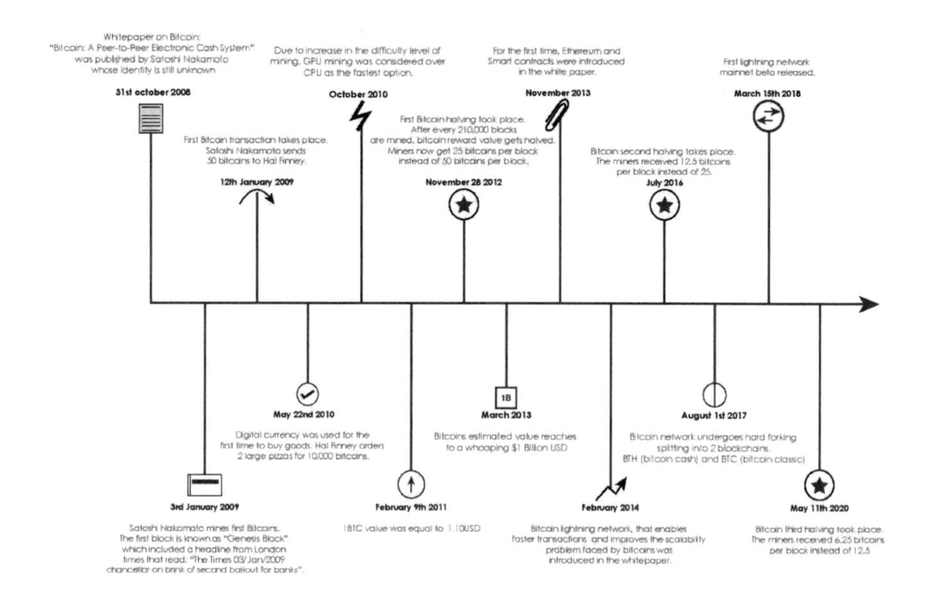

Fig. 2.5 Milestones of bitcoin (2008–2020)

- 2011, June to November—93%
  On the 11th of June in 2011, a trading firm called Mt.Gox [22] in Shibuya, Tokyo was responsible for the fall of the bitcoin price by 68% and the price of a single bitcoin came down to $0.01. Mt.Gox was responsible for over 70% of all the bitcoin trading in the world. This trading firm was hacked by using stolen credentials, hundreds of thousands of dollars were stolen from the Mt.Gox crypto wallet.
- 2012, August—57%
  A Ponzi scheme by a firm called bitcoin savings and trust, Caused the price of a single bitcoin to fall by 57%. The man behind the Ponzi scheme, namely Trendon Shavers used bitcoin as an investment for high returns at a later time. As he collected bitcoin he gave the older investors the bitcoin he collected from newer investors. Using the Ponzi scheme he had acquired about 146,000 bitcoins.
- 2013, April—87%
  After the initial bitcoin crash in 2011 due to Mt.Gox's carelessness, the bitcoin trading had taken a halt. As it started gaining popularity again in 2013. Due to the increased amount of trading, it was too much pressure on Mt.Gox, and yet again on April 10th, they were responsible for a 52% drop in the price of bitcoin in the time of 6 hours. However, the only reason for the decline of bitcoin wasn't that alone. A DDoS attack was launched against Mt.Gox, As the trading firm was handling over 70% of the bitcoin transactions in the world. It caused a widespread fall in the usage of bitcoin.
- 2013, December-2015, January—85%
  On the 28th of February in 2014 Mt.Gox announced that they lost about 850,000 bitcoins, this caused the firm Mt.Gox to file for bankruptcy. This was due to Mt.Gox's responsibility again. The amount of the bitcoin as per conversion rates at that time was about 450 million dollars.

One of the main reasons for the value of bitcoin to be at its lowest in its history is Mt.Gox. A list of all the incidents is listed below.

- Liberty Reserve withdrawal exploit: This occurred on the 20th of January 2011. The loss incurred in this incident is about 50,00 USD.
- 1st Wallet theft: This occurred on the 1st of March 2011. This incident was caused by the hackers obtaining a file known as the wallet.dat file. The hackers sole about 80,000 bitcoins.
- 2nd Wallet theft: This occurred on the 22nd of May 2011. An unknown person is believed to have accessed a wallet which contained 300,000 bitcoins. This wallet was stored on a public drive in an unencrypted manner. The person decided to return most of the stolen money. He only kept 1% of the total amount of bitcoins moved, this is about 3000 bitcoins.
- Major crash: This event is responsible for the biggest drop in the price of bitcoin. A hacker who gained access to Jed Mcalebs account sold many bitcoins for the price of the bitcoin to crash. This event occurred in June of 2011.
- Bitomat: Mt.Gox took over the debt of a company called bitomat after they deleted all their private keys. This event gave a loss of 17,000 bitcoins to Mt.Gox.

- Database hack: a hacker who gained permission to overwrite the database, increased the balances in the accounts, and withdrew the funds. The loss caused by this event is 77,500 bitcoins.
- 3rd Wallet theft: A hacker gained access to the wallet.dat file again and stole 603,000 bitcoins, this occurred in October of 2011.
- Incorrect deposits: The changes in the account balances from the 3rd wallet theft were updated in a wrong manner. The total error in the new balances resulted in about 44,300 bitcoins. Some of these errors were corrected at a later point in time. This error caused a loss of 30,000 bitcoins. This event occurred in October of 2011.
- Destroyed bitcoin: A bug in the software involved in the transfer of bitcoin. This bug caused a transaction in such a way that it was not redeemable. This resulted in a loss of 2,600 bitcoins. This event occurred on the 28th of October 2011.
- Law enforcement seizures: In May 2013 until August 2013 the US government sized an amount of 5,000,000 USD from the Dwolla account of Mt.Gox. This was done as the exchanges made by this account were not compliant with the regulations.
- Coinlab dispute: The firm Coinlab sued Mt.Gox for a dispute in a license agreement. The amount sued for was 5,000,000 USD. This event occurred in May 2013.
- Willy Bot: A trading program made to earn some bitcoins for all the above losses was implemented by Mt.Gox, this program didn't earn anything for the company, but it ended up incurring even more losses for the company. The total loss caused by this program was 51,600,000 USD and 22,800 bitcoins. This happened for two years from 2011 to 2013.

2.9 Challenges

The value of bitcoin was really low, almost considered next to nothing until July of 2010 where the spike in the value was from $0.0008 to $0.08. From then on, this cryptocurrency has gone up to the value of $20,000 at its peak during late 2017. Currently, the value of the bitcoin varies by the minute, its approximate value is $9178 [23].

Various reasons for the vast variety of the bitcoin price is as follows:

- **Regulations governing its sale**:
  If any major entity or country has initiated the usage of cryptocurrencies, this will increase the usage of multiple cryptocurrencies which will consequently increase the price of bitcoin. As major investors, use bitcoin as an investment, the usage of bitcoin will increase rapidly with the grant of usage.
- **Cost of producing or mining for bitcoin**:
  With the swift advancement of technology, computing speeds increase by multiple folds. The cost of these technologies increases at the same rate as the demand increases. Since the processing speeds of the mining system highly matters in the

mining of bitcoin, the price of bitcoin is affected by the expenses involved in the mining of bitcoin. A major cost incurred by miners is the electricity charges, the bitcoins are procured by solving a complicated mathematics problem. For this the computing speed of the miner's system mattes.
- **The number of competing cryptocurrencies**:
The price of any crypto currency is decided by the number or the rate of transactions taking place. Other than bitcoin if any other cryptocurrency has gained popularity, this will cause the price of bitcoin to fall or decline. Furthermore, with multiple ICOs (initial coin offering) constantly entering the crypto-market, as there are not many barriers for creating a cryptocurrency, there are hundreds of cryptocurrencies other than bitcoin. The reason for bitcoin staying in the market for so long is that it has high visibility over other cryptocurrencies.
- **The supply and demand for bitcoin**:
The number of bitcoins that ever be made or mined is 21 million, with the limit of bitcoin capped at that value after all the bitcoins are mined there will be no entry of new bitcoin into the market, this will cause the variation of the price of bitcoin. As miners will not be able to trade bitcoin this will cause the transactions of bitcoin to fall and consequently decrease the price of bitcoin.
- **Forking and halving of blocks**:
Since the halving of rewards takes place one every 210,000 blocks are mined, the incentive for the miner to mine for bitcoin reduces, this causes all in the mining in bitcoin which reduces the rate of introduction of bitcoin into the market. This will, in turn, decrease the trading of bitcoin which will cause the fall of the price of bitcoin.
- **Availability of bitcoin on currency exchanges**:
Bitcoin can be used as an investment just as NIFTY, SENSEX, and similar shares. When bitcoin is available on multiple currency exchanges, it opens up the possibility of trading with bitcoin to multiple investors.

Now that we have understood the reasons behind the variation in the price of Bitcoin, let's have a look at the reasons behind the volatility [24] of the Bitcoin price. The reasons for bitcoin being volatile are as follows:

- **Uncertainty of future value**:
The Price of each bitcoin could highly vary in a short time, this is one reason for only a small number of investors are in bitcoin trading.
- **High inflation affects bitcoin**:
The use of bitcoin as a daily usage currency will increase the activity in the bitcoin transactions, which results in increased inflation. This results in an offset of other currencies. This increases the volatility of bitcoin.
- **High profile losses**:
The wide variation in the price of bitcoin is the news event that scares investors which may cause them to suddenly buy or sell a large amount of bitcoin at any given time. This happens due to the fact there is high anonymity and there was a major loss for some of the original investors for bitcoin, this always decreases

the trade for the currency and in turn, the value of the currency goes down. Throughout the world, multiple scams and frauds are being pulled off using various cryptocurrencies, this is one of the reasons for not using bitcoin.

- **Tax treatment on bitcoin**:
  As bitcoin is considered an asset for taxable purposes. As the transactions of bitcoin have to be recorded by each party and the transactions have to be submitted to the Internal Revenue Services. This can be tedious for the sending and the receiving party. This will cause fewer people to use bitcoin. As any firm or country has to allow trading in any cryptocurrency, it has to set a strong set of rules and regulations for the trading, this will have a big toll on all the bitcoin traders. This may even cause the adoption rate of bitcoin to go to nil.
- **Security vulnerabilities**
  One of the ways the bitcoin community solves problems [25] is by presenting the problem to the public and getting a solution when each problem is presented for mass solutions. The investors in bitcoin get scared of investing in bitcoin and this changes the market on a big scale.

Although Bitcoin is not the only currency out there in the crypto-market, this digital coin offers some advantages that make it more unique than others. But like every coin has two sides, Bitcoin also has some disadvantages. Let us first look at the advantages:

> Generally, transactional information is subject to malware attack and may result in losing conventional financial or personal data of the sender or the person who is receiving the coins. But Bitcoin users are risk-free at the time of the transaction as any kind of data breach requires the hacker to know the private key of the users which is next to impossible. Another advantage of bitcoin is its low transaction fees. At present within bitcoin payments, no commissions are charged. If the transaction takes place faster, the user can include the fees at the time of the transaction process. Higher the fess, the higher the priority inside the network [26, 27].
>
> One of the most important advantages of bitcoin is that the information is transparent. All the completed transactions are visible to everyone inside the network, thanks to blockchain. The public address is visible, but it does not hint personal information of any kind, keeping the user completely anonymous. Since bitcoin is cryptographically safe, manipulating bitcoin protocol is far from possible by any organization, government, or individuals.

Well now that we have listed some of the advantages of bitcoin, let's see the drawbacks that raise concerns regarding the usage of bitcoins.

> One of the major disadvantages is the lack of information and understanding of bitcoin. Even today, many people don't know about the existence of bitcoin or other cryptocurrencies and the required transactional procedures. People should be informed about what bitcoin is and how it can be used.
>
> Bitcoins are risky and volatile. It is risky because bitcoin is still in its beginning stage with emerging functions, but still incomplete as bitcoin needs to grow up to reach its full potential. It is volatile because there's a limit on the number of bitcoins available in the market, resulting in increasing demand, day by day. It's difficult for people, especially the older generations, who are used to traditional classic money i.e. coins, notes, cards with their savings/earnings to trust and get used to the digital currencies since cryptocurrency is relatively new and it takes time for the people to understand what it is, it's working mechanism, etc.

The battle for the acceptance of bitcoin has been going on ever since the success stories about cryptocurrencies started surfing the internet. The problem with the cryptocurrencies is that they don't have a fixed procedural system, which is why many governments across the world are finding difficulty in legalizing these currencies, although a few governments have shown a positive attitude towards cryptocurrencies. Let us have a look at the countries where the use of cryptocurrencies is legalized and where they are strictly banned.

The past four years have visually perceived cryptocurrencies become ubiquitous, prompting more national and regional ascendant entities to grapple with their regulation. The expansive magnification of cryptocurrencies makes it possible to identify emerging patterns. Australia's ex-treasurer and current Prime minister Scott Morrison in 2016 released a statement saying that citizens are allowed to use digital currencies like bitcoin freely and that they no longer will be subject to GST. Similarly, the Canada Revenue Agency (CRA) has characterized digital currencies like bitcoin as a product and not a government-issued currency. Bitcoin is used for buying and selling goods or services that include trade transactions. As per the reports in 2013 US Treasury Department's Financial Crimes Enforcement Network claimed that purchasing goods and services using bitcoin is no longer treated as illegal and that bitcoin is a fully convertible decentralized virtual currency. On 11th of July 2014, the government issued a regulation note by stating that the country has legalized the operations of cryptocurrencies such as bitcoin, along with virtual currency exchanges, taxation and gave authority to those who were involved in trading and use of such currencies. Back in 2017, Estonia became the first jurisdictions in Europe to legalize crypto-related activities. Within 3 years, the number of licenses issued for companies to operate cryptocurrency has surpassed approximately 1400. According to Wikipedia 2017, South Korea made use of cryptocurrencies like bitcoin legal by not having any regulations but later in 2013 Bank of Korea suggested having regulations in the future for the use of bitcoin and other virtual currencies. In the year 2018, the Sweden government has legalized bitcoin as a mode of payment method though certain types of companies have a license to use it. In the same year, the Dutch court pronounced that bitcoin is a legitimate "transferable value". In the United Kingdom (UK), There's no specific cryptocurrency law but for exchanging purposes there are certain norms to follow which includes registration with the Financial Conduct Authority (FCA).

While Bitcoin is welcomed in many components of the world, a few countries are wary because of its volatility, decentralized nature, perceived threat to current monetary systems, and links to illicit activities like drug trafficking and money expurgating. El Banco Central de Bolivia, the central bank of Bolivia issued orders for banning bitcoin and other cryptocurrencies not regulated by a country or economic zone in 2014. Chinese regulatory authorities imposed a ban on virtual currencies and have termed it illegal in September 2017. All banks and other financial institutions are prohibited from trading or dealing in bitcoins. In the very same year, Nigeria banned all kinds of transactions in bitcoin and other cryptocurrencies. In 2018, the reserve bank of India (RBI) made a statement on banning the sale and purchase of cryptocurrencies for entities. The government doesn't recognize altcoins as legal tender. Iceland, the country which is known as one of the largest bitcoin mining facilities in the world but, it is prohibited to engage in any kind of foreign exchange trading with the cryptocurrency bitcoin. The Central Bank of Thailand issued a statement that the use of cryptocurrencies may include illegal activities such as money laundering or supporting terrorism because of which cryptocurrencies are not legal tender in Thailand. The National Assembly of Ecuador banned bitcoins and other altcoins due to the establishment of a new state-run electronic money system. The Central Bank of Thailand issued a statement that the use of cryptocurrencies may include illegal activities such as money laundering or supporting terrorism because of which cryptocurrencies are not legal tender in Thailand. The Finance minister of the Russian Federation stated that it's probably illegal to accept cryptocurrency payments. The state bank of Vietnam has declared the supply and use of bitcoins are illegal as a means of payment and have to pay a fine ranging from 140 to 200 million VND as punishment.

2.10 The Road Ahead

From the beginning stages of Bitcoin, the consensus algorithm that has been in use is the proof of work algorithm. Proof of work is an algorithm that allows trustless and distributed consensus. This algorithm requires participants to perform expensive computer calculations, also known as mining and these participants are known as miners. But recently, there have been many discussions that reveal the possibilities of using proof of stake in Bitcoin instead of proof of work as this algorithm has quite a few disadvantages like excessive energy usage. Proof of stake offers a virtual consensus mechanism. Although proof of work and proof of stake have a common goal, the method of reaching this goal is different for each algorithm.

With proof of work, the miners solve hard cryptographic puzzles using their computational resources. Whereas in proof of stake, there are validators instead of miners and these validators lock some of their cryptocurrency as a stake in the ecosystem. According to the founder of Swiss crypto broker Bitcoin Suisse, Niklas Nikolajsen's prediction, "Bitcoin (BTC) will move to Proof-of-Stake (PoS) once the Ethereum (ETH) network has proved the algorithm's success" [28]. Irrespective of the consensus mechanism that will be adopted in the future, many renowned personalities have made many predictions on the price of Bitcoin (see Exhibit V).

Although there has been a dramatic surge in the bitcoin acceptance rate worldwide, the rate of bitcoin adoption remains quite low. Governments all over the world need to take bold measures to regulate it and make it easier to buy/sell/hold Bitcoin. Till that happens, the crypto industry will continue to struggle amid all the uncertainty and will affect the rate of bitcoin adoption.

Exhibit I: Approximate Cost of BTC

Year	Jan–Jun	Jul–Dec
2010	1 BTC ~ 0.003 USD	1 BTC ~ 0.125 USD
2011	1 BTC ~ 1 USD	1 BTC ~ 2 USD
2012	1 BTC ~ 7.20 USD	1 BTC ~ 15 USD
2013	1 BTC ~ 45 USD	1 BTC ~ 800 USD
2014	1 BTC ~ 900 USD	1 BTC ~ 350 USD
2015	1 BTC ~ 250 USD	1 BTC ~ 450 USD
2016	1 BTC ~ 500 USD	1 BTC ~ 800 USD
2017	1 BTC ~ 1000 USD	1 BTC ~ 20,000 USD
2018	1 BTC ~ 7000 USD	1 BTC ~ 4000 USD
2019	1 BTC ~ 4000 USD	1 BTC ~ 8000 USD

* Above mentioned are only approximated values, actual values may vary within the range [29, 30]

Exhibit II: List of Some Well-Known Cryptocurrencies [31]

Release year	Currency	Founder
2009	Bitcoin (BTC)	Satoshi Nakamoto
2011	Litecoin (LTC)	Charlie Lee
2013	Ripple (XRP)	Chris Larsen and Jed McCaleb
2014	Monero (XMR)	Monero Core Team
2014	Neo (NEO)	Da Hongfei and Erik Zhang
2014	Stellar (XLM)	Jed McCaleb
2015	Nano (NANO)	Colin LeMahieu
2015	Ethereum (ETH)	Vitalik Buterin
2015	Tether (USDT)	Jan Ludovicus van der Velde
2016	ZCash (ZEC)	Zooko Wilcox
2017	Bitcoin Cash (BCH)	Fork of Bitcoin

Exhibit III: Blockchain Architecture Layers [32]

Layer	Function
Application layer	Composed of blockchain-based applications
Contract layer	Introduces programmability
Incentive layer	Is the driving force for the network
Consensus layer	Consists of different consensus algorithms
Network layer	Composed of various networking mechanisms
Data layer	Encapsulates the block-level information

Exhibit IV: Bitcoin Accepting Companies

Companies	Stores
Microsoft	Bitcoin.Travel
AT&T	Pembury Tavern
Burger King	Old Fitzroy
KFC	Zinga
Overstock	Helen's Pizza
ExpressVPN	Shopify.com
Subway	Amagi Metals

* Above mentioned are only a few examples [33]

Exhibit V: Bitcoin Price Predictions [34]

Personality	Predicted price
Anthony Pompliano is a well-known Bitcoin personality and he is a founder and partner at Morgan Creek Digital	**$100,000 by 2021**
Kay Van-Petersen is an analyst at Saxo Bank, a Danish investment bank that specializes in online trading and investment	**$100,000 by 2027**
Wences Casares is the Founder and CEO of Xapo, a Bitcoin wallet startup, and a board member at PayPal, among various other roles	**$1,000,000 by 2027**
Mike Novogratz is a former hedge fund manager who's been investing in Bitcoin and blockchain technology for a long time	**$7.5 trillion market cap by 2029**
Jeremy Liew is a partner at Lightspeed Venture Partners, famed as the first investor in the social media app Snapchat	**$500,000 by 2030**
Andy Edstrom is a wealth manager for a California-based investment advisory firm called WESCAP	**$8 trillion market cap by 2030**
Chamath Palihapitiya is the Founder of Social Capital and Co-Owner of the Golden State Warriors	**$1,000,000 by 2037**
Winklevoss twins—the famous Bitcoin billionaires have said Bitcoin has the potential to of thirty to forty times its current value	**$5 trillion market cap by 2028–2038**

2.11 Teaching Objectives and Target Audience

This case is designed to enable students to,

- Understand the foundations of Bitcoin and its significance in business.
- Understand the role of Bitcoin in reinventing the business models.
- Understand the factors that contribute to the success of Bitcoin.
- Understand the efficiency in Bitcoin markets.
- Understand the risks of using Bitcoin.
- Understand why business is accepting Bitcoin.
- To understand the legal status of Bitcoin.

This case is meant for MBA students as a part of their Information Technology/Management Information Systems/Business Analytics curriculum. It can also be used in the Strategic Management curriculum.

2.12 Immediate Issues

1. To protect themselves from Bitcoin scams, what steps should consumers take?
2. How is Bitcoin treated by RBI/Banks?
3. Is it possible for customers to make duplicate as it is a virtual currency?
4. Will the role of the traditional financial system decline due to the development of cryptocurrencies?
5. How would Bitcoin exchanges contribute to taxation and prohibit individuals from bypassing tax payments on transaction?
6. Is Bitcoin tangible like gold?
7. What are the risks involved?

2.13 Basic Issues

- Transaction Delays
- Regulation
- Volatility
- Tax issues
- Scalability.

2.14 Teaching Approach and Strategy

In the classroom mode, the moderator can initiate the discussion by giving brief introduction distributed ledger technologies and cryptocurrencies. The briefing can be extended by illustrating how they influence the global economy. The discussion can further include the working model of Bitcoins and its competition with other cryptocurrencies. This can be followed with the factors that contribute to the growth of cryptocurrencies and how the role of Technology, Economic and Regulatory factors contribute majorly to the wide acceptance of cryptocurrencies. Besides this, the students can discuss the Legal and social issues revolving around Bitcoin and can explore the opportunities and threats Bitcoin brings in. The discussion can be further extended through the following questions by moderator.

1. *Critically analyze the benefits of Bitcoin and discuss how different are Bitcoin transactions from Credit Card transactions.*
2. *How do you safeguard your Bitcoin from theft? Make few suggestions.*
3. *Is Bitcoin a medium of Exchange or a digital asset? Can it become a digital gold?*
4. *Can Bitcoin be the future of money? Discuss how Bitcoin is changing the e-commerce business.*

2.15 Suggested Session Plan

	Discussion pastures	Time (min)
Introduction		5
Discussion on Q1	*Critically analyze the benefits of Bitcoin and discuss how different are Bitcoin transactions from Credit Card transactions*	20
Discussion on Q2	*How do you safeguard your Bitcoin from theft? Make few suggestions*	20
Discussion on Q3	*Is Bitcoin a medium of Exchange or a digital asset? Can it become a digital gold?*	20
Discussion on Q4	*Can Bitcoin be the future of money? Discuss how Bitcoin is changing the e-commerce business*	20
Summary		5
Total		90

* *Distribution of case 2–3 days before the class*

2.16 Analysis

Bitcoin offers several advantages but the probable benefits can be, (i) low transactional costs: there are no intermediaries involved unlike traditional payment systems where a trusted third party charges the customer a significant fee to authenticate the digital transaction. Furthermore, sales of Bitcoin are irreversible, which eliminates the Likelihood of exploitation of customer pay-backs, which are costly to vendors. (ii) Increased Privacy: Customers who are looking for high end privacy for their financial and transactions are happy using Bitcoins Also, the probability of identity theft could be lower, and some might consider it desirable to exclude the government from the economic model. (iii) Inflation: No scraping of purchasing price: The mining activity lets the supply of bitcoin to grow at a consistent manner after which a standard amount is capped. If the demand for bitcoin decreases with respect to the fixed supply then inflation may occur. The supply of bitcoin may be increases in case of fractional reserve banking where the bitcoin network keeps a fraction of deposit in reserves and lends the remaining. If the held reserves get stabilises then the inflation concept might be eroded in Bitcoin.

1. *Critically analyze the benefits of Bitcoin and discuss how different are Bitcoin transactions from Credit Card transactions.*

Back in 2008, during the eve of Halloween, a revolutionary network model was revealed by a mysterious group or individual named Satoshi Nakamoto. Bitcoin is an electronic peer-to-peer cash system with no third party involved. It is one of the most popular cryptocurrencies backed by distributed ledger technology and modern cryptographic techniques to generate and regulate the currency units. The virtual currency Bitcoin is self-reliant that it does not need any financial institution to store or transfer the money. This digital currency can be used to buy any goods or avail any online services. Bitcoin transactions happens between Bitcoin wallets. Bitcoin wallets can be installed on your mobile phones or computer and can be identified using a unique Bitcoin address. For every transaction a new bitcoin address will be generated.

How bitcoin works?

A transaction is made whenever a value gets transferred between Bitcoin wallets and is included in the public ledger blockchain. To ensure secure transactions it has to be signed using the private key of the sender and will be broadcasted to the network. The transaction is considered to be valid when maximum participants validate and approves it. This process is called as Mining. This mining process prevents any individual from adding the block directly into the blockchain. The first miner to validate the request and add in the blockchain will receive rewards. (*see* ***TN Exhibit*** I).

Now that we know how the decentralized, peer-to-peer payment system works. Now, lets see the potential benefits of the system to the users.

A. *User Independence*

One of the primary factors that drew the attention of public is user independence. Users can handle the transactions independently and successfully without the help of trusted third parties like banks or any other financial institution.

B. *Pseudonymous*

Unlike real currency transactions where user identity is attached to every transaction, Bitcoin transactions are anonymous in nature. For every Bitcoin transaction a unique

Bitcoin address is generated thereby hiding the identity of the user. However, Bitcoin transactions are not completely anonymous, it is possible to track down the user based on the address used. Thus, it is considered as pseudonymous.

III. *Peer-to-Peer*

Bitcoin is a P2P network system which means any one can connect to any other person in the globe if only they are part of the network. And they do not require any consent from any external party.

IV. *Less Transaction and No maintenance Fee*

In traditional banking systems, a maintenance fee or minimum balance fee has to be paid to the central party but no such cost involved in Bitcoin system. For international transactions the exchange rates are high in fiat currencies but it is very less in bitcoin system as no government or external party involved.

E. *Accessibility*

Bitcoin payments can be made from any place provided internet access and completely avoids traveling to banks to make the transaction.

With so much discussed about Bitcoin and its benefits let's see how different are they from credit card transactions. Credit card transactions follow a centralized system and obviously provides several benefits like incentives, possibility to lend money, cash backs, less fraudulence but has its own disadvantages (*see* ***TN Exhibit*** II).

F. *Zero-Counterfeit*

It is not possible to counterfeit Bitcoins as every transaction will be recorded in the distributed ledger. Every node in the network has the copy of the transactions and it is impossible to modify anything.

G. *Inflation*

There's a limited supply of Bitcoin. There are around 18 million in circulation and the total supply is around 21 million. An increased demand for a limited supply item obviously will increase the unit price of Bitcoin.

TN Exhibit I: How Bitcoin System Works?

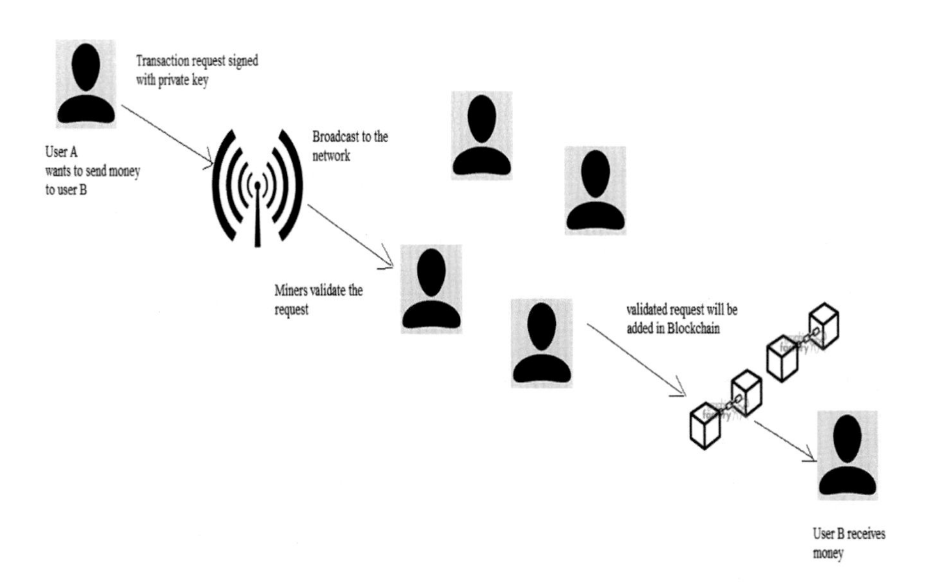

TN Exhibit II: Bitcoin Transaction Versus Credit Card Transactions

Bitcoin transaction	Credit card transaction
Direct transaction between the users without any intermediary	There is a central authority to do the transaction
Transactions are validated by nodes in the network	Validated by authorized centralized system
No personal information to be shared	Need to provide details like name, address, place etc.
Transactions done with alpha numeric codes (using cryptography)	No cryptographic techniques used
E-wallets used for transactions	Tangible credits are placed in physical wallets
Transactions cannot be reversed	Can be reversed
Not so widely accessed	Wider acceptance
High possibility for fraudulence	Comparatively less fraudulence
	High foreign transactional fee
Transaction cost is less than 1%	Costs associated transactions are 3–5%
No chargebacks or disputes	Disputes and chargebacks are high
Impossible to hack	Can be easily hacked
Accessed across the globe	Restrictions in few parts of the globe

2. *How do you safeguard your Bitcoin from theft? Make few suggestions. Analyze and discuss whether Bitcoin is a challenge or an opportunity*

With the increased demand and limited supply of Bitcoin, the possibility of cyberattacks on Bitcoin wallets are high. Based on its built, i.e. the underlying technology Blockchain provides a tamper proof mechanism which assures nearly impossible to tamper the data. Every user's wallet has their unique and secret private key. If the private key is lost, the user no more has control over his account. Wallets are also used to store Bitcoins. But there are no government regulations for Bitcoin and usually no insurance guaranteed. So, the Bitcoins are also to be saved and protected like fiat currencies. As the prices of Bitcoins are shooting up the number of attacks to hack the Bitcoin wallets are becoming high too. Some of the suggestions to ways to secure Bitcoins are,

- Bootable USB can be used to installBitcoin wallet. This assures a virus free OS and no log or cache can save the keys anywhere.
- It is advisable to use two wallets an offline one with more balance and an online with small amount of coins.
- Periodic backup of wallets will prevent from system failures or theft.
- Use a unique and strong password to protect it.
- Not advisable to store the bitcoins online.
- Use the latest version of bitcoin software.
- Can use paper wallet to store the coins physically and away from internet.
- Never share the private key to anyone.

Opportunities ahead

Secured Technology

The technology backing bitcoin system is considered to be more secure and it completely eliminates the huge double spending problem. The system provides complete transparency at the same time privacy of individuals. It is not possible to hack the data in Bitcoin system as it requires a large amount of stake in the network. It enables secure transactions with low transactional fees. There's no control with the single entity as the system is distributed in nature.

Costs

Bitcoin and other cryptocurrency transactions are considered to be low as compared with the fiat currencies. Besides cost, it can be operated round the clock across the globe with just having their internet.

More Returns

Due to the limited supply of Bitcoins, the demand for the cryptocurrency creates a huge return for the investors. Studies show that, prices of bitcoin is growing in par with the search queries on google trends. This creates more awareness among the public about its existence and it could benefit the users.

Even though Bitcoin offer lot of benefits it still has lot of challenges to overcome. The major challenges are summed below (*see* **TN Exhibit** III).

Robustness

Dealing with Bitcoins sometimes causes temporary shutdown due to high traffic. This shutdown causes hardware failure which may lead to lose one's hardware wallet. Because of this reason many people keep the coins in multiple places across the network. This pushes the strong need to keep a check on the robustness of the system.

Volatility

Since its origin Bitcoin has always been viewed as a part of scandalous activities. The infamous Silkroad made the currency to be highly suspicious and believed to be involved in illegitimate transactions Bitcoin's extreme inflation poses significant questions about its stability as a currency. Unlike the fiat currencies bitcoin is used by a small community and any small mischievous act also will affect Bitcoin value.

Lack of Trust

Though bitcoin started getting acceptance from mainstream users and businesses for the past couple of years it still is not accepted by many countries and its government. The reason being not having a centralized authority to monitor the bitcoin transaction and exchanges. This makes people think there's lack of security in bitcoin systems and is the reason for several cyber-attacks. Regulating the Bitcoin industry without any regulations or code of conduct will result in an unstable financial system.

Legal Issues

Bitcoin is a digital currency and is not regulated by any financial institution or government but people are using it as a monetary exchange. The users of bitcoin as of now receives no protection as the government is finding it difficult to regulate the use of bitcoins. Due to this reason several black markets and money laundering activities are popping up. Bitcoin offers plenty of opportunities for people to misuse it like theft, black market, money laundering etc. Especially the regions where internet usage is high the probability of such illegal activities. Since bitcoin claims to keep the identity of the user anonymous many people fall prey for scammers.

Even though we call bitcoin as a currency, it has to regulated i.e. it has to issued, accepted and used by a nation which is a major challenge for Bitcoin. While some country already have approved and regulated the use of bitcoin some have named the currency as illegal.

Less Adoption

In spite having several benefits the velocity of users wanting to use bitcoin system is very less. This could be due to one or all of the issues mentioned above. Some of the high-profile scams in Initial Coin Offering (ICO) markets are quite scary for even people who are interested in the system. At the end of the day all the users want their money to be safe and secure in bank. At the same time, users are expected to have some knowledge on bitcoin system which is not the case in traditional banking systems.

Technology still not matured

Besides the legal and regulatory changes, the underlying technology is still not matured enough and needs to be explored more to reap much benefits. A few technological issues are summed below.

Interoperability

One of the major challenges faced by Blockchain is its inability to exchange and apply information. The platform for the Internet devoted to Blockchain and crypto exchange needs to be made interoperable. By then, as long as there are people to abuse it through illegal and incorrect means, this technology is a danger to the economic structure, opening the doors to the adoption of virtual currencies.

Usability

It is not an easy task to buy and sell bitcoins in the crypto market. Since its origin it is still tedious to use. The procedures involved in preserving the data are complex which limits the system easy to adopt. The bitcoin system needs lot of user friendliness to buy, sell and use the cryptocurrency.

Scalability

Scalability is one major concerns since beginning. The bitcoin system is considered to be slow and faces transactional delays.

Conversion

When it comes to bitcoin to fiat currency conversion not many merchants come forward to do. Still payment method in local currency or dollars is preferred when compared with Bitcoins.

TN Exhibit III: Bitcoin: A Challenge or an Opportunity

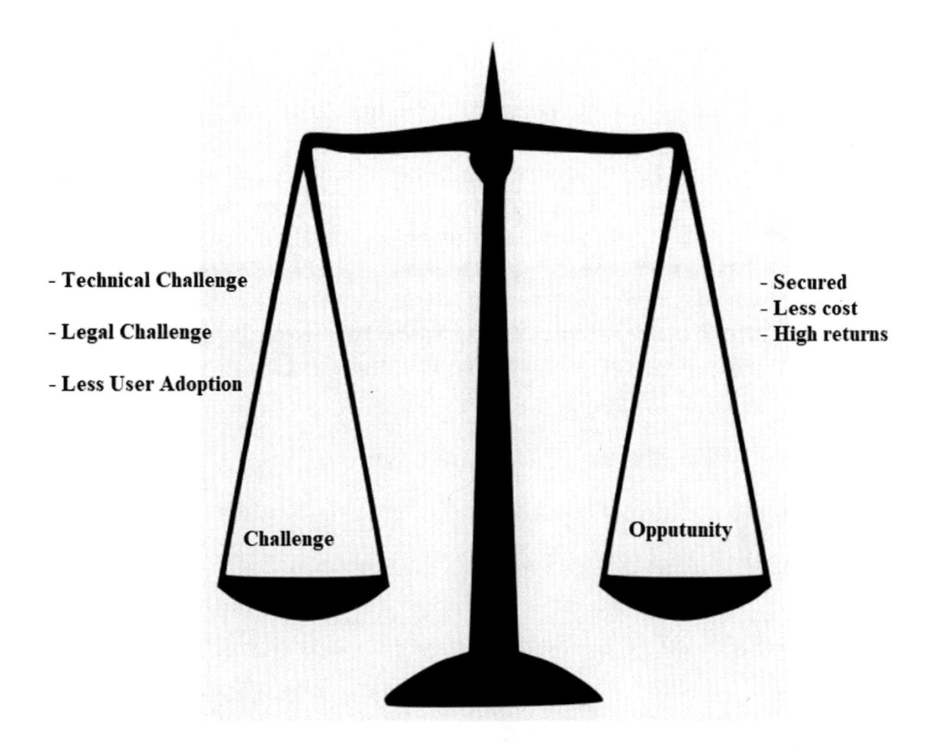

1. *Is Bitcoin a medium of Exchange or a digital asset? Can it become a digital gold?*

In recent years, cryptocurrencies are considered as digital assets and are quite popular [35]. Bitcoin and other related cryptocurrencies are presumed to do more than just termed as currencies. Bitcoin can be understood as a combination of commodity and fiat currency with no intrinsic value and not dependent on any regulating agencies. It is important to analyze both how assets can be used as replacements for currencies and to consider the distinguishing characteristic of a fiat currency that separates it from other assets. It is known that resources such as precious metals (e.g. gold, silver etc.) may be used to accomplish the same aims as those of currencies. Until very recently, it was traditional to fix currency values to gold to improve confidence in them and thereby preserve their value. However, precious metals are not the only substitutes for currencies. In fact, alternatives for paper currencies occur mostly because individuals do not trust a currency as a store of value, or if the physical representations (bank notes, coins) used to describe them are not generally available. Fiat currencies are regulated value-exchange receipts with all the inferred capacity for government interference implied by that definition. The main differentiator between currencies and assets is the dynamic aspect of a currency and the ability to control its worth. Simply put: currencies are a matter of opinion; assets are empirical matters. Bitcoin and digital assets backed by cryptography may not have the same qualities as fiat money since the number in existence cannot be directly modified, but they are similar to assets or resources. 'Digital assets' is a fitting term, as they are purely digital [36].

(*See* ***TN Exhibit*** IV) shows price of Bitcoin from 2010–2015 in US dollars and latest price movement till 2017. The price has increased so high from the beginning $5.28 to $2500 until 2017. The prices also have fallen down largely over certain time period from a huge value. Bitcoin is even referred with gold due its price growth, demand in supply and non-dependence on bank or any central authority. Because Bitcoin is separated from the existing fiat monetary system, despite it is volatile nature compared with other currencies it is still considered to be the safe place against any economic chaos. It is considered as safe because Bitcoin is not playing any significant role in any financial system and can therefore be unrelated to extreme negative stock returns.

TN Exhibit IV: Bitcoin Price (2010–17 in USD) [37]

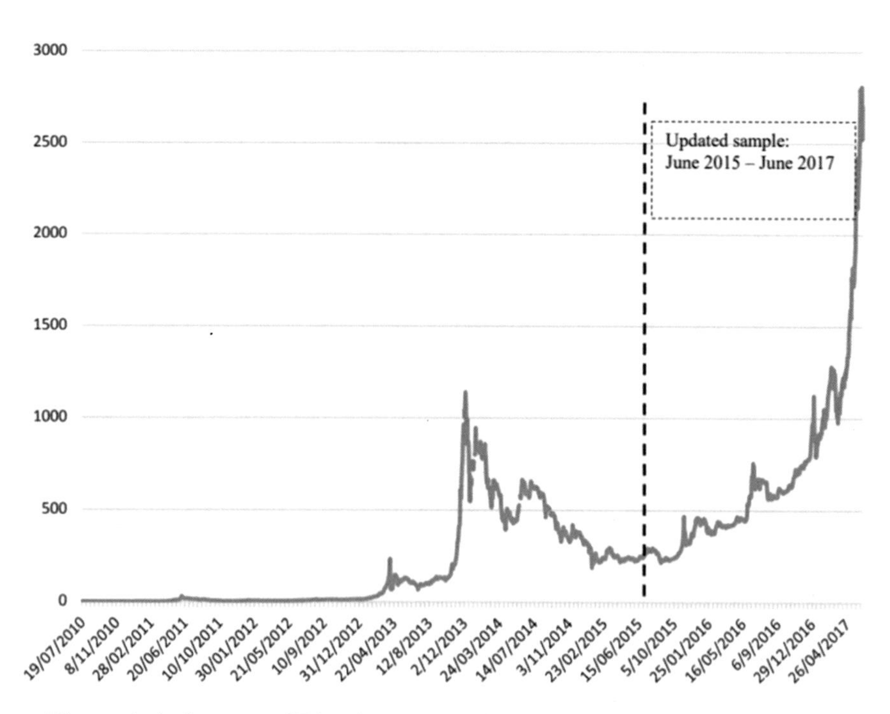

The statistical report of Bitcoin returns with other assets are tabulated (*see* ***TN Exhibit*** V). Based on the data Bitcoin shows higher returns compared with other returns. At the same time Bitcoin is showing highest volatility compared with other assets. It is also found that the monthly average volatility of Bitcoin is greater than gold and other similar assets. Similarly, high kurtosis and negative skewness is also exhibited by bitcoin. This skewness can be similar to the high yield silver, gold or corporate bond's skewness. This skewness also indicates that bitcoin exhibits an asymmetric return distribution. The really high kurtosis of bitcoin shows bitcoin returns are associated with large number of tail events. With all the ups of stock markets and pullback of Bitcoins, early investment on bitcoin was strategically a brilliant move (*see **TN Exhibit*** VI). The returns of Bitcoin outperformed even the global giants of the market.

TN Exhibit V: Statistics Made Based on Daily Data from 2010–2015 [38]

	bitr	*sp5r*	*sp6r*	*gldr*	*silvr*	*Eurr*	*audr*	*jpyr*	*gbpr*
Mean	0.65%	0.05%	0.06%	0.00%	-0.01%	-0.01%	-0.01%	0.03%	0.00%
Stdev	7.60%	0.95%	1.27%	1.09%	2.20%	0.60%	0.71%	0.58%	0.47%
Skewness	-1.01	-0.49	-0.24	-0.89	-0.89	-0.32	-0.19	0.38	-0.06
Kurtosis	17.04	8.25	7.64	10.85	12.94	4.79	4.85	8.22	3.63

	cnyr	*hufr*	*twus*	*wtir*	*hhr*	*cbr*	*tbr*	*hbr*
Mean	-0.01%	0.02%	0.01%	-0.03%	-0.04%	0.01%	0.01%	0.03%
Stdev	0.13%	0.93%	0.29%	1.23%	2.27%	0.05%	0.27%	0.18%
Skewness	0.05	0.17	0.29	-0.64	0.00	-0.31	-0.17	-1.92
Kurtosis	13.56	4.36	5.97	9.16	3.87	4.98	3.77	17.58

TN Exhibit VI: Bitcoin Versus Other Global Giants [39]

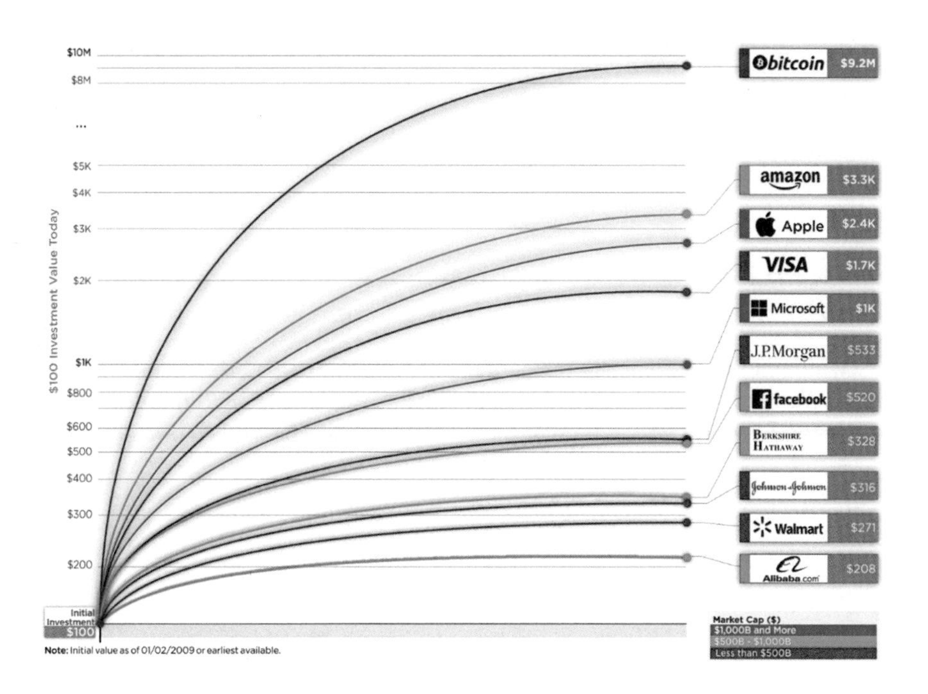

The prices in the graph was plotted using data from Yahoo finance. The cryptocurrency has still produced an incredible 9,150,088% return, even with Bitcoin's current price implosion. Other businesses are also amazing, such as Amazon (+3,156%) and Apple (+2,345%), but nothing really comparable to the world's crypto king [40].

TN Exhibit VII: Bitcoin Versus Gold

Asset type	Pros	Cons
Gold	Stable: Considered as a great investment on a longer run. The price goes high during recession and is always a stable investment	In case of financial crisis where the paper money is not viable gold cannot be used as a currency
	Longevity: The beauty, utility and scarcity of gold made a perfect foundation for trading and associated value. Since longer time it is being used as a currency across the world and has proven its value in every testing time	Even in cases where gold coins are considered as currency the worth of gold coins are much higher than fiat currencies
	Safe: It can be safely stored in a vault and can also be insured to certain amount	Has to be stored physically and should be extremely careful against thefts
Bitcoin	Easy: Bitcoin are digital and extremely easy to transfer it from one user account to another	Since it is digital, it may be subjected for theft and fraudulent activity
	Used as currency: You can pay through bitcoin at any place that accepts bitcoins	Based on demand and supply the value of bitcoin might get increased. If any other cryptocurrency exceeds the demand then there could be decrease in bitcoin value
	New: Since it is new it has the capability to have an increasing value in the future	Its new. At the same time highly volatile. It can give high returns and can crash heavily as well

1. *Can Bitcoin be the future of money? Discuss how Bitcoin is changing the e-commerce business.*

According to Deutsche bank research report "Imagine 2030", the current monetary system is fragile and the use of digital currencies will increase up to 200 million users by 2030. Though at present bitcoin acts an addition to the current monetary system but eventually it could even possibly replace the paper currency because of its unique features like decentralization, anonymity, etc. The bank also predicts that users of cryptocurrency can be increase up to four folds in the next ten years i.e. around 200 million. This is almost similar to what internet was in its first twenty years (*see **Exhibit*** VIII).

TN Exhibit VIII: Cryptocurrency Versus Internet Adoption Rate

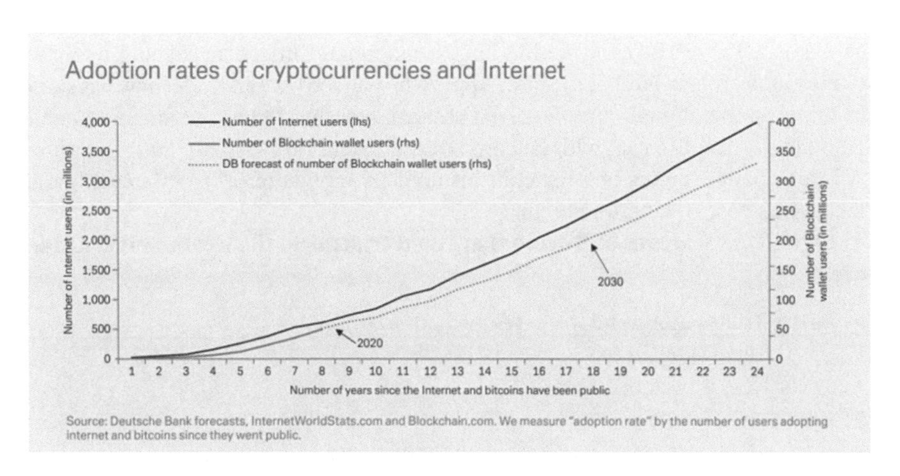

The process of regulating bitcoin is in progressing and once it outgrows the hurdles around it may be a legal substitute for fiat currencies. If cryptocurrencies become legitimate the actual victims are the cards and not the fiat money. If crypto acceptance increases the possibilities of credit and debit cards to become obsolete is high. All over the world financial markets have taken a toll due to covid-19. Since the outbreak of coronavirus there's an increased association between stock market and Bitcoin. For example, in March 2020, after a sharp reduction in S&P index the price of Bitcoin also fell below $4000. The prices though have increased around $7000 few crypto investors were appalled by the fall. But Bitcoin manages to rise with a roar.

Ecommerce

Bitcoin being the most popular cryptocurrency it is revolutionizing the E-commerce business big time. The struggle with current mobile payment is, it involves a complete process of adding receiver's details, sender's details and then waiting for transaction completion. Some of the downsides of this process is, the recovery process is not easy if a wrong input is given. If any fraudster gets access to the sender's PIN details the money is at risk. Transaction fee is charged for every transaction made. And for every transaction individual identity has to be verified. Thus, no privacy in tradition payment methods [41, 42].

Bitcoin reforms the overall shopping experience. All that a user needs to do is buying some bitcoins and having them in their wallet. When the shopping is complete the user just has to open the wallet app and scan the QR code of the store's device. The engraves secret code gets unlocked and informs the bitcoin network about the transaction. The transaction can be completed with less time compared with traditional methods. The relationship between bitcoin and ecommerce is secure, cheap and instant. Bitcoin based payment methods help individuals to send money to their families directly via mobile phones. Unlike credit cards where international payments are risky, having bitcoin-based payment system assures the payment cannot be undone once it is complete. This discards the fraudulent risks and make the vendors to sell worldwide.

In recent years, there's a gradual increase in Bitcoin's mainstream process. John Donahoe, CEO of eBay, the world's largest ecommerce merchant stated, Bitcoin will play a crucial role in PayPal's future. Apple who banned bitcoin wallet now has started to bring up this digital currency. And amazon, another largest ecommerce retailer who did not want to risk with Bitcoin holds a patent on Bitcoin-Cloud computing. All this change shows how Bitcoin restored its reputation after infamous money laundering case of Charlie Shrem.

Significant features of Bitcoin that could transform the Ecommerce business are,

- Faster Transactions and instant Satisfaction.
- Easy overseas business.
- No or Less transaction fees.
- Builds user Trusts.
- Accountable access of funds or finances by users.
- Highly secured transactions.
- Inspiring innovations in digital currencies.

References

1. Bitcoin: A Peer-to-Peer Electronic Cash System
2. https://medium.com/block-journal/digital-currency-a-brief-history-98be6f6f0f10
3. https://www.businessinsider.in/tech/everything-you-need-to-know-about-bitcoin-itsmysterious-origins-and-the-many-alleged-identities-of-its-creator/articleshow/61895890.cms
4. Varaprasada Rao, K., Panda, S.K.: Secure electronic voting (E-voting) system based on blockchain on various platforms. In: Satapathy, S.C., Lin, J.C.W., Wee, L.K., Bhateja, V., Rajesh, T.M. (eds.) Computer Communication, Networking and IoT. Lecture Notes in Networks and Systems, vol. 459. Springer, Singapore (2023). https://doi.org/10.1007/978-981-19-1976-3_18
5. Grinberg, R.: Bitcoin: an innovative alternative digital currency. Hastings Sci. Technol. Law J. **4** (2011)
6. https://www.wired.com/2014/03/bitcoin-exchange/
7. https://www.technologyreview.com/2018/03/29/67091/how-network-theory-predicts-the-value-of-bitcoin/#:~:text=Metcalfe's%20Law%20states%20that%20a,the%20number%20of%20its%20users.&text=With%20these%20parameters%2C%20the%20generalized,when%20Bitcoin%20has%20been%20overvalued
8. https://ethereum.org/en/whitepaper/
9. Panda, S.K., Mohammad, G.B., Nandan Mohanty, S., Sahoo, S.: Smart contract-based land registry system to reduce frauds and time delay. Secur. Priv. e172 (2021). https://doi.org/10.1002/spy2.172
10. Panda, S.K., Satapathy, S.C.: Drug traceability and transparency in medical supply chain using blockchain for easing the process and creating trust between stakeholders and consumers. Pers. Ubiquit. Comput. (2021). https://doi.org/10.1007/s00779-021-01588-3
11. Niveditha, V.R., Sekaran, K., Amandeep Singh, K., Panda, S.K.: Effective prediction of bitcoin price using wolf search algorithm and bidirectional LSTM on internet of things data. Int. J. Syst. Syst. Eng. **11**(3–4), 224–236.

12. Sathya, A.R., Panda, S.K., Hanumanthakari, S.: Enabling smart education system using blockchain technology. In: Panda, S.K., Jena, A.K., Swain, S.K., Satapathy, S.C. (eds.) Blockchain Technology: Applications and Challenges. Intelligent Systems Reference Library, vol. 203. Springer, Cham (2021). https://doi.org/10.1007/978-3-030-69395-4_10
13. Lokre, S.S., Naman, V., Priya, S., Panda, S.K.: Gun tracking system using blockchain technology. In: Panda, S.K., Jena, A.K., Swain, S.K., Satapathy, S.C. (eds.) Blockchain Technology: Applications and Challenges. Intelligent Systems Reference Library, vol. 203. Springer, Cham (2021). https://doi.org/10.1007/978-3-030-69395-4_16
14. Panda, S.K., Daliyet, S.P., Lokre, S.S., Naman, V.: Distributed Ledger Technology in the Construction Industry Using Corda, The New Advanced Society: Artificial Intelligence and Industrial Internet of Things Paradigm. https://doi.org/10.1002/9781119884392.ch2
15. Panda, S.K., Satapathy, S.C.: An investigation into smart contract deployment on Ethereum platform using Web3.js and solidity using blockchain. In: Bhateja, V., Satapathy, S.C., Travieso-González, C.M., Aradhya, V.N.M. (eds.) Data Engineering and Intelligent Computing. Advances in Intelligent Systems and Computing, vol. 1. Springer, Singapore (2021). https://doi.org/10.1007/978-981-16-0171-2_52
16. Panda, S.K., Rao, D.C., Satapathy, S.C.: An investigation into the usability of blockchain technology in internet of things. In: Bhateja, V., Satapathy, S.C., Travieso-González, C.M., Aradhya, V.N.M. (eds.) Data Engineering and Intelligent Computing. Advances in Intelligent Systems and Computing, vol. 1. Springer, Singapore (2021). https://doi.org/10.1007/978-981-16-0171-2_53
17. Panda, S.K., Dash, S.P., Jena, A.K.: Optimization of block query response using evolutionary algorithm. In: Bhateja, V., Satapathy, S.C., Travieso-González, C.M., Aradhya, V.N.M. (eds.) Data Engineering and Intelligent Computing. Advances in Intelligent Systems and Computing, vol. 1. Springer, Singapore (2021). https://doi.org/10.1007/978-981-16-0171-2_54
18. https://litecoin.org/
19. Nanda, S.K., Panda, S.K., Das, M., Satapathy, S.C.: Automating vehicle insurance process using smart contract and Ethereum. In: Chakravarthy, V.V.S.S.S., Flores-Fuentes, W., Bhateja, V., Biswal, B. (eds.) Advances in Micro-Electronics, Embedded Systems and IoT. Lecture Notes in Electrical Engineering, vol. 838. Springer, Singapore (2022). https://doi.org/10.1007/978-981-16-8550-7_23
20. Panda, S.K., Elngar, A.A., Balas, V.E., Kayed, M. (eds.): Bitcoin and Blockchain: History and Current Applications, 1st ed. CRC Press (2020). https://doi.org/10.1201/9781003032588
21. Panda, S.K., Jena, A.K., Swain, S.K., Satapathy, S.C. (eds.): Blockchain Technology: Applications and Challenges. Springer, Intelligent Systems Reference Library. https://doi.org/10.1007/978-3-030-69395-4
22. https://www.coindesk.com/company/mt-gox
23. Navon, J.: Econsultancy. https://econsultancy.com/blog/64542-bitcoin-or-bitcon-the-challenges-facing-the-crypto-currency-sector/
24. https://www.fxcm.com/markets/insights/what-causes-volatility-in-bitcoin/
25. https://news.bitcoin.com/bitcoin-history-25/
26. Irrera, A., Wilson, T.: Is PayPal's Crypto Move a Game-Changer for Bitcoin? Probably Not, Say Experts. https://bfsi.economictimes.indiatimes.com/news/fintech/is-paypals-crypto-move-a-game-changer-for-bitcoin-probably-not-say-experts/78839918
27. Varaprasada Rao, K., Panda, S.K.: A design model of copyright protection system based on distributed ledger technology. In: Satapathy, S.C., Lin, J.C.W., Wee, L.K., Bhateja, V., Rajesh, T.M. (eds.) Computer Communication, Networking and IoT. Lecture Notes in Networks and Systems, vol. 459. Springer, Singapore (2023). https://doi.org/10.1007/978-981-19-1976-3_17
28. https://cointelegraph.com/news/bitcoin-will-follow-ethereum-and-move-to-proof-of-stake-says-bitcoin-suisse-founder
29. Source: https://www.statista.com/statistics/326707/bitcoin-price-index/
30. https://en.wikipedia.org/wiki/History_of_bitcoin#Prices_and_value_history
31. https://www.investopedia.com/tech/most-important-cryptocurrencies-other-than-bitcoin/, https://en.wikipedia.org/wiki/List_of_cryptocurrencies

32. https://subscription.packtpub.com/book/data/9781789804164/1/ch01lvl1sec06/layered-structure-of-the-blockchain-architecture
33. https://99bitcoins.com/bitcoin/who-accepts/
34. https://www.bitcoinprice.com/predictions/
35. Axelrod, A.: The Future of Digital Assets: Stepping into the World of Main-stream Crypto. https://cointelegraph.com/news/the-future-of-digital-assets-stepping-into-the-world-of-mainstream-crypto
36. Windsor, R.: Why Bitcoins Are Digital Assets Rather than Cryptocurrencies. https://digitalassetmanagementnews.org/features/why-bitcoins-are-digital-assets-rather-than-cryptocurrencies/
37. Baur, D.G., Hong, K., Lee, A.: Bitcoin: medium of exchange or speculative as-sets? J. Int. Financ. Mark. Inst. Money **54** (2017). https://doi.org/10.1016/j.intfin.2017.12.004
38. BTC to USD: WinkDex; Other: Bloomberg
39. https://howmuch.net/articles/biggest-companies-vs-bitcoin-last-decade-performance
40. Hatzis, I.L.: Is Cryptocurrency the Future of Money? https://dailyfintech.com/2019/12/16/473
41. Rampton, J.: How Bitcoin Is Changing Online eCommerce. https://www.forbes.com/sites/johnrampton/2014/07/02/how-bitcoin-is-changing-online-ecommerce/?sh=1f8c005c5e72
42. Rampton, J.: How Bitcoin Is Changing Online eCommerce. http://www.forbes.com/sites/johnrampton/2014/07/02/how-bitcoin-is-changing-online-ecommerce/#3cb7a2c3493c

Chapter 3
HR Digital Transformation: Blockchain for Business

Sujata Priyambada Dash

Abstract India is striving very hard to establish in today's competitive world. We need to be ahead in our socio-cultural, socio-psychological, political and economic business scenario. We need to act very fast in digitalization of business processes and practices. People who are regarded as an employee are the human resources and are also the epitome of both of private and public sector organizations. By the passage of time, these human resources act in many fields such as in e-governance, insurance sector, sourcing and procurement, healthcare, education, supply chain, banking, real-estate, logistics, tourism and hospitality, and cyber security. Human resources showed tremendous efforts and contributions in attainment of the organizational goals and objectives. Simultaneously received tremendous pain and sorrow when betrayed, cheated, disturbed by the corrupted system. When the evolution of blockchain technology is being hailed as the second-generation internet which is a game-changer have changed the entire business scenario being completely decentralized and totally transparent. The motivation behind the study is that people for the first time in human history can trust and transact with each other in a much-secured manner. The emergence of blockchain technology could further transform the world of HR. This study has explored the concept of digitalization, human resource, and blockchain technology from an extensive literature survey. The purpose of this study is to explore the concept of blockchain technology and its relevance in today's competitive businesses. The study also emphasizes on different characteristics and the importance of blockchain technology in varied HR practices. This chapter has attempted to discuss certain research questions such as (a) what is human resource and why human resource? (b) What is digital transformation? (c) What is blockchain? (d) Why blockchain? (e) What are the various types of blockchain? (f) Why blockchain in HR? (g) How blockchain has transformed HR? (h) How far blockchain is relevant for HR? and (i) HR digital transformation vis-à-vis blockchain is examined in this study along with the challenges and criticisms. This paper highlights the countries and companies who have already formulated, implemented, and executed blockchain technology in their work places for higher productivity gains and to be recognized

S. P. Dash (✉)
Department of Management, Birla Institute of Technology, Mesra, Ranchi 835215, Jharkhand, India
e-mail: spdash@bitmesra.ac.in

S. K. Panda et al. (eds.), *Recent Advances in Blockchain Technology*,
Intelligent Systems Reference Library 237, https://doi.org/10.1007/978-3-031-22835-3_3

as digitalized country in global scenario. Hence, the main results of the current study focused on the HR digital transformation considering blockchain for businesses. This study will help HR practitioners to think out of the box and thereby implement blockchain technology for the beneficial of the organization as well as to all the internal and external stakeholders and society at large. This study will also help in cost reduction in an effective manner. Therefore, it becomes inevitable to focus on the HR digital transformation considering blockchain for businesses to remain sustainable.

Keywords Blockchain · Blockchain technology · Business · Digital transformation · Human resource

3.1 Introduction

We the people have built the country can never see it deteriorating anyway. We are the strong workforce with high morale, responsiveness, creativity and innovation, work experience, commitment, and competency can strive the world with their performances. Human resource is the epicenter of any organization full of dynamism and enthusiasm. Sentiments, emotions, feelings are the main elements of human resources. Human resources are the social animals who act, react, likes, dislikes, cry, and laugh. These human resources too have their own tastes and preferences. These human resources need to be handled properly and diplomatically by providing a well-designed work environment thereby boosting their morale. Higher the satisfaction higher is the morale, higher the morale higher is the productivity of an organization. Better human relations need to prevail in an organization to get rid from the grievances and conflict between human resources. Human resources are of paramount importance due to their dynamic activity in an organization. Therefore, they need to be treated in a fair and equitable manner. We are the social animals and socialization process never occur in vacuum. We treat each other with just and fair, respect with dignity, participate in decision-making processes, introduce incentive scheme to motivate human resources. Cooperation, coordination, integrity, cohesiveness, and collectivism of human resources are very vital for the prosperity of an organization. SMART (Smart, Measurable, Achievable, Realistic, and Time bound) people are objective oriented, meet the deadlines, and become success in their present and future endeavors. Sometimes the biased, corrupted system with fraudulent practices, hacking and ransomware embarrassed human resources in a larger way. The emergence of blockchain technology promulgated by Satoshi Nakamoto whose identity remained mystery till date formed the foundation of the cryptocurrency, Bitcoin, in 2009. Blockchain technology is 'an incorruptible digital ledger of economic transactions that can be programmed to record not just financial transactions but virtually everything of value' is indicated by Blockchain Revolution Authors—Don and Alex Tapscott. Individual data, such as education and career history, can be safely kept and interacted with on a real-time digital ledger. Suddenly, blockchain appears to be quite relevant to the field of human resources.

According to Gartner Spectrum [1], the evolution of blockchain technology takes place up to four archetypes. Firstly before 2010, the technology used is termed as *enabling technology* which provides *Internet of information.* Cryptography, peer-to-peer (P2P) networking, and distributed computing and messaging are the building blocks of blockchain-enabling technology improves operational efficiency. Second phase of the technology starts between 2010 to 2020 is *blockchain inspired* which provides *internet of content*. Blockchain-inspired used only three characteristics namely distribution, encryption, and immutability to further the efficiency of the existing processes. Alibaba, for example, tracks and traces food products everywhere using this technology. Third phase is termed as *blockchain-complete* beginning around 2023 regarded as "*The Internet of Value*". A blockchain-complete enterprise-ready solutions will emerge with all the five elements of blockchain technology namely decentralization, distribution, immutability, encryption, and tokenization. However, these components are the *catalyst* for the introduction of new business models. Fourth phase moving into 2025 will be the *blockchain-enhanced*. The artificial intelligence (AI), internet of things (IoT), and decentralized self-sovereign identity (SSI) will be incorporated as complementary technologies during this phase. This allows people to share their identities and data in a traceable and trackable digital wallet. These methods increase the number of micro transactions possible by smart contracts and expand the types of value that can be tokenized.

This study reveals certain questions that need to be answered in a very systematic manner. The questions are (i) what is human resource and why human resource? (ii) what is digital transformation? (iii) what is blockchain? (iv) why blockchain? (v) what are various types of blockchain? (vi) what benefits blockchain bring in to the business? (vii) why blockchain in HR? (viii) how blockchain has transformed HR? (ix) how far blockchain is relevant for HR? (x) HR digital transformation vis-à-vis blockchain, and (xi) challenges and criticisms.

The study purpose is to explore the concept of blockchain technology, its relevance in today's competitive businesses. The study also emphasizes on different characteristics and the importance of blockchain technology in varied HR practices. Therefore, it become inevitable to focus on the HR digital transformation considering blockchain for businesses to remain sustainable.

3.2 What Is Human Resource and Why Human Resource?

Resources in an organization are physical resources (men, machine, money, raw materials, and market), financial resources (capital/funds/budget), human resources (employees/personnel/workforce/manpower), and technological resources (system, computers, and technical know-how). This human resource are the employees of the organizations who comes under the payroll with proper identity. Human resources are highly intangible in nature embodied with knowledge, skills, abilities, and potentialities. All other resources such as physical, financial and technological resources are tangible in nature. Optimum utilization of tangible resources is done by the intangible that is human resources. Once Peter F Drucker said "leave my men, I will create another organization". Human resources are the lifeblood to any organization. Human resource makes its entry and exit from the organization. Human resource gives life to the workplace by switching on the lights, fan, machineries and system and start working after making entry into the office premises. Exit also takes place of human resources at the evening hours meaning takes away all his knowledge, skills, and abilities, experiences and qualification along with him. Therefore, human resources play a pivotal role in both public and service sector organizations.

3.3 What Is Digital Transformation?

Digitization or digital transformation is something that all organizations have to bring in order to be effective, relevance and sustainable in future. The pillar of digital revolution has been dubbed blockchain. Many scholars believe that blockchain and its derivative technologies are fundamentally disrupting the existing economic and corporate landscape [2], particularly now that it appears to have found a suitable partner. Institutions in the corporate, insurance, banking, and other financial and non-financial service sectors are also undergoing significant digital transformation. Digital technology is a new disruptive force that is disrupting business practices and becoming a more important factor around the world [3]. The way we govern and retain administrative control in a digital age must alter. Blockchain is the answer to this problem. This technique guarantees that such problems will be resolved. Blockchain technology is an open distributed ledger that can be used to record productive and long-lasting trades between two parties [4].

3.4 What Is Blockchain?

Blockchain technology was promulgated by Satoshi Nakamoto in 2008 also regarded as digitalized ledger technology (DLT) for cryptocurrency [5]. Blockchain technology has become a trend as it uses *decentralized* network which means removal of middlemen. It is considered as digital ledger as it comprised of list of records and keep track of transactions. Blockchain technology is a proven disruptive technology has gathered the attention of HR practitioners, researchers and organizations as well across the globe [6]. It is an open, decentralized, distributed, public peer-to-peer technology with higher security. Blockchain has evolved from Bitcoin which is a peer-to-peer electronic cash system, a virtual cryptocurrency which do not have physical existence. It is *distributed* in that is a mess network where there is no server concept. Every node or peer is connected to other node in peer-to-peer network where data is not lost and it delivers to its destination properly. All nodes are both clients and servers which provides and consume data. It is a distribution system architecture and all clients are heterogeneous in nature. Data is recorded in a secure and semi-anonymous manner using technology. Participants have complete regulator over their personal identification and other information, and provide only what is necessary for a transaction is one of the best features of blockchain technology-*encryption.* It is *immutable* in that the finalized transactions are signed cryptographically, timestamped, and appended to the ledger in a sequential order. Changes to records are only possible if all parties agree. Tokens are used to trade value and can signify a varied range of asset kinds, with 'money,' data units, and a user's identity. The generation of tokens termed as *tokenization* is how a blockchain reflects and allows for the trading of a 'native value' ('currency'). All these five elements create a true blockchain.

3.5 Why Blockchain?

The traditional name of blockchain technology was trust-machine. Vast number of data and information are used in a secured manner stored in blocks. Organizations use blockchain technology widely (a) to view the history of any transaction at any time as data is non-modifiable in nature, (b) to keep the data and information in a more secured and safe manner as blockchain is tamper-evident and tamper-proof in nature [7], (c) to reduce the cost for financial transactions for all kinds of applications since there is no intermediaries [8], (d) allows trust-free business services [9], (e) data are stored in a confidential manner and transactions made are time stamped [10]. The extant literature survey mapped the following blockchain properties to possible benefits in strategic, organizational, economic, informational, and technological domains [2]: (a) transparency, fraud, and corruption were the strategic categories, (b) accountability, trust, traceability, and auditability were the organizational category, (c) cost and resilience is the economic category, (d) availability, scalability, distributed or decentralization, no intermediaries, sharing, reliability, privacy, quality, and integrity of data are the informational category, and (e) tamper-resistance, security, authentication, efficiency, and immutability are the technological category.

3.6 What Are the Types of Blockchain?

The types of blockchain are of paramount importance. Basically, blockchain is categorically of three types namely.

3.6.1 Public Blockchain

On the network, anyone can manage transparent transactions anonymously. It's primarily used for Bitcoin, and it's completely decentralized and based on user consensus. Hackers can duplicate all the blocks of information's that have been modified without user's identification.

For example: Bitcoin, Ethereum, Bitcoin Cash, Litecoin, Monero.

3.6.2 Private Blockchain

Although the users are recognized, the transactions are anonymous and the data is not accessible to the public. A user cannot publish or read any information in the blockchain network without first receiving an invitation to join. Typically utilized by large organisations with clearly established permissions among the blockchain enterprise's stakeholders.

For example: Multichain.

3.6.3 Consortium Blockchain

This is a hybrid model that combines public and private elements. The organizations may have their own private blockchain network to share information with the consortium's users, which might include banks, businesses, and corporations.

For example: Hyperledger, Corda, Quorum.

3.7 What Are the Benefits Blockchain Can Bring to the Business?

Benefits that blockchain can bring to the business are described below.

3.7.1 Trust

Allows persons who are unfamiliar with one another to trust one another.

3.7.2 *Decentralized Structure*

Reduces weak places while enabling real-time data sharing between companies like distributors and suppliers.

3.7.3 *Improved Security and Privacy*

End-to-end encryption create a permanent record of transactions, reducing fraud and unlawful behavior.

3.7.4 *Visibility and Traceability*

Track the provenance of a range of products, including pharmaceuticals, to ensure they are genuine and not counterfeit and organic products, to ensure they are actually organic.

3.7.5 *Immutability*

Prevents the modification or deletion of transactions.

3.7.6 *Speed*

Transactions are processed more quickly than with traditional techniques since intermediaries are eliminated.

3.7.7 *Tokenization*

Converts an asset's value into a digital token that is stored on a blockchain and shared among users. The exchange of digital art is done with non-fungible tokens.

3.7.8 *Individual Control of Data*

Allows for the choice of what digital data is shared, with whom, and for how long, with smart contract restrictions.

3.7.9 *Reduced Costs*

Enhances reporting and auditing while decreasing the need for manual processes like data aggregation and editing.

3.7.10 *Innovation*

Leaders in a variety of industries are investigating and putting into practice blockchain-based systems to solve insurmountable issues and enhance time-consuming, clumsy procedures like confirming the facts on a CV.

3.8 Why Blockchain in HR?

Blockchain in HR is of significant importance in that it reduces the risk of insufficient hiring, and providing protection while verifying the source of transactions. The lists of importance are mentioned below:

(i) To avoid fraud, hacking, and ransomware; (ii) to prevent totally manufactured papers such as school credentials or past work records from being faked or tampered with; (iii) to protect the foreign documents mistranslated, misrepresentation of originals or changing the context together; (iv) to prevent from skills gap between company needs and competencies occupied by the workforce; (v) to eliminate the possibility of candidates providing false information during job interviews; (vi) to avoid selecting the wrong individual, which can have major consequences for businesses; and (vii) to avoid paying intermediaries when forming contracts.

3.9 Relevant Literature Review

Secondary data has been explored with different terminologies in Google engine, Scopus, Springer Link, Taylor & Francis, and IEEE Xplore databases with varied terminologies such as Digital, Digital transformation, Blockchain, Blockchain Technology, Digital Transformation and blockchain. From the extensive literature survey researchers name, year of publication, paper description, type of paper, and area of research is depicted in the below mentioned Table 3.1 [3, 6, 11–33].

3.10 Transformation of Blockchain Technology in Human Resource in Three Waves

Transformation of blockchain technology in human resource has taken place in three waves by the passage of time and performance made by the organizations to keep abreast of the digital technology, competitive market scenario, and sustainable development.

3.10.1 First Wave

HR professionals have excelled in utilizing cloud-based HR systems and outsourcing recruitment, benefits, and payroll; nonetheless, the core economic fundamentals have remained unchanged. Despite the fact that there is an oversupply of skilled individuals in the global workforce, businesses face a talent supply challenge known as the "talent management conundrum." Although some HR groups utilize people analytics to anticipate which factors lead to more successful recruits, the majority of HR departments are either not collecting the necessary data or lack the necessary skills and procedures. Approximately 20–50% of job applicants exaggerate or lie about their qualifications. The preservation of confidential customer and employee data is the touchstone of all firms, and trust is the ultimate currency. In a matter of days or even hours, data security failures can devastate a brand that has been meticulously developed over years. The CHRO may be interested in any solutions that claim to minimise the cost (and bureaucracy) of contracting and coordinating. The initial wave of blockchain would start to address some of the most fundamental concerns with recruitment, such as identity management and credential verification. The initial wave of blockchain technology would start to address some of the most basic difficulties in recruitment, such as identity management and credential verification. Individuals could have full control over their data by owning and administering their own secure digital identity, allowing them to avoid misrepresentation. The solution is to build decentralized networks in which people own their own professional data, choose who they share it with, and monetize it equitably. Ensure that candidates are who they say they are and have verifiable credentials is a major concern for recruiters. The worker is not required to present the original document because the proof of validation serves as confirmation of credential legitimacy. The cost of third-party background checks may be eliminated if candidate credentials could be verified efficiently, allowing for speedier and more effective onboarding of new recruits. In a competitive labour

Table 3.1 Relevant Literature Review

Sl. No.	Researchers'	Year	Paper description	Paper type	Area
1	Michailidis [11]	2021	Blockchain usage in terms of security, fraud prevention, and productivity gains	Theoretical	Blockchain and human resource management
2	Li et al. [12]	2021	Blockchain technology applications in HRM	Theoretical	Blockchain and human resource management
3	Jiang et al. [13]	2021	Design of college student information sharing system	Theoretical	Blockchain
4	Rhemananda et al. [14]	2020	Blockchain technology applications in HRM	Theoretical	Blockchain and human resource management
5	Ferri et al. [15]	2020	Blockchain in auditing services intent to use	Empirical	Blockchain
6	Ingold and Langar [16]	2020	Different types of resumes	Empirical	Resume fraud
7	Serranito et al. [17]	2020	Verification of digital documents ecosystem	Empirical	Blockchain
8	Tang et al. [18]	2019	Blockchain ethicality using a framework	Research Framework	Blockchain
9	Hughes et al. [6]	2019	Application and usage of blockchain	Conceptual	Blockchain
10	Koncheva et al. [19]	2019	Blockchain and managing labour relationships	Conceptual	Blockchain and human resource management
11	Sakran [20]	2019	Fraud resumes identification	Conceptual	Fraud resume
12	NASSCOM Avasant [21]	2019	Opportunities of blockchain in India	Report	Blockchain
13	Schmitz and Leoni [22]	2019	Issues such as governance, transparency and trust with blockchain is addressed	Review of literature	Blockchain

(continued)

Table 3.1 (continued)

Sl. No.	Researchers'	Year	Paper description	Paper type	Area
14	Linda et al. [23]	2019	Employees survey regarding Blockchain	Report	Blockchain
15	Coita et al. [24]	2019	Marketing and human resource management in the area of blockchain	Conceptual	Human resource management and blockchain
16	Clohessy and Acton [25]	2019	Factors (organizational) affecting blockchain adoption	Theoretical	Blockchain and human resource management
17	Spence [26]	2018	Blockchain applications in human resource management	Whitepaper	Blockchain and human resource management
18	Onik et al. [27]	2018	Blockchain recruitment framework	Research framework	Blockchain and human resource management
19	Murray et al. [28]	2018	Blockchain smart contracts	Conceptual	Blockchain in organizations
20	Li et al. [29]	2018	Application of blockchain in various functions of organizations	Conceptual	Blockchain in organizations
21	Gatteschi et al. [30]	2018	Merits and demerits of blockchain technology in insurance sector	Conceptual	Blockchain in organizations
22	Bhattacharya et al. [31]	2018	Human resource management framework using blockchain	Model research paper	Blockchain and human resource management

(continued)

market, the transaction cost of human resources should decrease, allowing companies to pay higher wages to workers.

Table 3.1 (continued)

Sl. No.	Researchers'	Year	Paper description	Paper type	Area
23	Paul et al. [32]	2018	Fraud prevention	Conceptual	Blockchain and human resource management
24	Deloitte ASSOCHAM [3]	2017	Blockchain application in India	Report	Blockchain
25	Wang et al. [33]	2017	HRIS framework using blockchain	Research framework	Blockchain and human resource management

Source Author's Representation

3.10.2 Second Wave

The second wave of blockchain technology could have a positive impact on the broader talent ecosystem, resulting in a larger, stronger, and more equitable gig economy while reducing the number of full-time job contracts. The procedure essentially eliminates conscious or unconscious bias and discrimination, which is another source of friction, by relying on easily verifiable credentials. In the market for business counsel, capable teams of freelancers may disrupt the consulting profession if they could form to address complicated business challenges. It may be even more crucial to attract the top people to an organisation in a flexible labour market with ever-higher turnover of projects and roles. The use of blockchain to create a distributed workforce could minimise the requirement for permanent contracted personnel and their accompanying benefits packages. Blockchain eliminates non-value-adding intermediaries.

3.10.3 Third Wave

Blockchain is, at its heart, a peer-to-peer technology that allows us to transact without the use of centralized intermediaries. Organizations will always require work and workers. In a more globalized and automated industry, what is the best method to deploy our workers? Organizational structures are evolving as professions become more specialized, people work in teams with cross-functional boundaries, and competence, rather than span of control, is used to define success. Many HR and people management practices are still focused around the person, not the team, which is why successful team dynamics are so important in developing enterprises. HR has struggled to convey the benefit of people management knowledge to the organization throughout its existence.

The initial wave of HR solutions is anticipated to start with systems that provide identity and qualification verification, real-time payment of contractors, smart employment contracts, and eventually electronic résumé standards. Blockchain could someday allow CHROs to gain more direct access to talent, dramatically altering HR's position and pave the road for completely new businesses.

3.11 How Blockchain Has Transformed HR?

Blockchain has transformed HR in various ways. Below mentioned are the few start-ups or organizations utilizes blockchain technology in a most effective and efficient manner [34].

3.11.1 Beowulf: Streamlining Communication in the Workplace

A decentralized cloud network is used by this 2019 start up to expedite internal communications in a variety of work contexts. Secure organizational communication, distance learning, remote worker healthcare, and ready to use software development kits (SDKs) for unique features of communication are just a few of the company's products.

3.11.2 The BeSure Network: Ensuring Workplace Safety Protocols Are Valid

BeSure Network is a firm founded in 2017 with the goal of removing unqualified individuals from hazardous/dangerous jobs. Its blockchain technology gathers auditable safety and enforcement data from a variety of sources, allowing businesses to demonstrate their commitment to workplace safety. Workers and regulatory agencies, as well as managers (for example, plant floor supervisors), have secure access to data. To aid in the maintenance of a secure workplace, BeSure Network provides features such as smart contracts and automatic data entry. The technology is now undergoing beta testing with the help of UK-based safety organizations.

3.11.3 Etch: Instant Payroll

Etch, a blockchain-based payroll system, was created in 2019 and allows employees to collect money anytime they choose. Credit is given in Etch tokens, which are then put in a digital wallet. Workers can use their Etch card to make purchases at millions of locations around the world, and the value of the token is decided by the cash provided by the employer.

3.11.4 *eXo Platform: Revolutionizing Employee Rewards and Recognition*

eXo Platform, which was created in 2003, isn't really a start-up; however, it has recently increased its blockchain investments and integrated distributed ledger technology into its rewards and recognition module. Connectivity, teamwork, employee empowerment, and incentive features are all included in eXo, which is built on open-source technology. eXo is known for its high level of interoperability, which is especially beneficial if one has in-house technology expertise. Given its history as an HR technology pioneer, eXo naturally serves to well-known companies like UCLA and HSBC Bank.

3.11.5 *Gospel Technology: Changing Human Resource Information Systems (HRIS)*

Gospel Technology was founded in 2016, and in 2017 CIO Review named it one of the "20 Most Promising Blockchain Technology Solution Providers." Although Gospel's data platform may be applied to any corporate situation, it is particularly useful in human resource management. It enables users to share sensitive data with internal and external stakeholders while maintaining control and safeguarding employee privacy rights. Despite its two-decade heritage, Work.com just relaunched in a new guise, utilizing artificial intelligence and blockchain to connect job seekers with the appropriate companies.

3.11.6 *Job.com: Changing the Face of Recruiting*

Despite its two-decade heritage, Work.com just relaunched in a new guise, utilizing artificial intelligence and blockchain to connect job seekers with the appropriate companies.

3.11.7 *Lympo: Employee Wellness Data Democratization*

In 2016, Lympo was founded as a platform for healthcare and employee well-being. It connects wearables and other data sources to build a large-scale data-sharing ecosystem. A blockchain-based reward/incentive is offered based on the user's fitness actions. Lympo's plan includes a health and fitness wallet, employee incentivization, a wellness marketplace, and healthcare company crowdfunding. A free trial of its human resources solution is available.

3.11.8 *Peoplewave: Employee Relationship Management (ERM)*

This 2016 business offers a variety of services. Wavebase, its most recent product, is fascinating. Wavebase employs blockchain technology to automate applicant sourcing, selection, and screening. Peoplewave's larger enterprise resource planning (ERP) strategy interacts smoothly with Wavebase, providing a centralized source of employee data for everything from recruiting to engagement and performance monitoring.

3.11.9 *Vault Platform: Increasing Accountability in Workplace Abuse Reporting*

Vault was founded in 2017 to give employees a safe place to document and correct misbehaviour. It employs blockchain technology to date and time stamp data records indefinitely, resulting in an accurate history of occurrences. It speeds up employment discrimination investigations so that workers can get answers as soon as possible.

Furthermore, Vault offers advanced analytics that give HR insight into the company's work culture.

3.11.10 *WurkNow: Job Advancement for Blue-Collar Workers*

WurkNow is a blockchain-based human resources firm created in 2018 with the goal of changing the way blue-collar job searchers interact with employers. Hiring, departments, human resource management, time and labour management, and enforcement are the four components. The enforcement module relies on blockchain technology to maintain a centralized collection of data records ranging from worker drug monitoring to performance assessments, creating trust and accountability.

3.12 How Far Blockchain Is Relevant for HR?

Blockchain relevance for HR is felt in the era of internet while dealing with a range of issues and challenges than ever before. To reduce the danger of ineffective recruiting, HR departments devote a significant amount of effort to linking, screening, and confirming candidates' applications, as well as completing validation tests and validating data. Recruiters' primary responsibility is to connect applicants' profiles to a variety of sources, including direct applications, recruiting agencies, and social networking sites. As a result, resume screening takes time [35]. Aishwarya [36] pointed out, the blockchain system has far-reaching and comprehensive effects on HR practices. Smart contracts may be used to set up blockchain network solutions for organizations to tackle some of the most major HR concerns which is described below.

3.12.1 Background and Employment-History Checks

In a distributed blockchain network, virtual credentials that is tokenizing applicants' identity can be obtained by permissioned applicants. It gives an immutable record of Smart Contract applicants' job history. Using this blockchain technology, prospective employers can remove the possibility of false applications.

3.12.2 Access and Data Security

Organizations have access to a wealth of personal information about their employees. Documents can be encrypted and immutably recorded on a blockchain, which is especially crucial for private records relating to medical issues or performance history. These records, however, can be shared in tokenized form with individuals who have granted confirmed permission as appropriate.

3.12.3 Smart Contracts for Temporary Workforce

A smart contract establishes enforceable and irreversible rights and obligations for all participants. Immutable contracts in HR, for example, can release money from escrow when workers complete tasks, smoothing our both workers' and employees' cash flow.

3.12.4 Compliance and Regulations

Employees can use legislation like the European Union's General Data Protection Regulation (GDPR) to exercise their 'right to be forgotten' rights by simply removing the encryption key and rendering personally identifiable information unrecoverable. Human resource will be able to leverage blockchain to ensure that employees have control over their own data as more stringent regulations becomes available parties to process payments, such as banks. Similarly, banks will no longer be able to manipulate money flows to the extent that they can now by trading- and so modifying the value of-fiat currencies (government-issued).

3.13 HR Digital Transformation vis-a-vis Blockchain

Digital transformation in HR evolved with a digital revolution, e-HR practicing HRM activities in a computer and web-based tools and applications. Instead of paper work these tools have eased the work of HR managers to perform the specialised task in a more strategic manner. Digital revolution in HRM has decentralised HR services such as recruiting, training and development, performance review, and compensation management through a comprehensive and integrative method. Considering some disadvantages in digitalization of HR functions blockchain technology becomes inevitable to carry out the work place task in a more effective and predominant way thereby satisfying the entire HR system. The areas of HR digital transformation vis-à-vis blockchain are described in a better manner.

3.13.1 E-Recruitment vis-à-vis Blockchain in Recruitment

Organizations started using quite popular system as E-recruitment system permitting HR managers to mark a large number of applicants at a lower cost [37]. According to [38], organizations do job posts publishing, namely open positions in their own organizations in virtual mode, having a virtual form to fill-in which is available for candidates. A web-based database existing to store the resumes of candidates. Candidates specifically face privacy risk while applying job online. Suspicions arises that their information may be shared without permission by the third party. People avoid using technological tools due to their technology bias. Whereas this disruptive technology can leverage to help untap fresh talents and also help recruiters to find out more information what is than written in the resume. Compliments or criticisms also can be available in their ledger which is immutable. This enables the recruiters to categorize competent HR talent pool. The likelihood of the proper selection enhances when the blockchain saves recruiters from the resume polishers who overstate or even create qualifications for themselves.

3.13.2 E-Compensation vis-à-vis Blockchain in Compensation

According to [39] the use of web and computer-based technology for employee compensation planning is known as e-compensation. E-compensation is well-defined as the use of computer-based software applications. Managers administer and disclose information about pay and benefits procedures and information in their organizations effectively as described by [40]. The growth in integrated human resource management systems enabled organizations to provide real-time and trustworthy data and information's to their employees. These technological developments gave employees the opportunity to handle HR processes. Members of the organization can use these systems to update their data, HR professionals can use them to accommodate wages and fringe benefits, most notably, bureaucratic duties are carried out using real-time data and e-compensation procedures. Web-based technology ensure that workflows are more up to date, rapid, and reliable and managers can use them to establish plans by generating reports [41]. E-compensation reduces the amount of mistake in the HR compensation process. Companies that use e-compensation, for example, have less errors when it comes to paying wages or other types of payments to employees. With the ability to control and report all of the pay in an organization at the same time, as well as the ability to analyze this massive data, a fair team of HR may take actions to ensure salary equality with the help of e-compensation. Inequalities that have been created in the past can be easily identified and remedied. During everyday procedures, if a subordinate makes a mistake in the compensation function, such as sending the wrong payroll to the wrong employee, the HR manager may readily discover it and take corrective action. Problem still exists in handling the HR payroll system. Whereas blockchain, in combination with the data mining tools, will be a godsend to the compensation department, allowing events, payments, and benefits to be processed quickly. When an employee completes his probationary period, blockchain will automatically trigger a raise in his wage. The function of intermediate banks and other financial organizations may be bypassed when using Bitcoin, and payments can be processed in a matter of hours. It also decreases the amount of paperwork and receipts processed for each transaction, resulting in lower back-office costs and more sustainability. Bitwage-a US based company provides frictionless invoicing assisting international payments by handling the conversion of bitcoin to local currencies. For example, many organizations such as Japan-based GMO internet, Dana Crowdfunding, Chrono bank, Fairley allow organizations to pay their contract workers without going to third party using blockchain-type technology. Similarly, IBM has projected a new blockchain banking solution for cross-border payments that permit financial bodies to go for last transaction.

3.13.3 E-Training vis-à-vis Blockchain in Talent Management

E-training is a type of distance learning that uses web-based technologies such as videos, e-books, emails, multimedia messages, sound tracks, and discussion groups to provide individuals with the necessary knowledge on specific selected themes or a specific specialty [42]. The use of technology to educate is prevalent in e-learning [43]. This type of training can take the form of face-to-face instruction, distance mediated instruction, or pure online instruction. Virtual, distance, online, and web-based training are some of the terminologies used to describe e-training [42]. Writing, communication, visualization, and storage technologies are all included in E-training systems [44]. E-learning is the use of Internet-based technology to provide a sufficient number of alternative methods for increasing knowledge acquisition [45]. E-learning also enables flexibility in the time period in which the student wishes to profit from the learning content, ease of access, just-in-time delivery, low costs, and great customer value [45]. Organizations favor e-training because it cuts down on the amount of money, they have to spend on education to teach their employees. Given the high cost of training personnel across many countries, this is a critical solution for reducing both wasted time and travel expenses. E-learning techniques enable businesses to access a vast number of educational materials at a reasonable cost [46]. It serves from diverse hierarchical levels and geographies due to the increased number and quality of employees. Whereas, a trustworthy and comprehensive blockchain record of the employees including their training, competencies, and performance can be constructed by the organizations. Organizations may map the right individuals for the right position using Artificial Intelligence and People Analytics tools, supporting a competency-based work culture within the organization.

3.13.4 E-Performance Appraisal vis-à-vis Blockchain Performance Management

E-performance systems are defined as mechanisms that use a company's web-based portals and applications to conduct an internet-based assessment of an employee's potential, knowledge, and performance [39]. All the processes are online in e-performance appraisal system. HR managers can use web-based performance management solutions to track the performance review process thereby reducing time through paperwork. In e-performance appraisal system the comfort and ease of the system can benefit all the employees of the performance assessment system with the employee whose performance is being reviewed, the line manager who is one of the accessors, and the HR manager who is responsible for the process success. The employee whose performance is being reviewed, the line manager who is reviewing the performance, and the HR manager who is coaching the process can all access the linked workflow, but each can only see and control the portion of the process that is related to his or her particular task. Whereas organizations can use block chain technology to establish performance blockchains in which each employee's assessments and reviews are recorded. This will assist prospective recruiters in determining the true potential of the candidate they are considering. The favorable impact of blockchain on employee productivity would benefit start-ups and small businesses that have to spend a lot of money to find the right people for their jobs. For employees and private contractors, blockchain also ensures fraud prevention and HR cyber security.

3.14 Blockchain Used in Other Industries and Countries

Many tech companies are using blockchain technology through proper investments otherwise they will lag behind from their competitors. The companies are IBM, Microsoft, and Accenture started building products and applications to be benefitted from blockchain technology. Patent filing every year is executed by the Fintech companies such as JP Morgan, Bank of America, and MasterCard. In particular, 45 patents in cryptocurrency related patents are owned by Bank of America. Handling of raw material and inventory management has become easier and more efficient in the fashion business with blockchain technology. Food safety using blockchain is regulated by the companies Nestle, Walmart, and Dole. Raw material handling and inventory management becomes simpler and more efficient using blockchain technology in fashion industry. A project is initiated by UN for human and child trafficking using blockchain technology for the greater good of the society. Countries such as South Korea, Estonia, and Dubai are using blockchain technology and also advancing these concepts towards storing government records such as health records, marriage certificates, business registrations and much more.

3.15 Blockchain Start-ups in India

In India, blockchain has grown in popularity, and both the government and businesses are seriously considering the possibilities for its use. In order to determine the areas where India can utilize blockchain, NITI Aayog began working on a national strategy for it in 2018 [47]. Banking, insurance, finance, and the public sector are major players who have jumped in and established blockchain pilot projects, according to the India Blockchain Report 2019 published by NASSCOM and Avasant jointly. Nearly half of Indian states have started blockchain projects, according to a report, and the public sector is leading the way. Additionally, it notes that the public sector is now carrying out more than forty blockchain initiatives, with 92% of them still in the pilot stage and 8% at the production stage. Blockchain start-up companies founded in India over the previous 18 years were featured in the Blockchain Technology Report India 2018 [48] along with the most common use cases these companies have evaluated.

(i) ***Year 2021***: By developing valuable business/incentive models for all ecosystem participants, Infosys is promoting the enterprise-wide adoption of blockchain-powered business networks across industries. For its blockchain development services, Infosys was rated as a top vendor in the Nelson Hall NEAT Vendor review.
(ii) ***Year 2019***: The creation of a blockchain platform for coffee trading has been announced by the Indian Ministry of Commerce and Industry with the goal of increasing product traceability and farmer income by removing middlemen.
(iii) ***Year 2018***: In order to stop the circulation of fake drugs, NITI Aayog announced a pilot project on a blockchain-based platform for India's domestic pharmaceutical supply chain services.
(iv) ***Year 2018***: The Telangana Government and TechMahindra collaborated to establish India's first Blockchain District, which supports and advertises blockchain start-ups and businesses.
(v) ***Year 2018***: To enable cross-border payments, ICICI Bank has joined JP Morgan's proprietary blockchain technology Interbank Information Network (IIN), a first live blockchain-powered Quorum.
(vi) ***Year 2018***: Blockchain will be used by the Kerala government's K-DISC (Kerala Development and Innovation Strategic Council) to handle the supply chains for milk, vegetables, and fish with the states.
(vii) ***Year 2018***: Microsoft and TechMahindra collaborated to develop a distributed ledger-based solution for a blockchain technology aimed at controlling spam calls.
(viii) ***Year 2018***: Blockchain technology was developed by Yes Bank using Hyperledger Distributed Technology (DLT) to give Bajaj Electricals a digital, automated vendor financing mechanism.
(ix) ***Year 2017***: IndiaChain, a blockchain project by NITI Aayog, has been announced for use by the nation for a number of reasons, including maintaining land records and allocating public goods.
(x) ***Year 2017***: SBI has joined up with other top Indian banks to form the BankChain consortium, which will work with tech giants like IBM and Intel to create financial industry solutions for services like KYC.
(xi) ***Year 2016***: In cooperation with Emirates NBD and EdgeVerve System, ICICI Bank, the first Indian bank, has successfully completed a trial transaction in international trade financing and remittance on blockchain technology.

3.16 Challenges and Criticisms of Blockchain Technology

Before the existing blockchain technology can concurrently guarantee scalability, privacy, and reliability at scales with billions of transactions every second, there are a number of issues that need to be resolved. The following are the main difficulties:

3.16.1 Scalability

The present blockchain mandates that all transactions be saved and made accessible for each new transaction to be validated. As a result, cryptocurrencies like Bitcoin can only complete a small number of transactions every second. After a certain record and network size threshold, newer systems stop scaling. To address the challenges, a number of sub-problems must be resolved, such as the optimization of storage for transactions, which necessitates intelligent methods to maintain only a minimal quantity of data to validate transactions. The difficulty of knowing when to archive and delete material is also present here. Another issue in this regard is how data should be split across several nodes to guarantee the greatest effectiveness and scalability. The consensus mechanism is another crucial factor to take into account when evaluating the scalability of blockchain networks. As a result, another crucial point to be addressed is load-balancing in terms of how many and which nodes should be employed to validate each transaction among participating nodes [49, 50].

3.16.2 Multi-blockchain Interoperability

Given the internet's very heterogeneous and distributed structure, it is conceivable that the ecosystem will contain a number of both private and public blockchains. These various blockchains should be able to communicate securely and transparently without compromising security in order to maintain a single state of the information. Before a blockchain validates the transaction of a user, for instance, it may query numerous other blockchains to determine that user's precise identity [51, 52].

3.16.3 Blockchain and AI

Although the number of nodes in the network is relatively modest, the majority of the current blockchain protocols are simplistic and require lengthy verification durations. This suggests that the efficiency of the present consensus protocols needs to be increased. Millions of nodes are expected at larger scales, which raises the possibility of malicious nodes attempting to compromise the system. By making various components of the blockchain 'smarter,' a number of AI algorithms can assist in resolving this issue as well as many others. In order to make wise decisions about whether a certain node is trustworthy or not, for instance, the behavior of nodes can be learned from their various activities and communication interactions. As a result, the cost of adding new blocks is decreased because unreliable nodes can be immediately removed from decision-making [53–55].

3.16.4 Energy

A secure system must be maintained, which costs money. It can be particularly demanding. It has been calculated that the existing Bitcoin ecosystem uses as much electricity as some small cities. Every usage becomes important when one takes into account Internet-scale networks with a variety of connection types and devices, such as mobile phones. Depending on the computing and battery capacity of the device, this necessitates efficient data and computation management [56–58].

3.16.5 Simulation and Testing

A variety of blockchain-based solutions have recently come into existence. They all assert that they have advantages over the competition. In order to compare various suggested consensus and data structures and assess security issues, there are currently no standardized simulation environments or benchmarks available. Simulation environments are crucial for future development as well as for testing the currently suggested solutions. Additionally, simulation environments are economical and enable repeatable findings.

3.16.6 Lack of Industry-Ready Resources

There is a shortage of skilled resources with experience in blockchain, i.e., there are only 45,000–60,000 trained resources who are industry-ready globally. This is due to talent upskilling as the demand for blockchain talent is growing at over 40% every quarter. To meet demand, reskilling and upskilling are absolutely necessary. According to [59] due to the nascency of the technology and the small number of real engagements, blockchain expertise and competencies, for both fundamental platform programming and blockchain application development, are extremely rare globally. Talent upskilling is the main difficulty facing providers in India and around the world, according to [25]. In India, service providers are having trouble locating staff members with knowledge and experience in proofs-of-concept, pilots, or implementations of Blockchain solutions, especially at the mid- or senior level. Through the use of current workforce, cross-training programmes, and partnerships between the private (service providers working with platform providers) and public sectors, the service providers are attempting to close the gap [60, 61].

3.16.7 Lack of a Clear Regulatory Direction

Different countries have adopted various regulatory strategies. India is taking a cautious stance when establishing the blockchain ecosystem. India is therefore still working to provide the Blockchain ecosystem a clear regulatory orientation. Despite having a negative opinion of cryptocurrencies, the Indian government (GOI) is supportive of Blockchain technology and is working to establish a national digital currency as well as find other fields in which it might be used. The BFSI industry's usage of the technology has been a major focus of the GOI's regulatory approach to blockchain technology. Most regulatory guidelines for blockchain in India have been set by the Reserve Bank of India (RBI). The RBI has issued a warning to the general public against dealing with cryptocurrencies and has cut off all legal banking channels to organizations that deal with them. Cryptocurrencies are not accepted as legal money by the GOI. Although there are no official regulatory structure governing cryptocurrency exchanges, the inability to access official banking channels has forced several of India's most well-known exchanges to close, as per [62–64].

3.16.8 Challenges Related to Human Resource

Hirsch stated that one significant barrier to implementing blockchain technology is that HR is frequently a bit of a slacker when it comes to technological adoption. The early stage of the technology may also be a barrier since blockchain isn't yet a widely accepted option for many corporate applications. Other barriers and resistance are present in addition to these strong opposing forces. According to Daher, the operational hazards can be divided into the following four groups cited by Baker [65].

(i) **Cyber security**: HR professionals who deal with sensitive personal data and financial transactions are at danger since blockchain is still susceptible to data flaws from endpoints that hackers can use to intercept data as it is being transmitted.
(ii) **Compliance risk**: Regional regulatory standards for blockchain are still lacking, putting businesses at risk of losing money and facing legal consequences if they don't protect employee data rights and follow regulations like the General Data Protection Regulation of the European Union.
(iii) **Counterparty risk**: To support blockchain transactions, third-party vendors must frequently be used. As a result, even though those suppliers' websites and applications might not be as secure as the blockchain, the confidence established by a blockchain is extended to them.
(iv) **Data privacy**: The human element is the main internal risk factor for HR. Personal data on a distributed ledger may not yet feel secure to employees.

3.16.8.1 The HR Blockchain Situation in Summary

Despite holding great potential for both businesses and employees, blockchain is still very much in its infancy. "Blockchain technology has the ability to fundamentally change the way that HR functions, potentially impacting everything from benefit administration to the management of sensitive employee data to how HR transactions are conducted," said Elissa Tucker, APQC's principal research lead for human capital management. Despite these advantages, she noted, "few HR leaders reported that their HR unit was already embracing blockchain," indicating that the technology is still in its infancy for HR.

There are *criticisms* of blockchain technology which are discussed very briefly.

Every HR department today needs blockchain technology because it is secure, immutable, and verifiable. But in order for that to occur, we must have a fully digitalized economy, in which every component of the economy is interconnected and digitized. The first stage in deploying HR Blockchain is the requirement of universality of digital certification by all colleges so that credential scanning can be quickly updated and confirmed. For instance, the academic background check process would be substantially sped up if every job applicant had a blockchain ledger where all of their credentials are kept, but if only a small number of colleges supply that data in blockchain, it doesn't significantly improve the procedure. Furthermore, it is crucial to guarantee that every employee always has access to an updated ledger. Other issues with blockchain include people's resistance to change, concerns about data security and people's reluctance to store their certificates online, a small percentage of people who are computer and internet illiterate, challenges with using blockchain, and problems with tokens that include theft, loss, and other issues. The lack of blockchain specialists to develop tools is one of the biggest issues that blockchain companies confront. Few people are skilled in creating the technology, despite the fact that demand for it is rising daily. While looking at candidates' social media profiles, companies occasionally find material they shouldn't have seen; the same might occur with blockchain data. Will we in the future be debating whether we ought to grant candidates access to blockchain. In addition to having the technology, Bridgers added, we also need to have the cultural readiness to make the change in how we think about sharing information. As soon as everything is in place, we will observe how blockchain actually works in HR and how it improves people's lives.

3.17 Conclusions

This study has revealed two aspects of the digital world. First aspect is HR digital transformation has evolved as e-HR and the other aspect is blockchain in HR. Internet and web-based technology applications are widely used in e-HR concept and their implications in human resource management function [66]. Blockchain is one such disruptive technology which can be used by every industry. Blockchain in human resource for businesses has benefitted the organizations in their current operations.

Right employees need to be placed at right job and at right time is the mantra of HR department to have a satisfied and contended workforce. Organizations with blockchain technology can regain momentum to attain competitive advantage and sustainable development which has never happened before. This technology aimed at reliability, privacy, confidentiality, security, integrity, transparency of data in a most immutable, tamper-evident and tamper-proof manner. This technology also used in scalability, trustworthiness, distributed, decentralized, quality, fraud free, cost reduction, availability, authentication, efficiency, and reduction in energy consumption. Therefore, HR practitioners along with bitcoin may open doors for business in the coming years. This study may help the HR professionals to handle all the data and information and smooth functioning of blockchain technology. Blockchain technology solutions play a vital role in transforming the world of work. The mainstream adoption of blockchain has become widely accessible. For the organizations to be successful, this disruptive technology will be utilized and greatly influence human resource management practices thereby become a necessary HR tool.

References

1. Panetta, K.: The 4 Phases of the Gartner Blockchain Spectrum (2019). https://www.gartner.com/smarterwithgartner/the-4-phases-of-the-gartner-blockchain-spectrum. Accessed 17 Jan 2021
2. Olnes, S.J., Ubacht, Janssen, M.: Blockchain in government: benefits and implications of distributed ledger technology for information sharing. Gov. Inf. Q. **34**(3), 355–364 (2017)
3. Deloitte ASSOCHAM: Blockchain Technology in India: Opportunities and Challenges. Deloitte, London (2017)
4. Pilkington, M.: Blockchain technology: principles and applications. In: Xavier, F., Olleros, Zhegu, M. (eds.) Research Handbook in Digital Transformation. Edward Elgar Publishing Limited (2016)
5. Nakamoto, S.: Re: bitcoin P2P e-cash paper 2008-11-17 16:33:04 UTC. Satoshi Nakamoto Institute. Archived from the original on 7 December 2016. (17 November 2008). Accessed 4 Dec 2016
6. Hughes, L., Dwivedi, Y.K., Misra, S.K., Rana, N.P., Raghavan, V., Akella, V.: Blockchain research, practice and policy: applications, benefits, limitations, emerging research themes and research agenda. Int. J. Inf. Manag. **49**, 114–129 (2019)
7. Yaga, D., Mell, P., Roby, N., Scarfone, K.: Blockchain Technology Overview, pp. 1–46. National Institute of Standards and Technology (2018). https://doi.org/10.6028/NIST.IR.8202
8. Makridakis, S., Klitos C.: Blockchain: current challenges and future prospects/applications. Future Internet MDPI **11**(12), 1–16 (2019)
9. Frémont, V., Jonathan, G.M.: Can the blockchain technology solve trust issues in industrial networks? In: BIR Workshops (2018)
10. Franciscon, E.A., Nascimento, M.P., Granatyr, J., Weffort, M.R., Lessing, O.R., Scalabrin, E.E.: A systematic literature review of blockchain architectures applied to public services. In: 2019 IEEE 23rd International Conference on Computer Supported Cooperative Work in Design (CSCWD), pp. 33–38. IEEE (2019)
11. Michailidis, M.P.: Blockchain Technology: The Emerging Human Resources Challenge. Preprints, 2021, 2021050035 (2016). https://doi.org/10.20944/preprints202105.0035.v1
12. Li, L., Zhang, H., Dong, Y.: Mechanism construction of human resource management based on blockchain technology. J. Syst. Sci. Inf. **9**(3), 310–320 (2021)

13. Jiang, P., Feng, Y., Dai, Y.: Design of college student information sharing system based on blockchain. In: 2021 IEEE 2nd International Conference on Information Technology, Big Data and Artificial Intelligence (ICIBA), vol. 2, pp. 568–572. IEEE (2021)
14. Rhemananda, H., Simbolon, D.R., Fachrunnisa, O.: Blockchain technology to support employee recruitment and selection in industrial revolution 4.0. In: International Conference on Smart Computing and Cyber Security: Strategic Foresight, Security Challenges and Innovation, pp. 305–311. Springer, Singapore (2020)
15. Ferri, L., Spano, R., Ginesti, G., Theodosopoulos, G.: Ascertaining auditors' intentions to use blockchain technology: evidence from Big 4 accountancy firmd in Italy. Meditari Account. Res. (2020). https://doi.org/10.1108/MEDAR-03-2020-0829
16. Ingold, P., Langer, M.: Resume5 resume? The effects of blockchain, social media, and classical resumes on resume fraud and applicant reactions to resumes. Comput. Hum. Behav. **114**, 1–13 (2020)
17. Serranito, D., Vasconcelos, A., Guerreiro, S., Correia, M.: Blockchain ecosystem for verifiable qualifications. In: 2nd Conference on Blockchain Research & Applications for Innovative Networks and Services (BRAINS), pp. 192–199. IEEE (2020)
18. Tang, Y., Xiong, J., Becerril-Arreola, R.: Iyer, L: Ethics of blockchain: a framework of technology, applications, impacts, and research directions. Inf. Technol. People **33**(2), 602–632 (2020). https://doi.org/10.1108/ITP-10-2018-0491
19. Koncheva, V., Odintsov, S., Leonid, K.: Block chain in HR. Advances in economics. Bus. Manag. Res. **105**, 787–790 (2019)
20. Sakran, T.: Educational tips for the detection of resume padding. Int. J. Policy Inf. **7**(2), 31–41 (2019)
21. NASSCOM Avasant: India Blockchain Report 2019. NASSCOM Avasant, New Delhi (2019)
22. Schmitz, J., Leoni, G.: Accounting and auditing at the time of blockchain technology. Aust. Account. Rev. **29**(2), 331–342 (2019)
23. Linda, P., Massey, R., Holdosky, J.: Deloitte 2019 Global Blockchain Survey: Blockchain Gets Down to Business. Deloitte, London (2019)
24. Coita, D.C., Abrudan, M.M., Matei, M.C.: Effects of the blockchain technology on human resources and marketing: an exploratory study. In: Kavoura, A., Kefallonitis, E., Giovanis, A. (eds.) Strategic Innovative Marketing and Tourism. Springer Proceedings in Business and Economics. Springer, Cham (2019)
25. Clohessy, T., Acton, T.: Investigating the influence of organizational factors on blockchain adoption. Ind. Manag. Data Syst. **119**(7), 1457–1491 (2019)
26. Spence, A.: Blockchain and Chief Human Resource Officer, Blockchain Research Institute, Mountain View, California (2018)
27. Onik, M.M., Miraz, M.H., Kim, C.S.: A recruitment and human resource management technique using blockchain technology for industry 4.0. In: Proceeding of Smart Cities Symposium (SCS-2018), Manama, Bahrain (2018). https://doi.org/10.1049/cp.2018.1371
28. Murray, A., Kuban, S., Josefy, M., Anderson, J.: Contracting in the smart era: the implications of blockchain and decentralized autonomous organizations for contracting and corporate governance. Acad. Manag. Perspect. **35**(4), 1–53 (2018). https://doi.org/10.5465/amp.2018.0066
29. Li, Y., Bienvenue, T.M., Wang, X., Pare, G.: Blockchain technology in business organizations: a scoping review. In: Proceedings of the 51st Hawaii International Conference on System Sciences, pp. 4474–4483 (2018). https://doi.org/10.24251/HICSS.2018.565
30. Gatteschi, V., Lamberti, F., Demartini, C., Pranteda, C., Santamar, V.: To blockchain or not to blockchain: that is the question. IT Prof. **20**(2), 62–74 (2018)
31. Bhattacharya, S., Dana, S., Banerjee, A.: Real time human resource investment management using blockchain based implementation. Int. J. Manag. Appl. Sci. **4**(5), 75–77 (2018)
32. Paul, S., Chowdhury, M.J.M., Colman, A., Kabir, M.A., Han, J.: Blockchain for fraud prevention: a work-history fraud prevention system. In: 17th IEEE International Conference on Trust, Security and Privacy in Computing and Communications/12th IEEE International Conference on Big Data Science and Engineering, pp. 1858–1863. IEEE (2018)

33. Wang, X., Feng, L., Zhang, H., Lyu, C., Wang, L., You, Y.: Human resource information management model based on blockchain technology. In: IEEE Symposium on Service Oriented System Engineering (SOSE), San Francisco, CA, pp. 168–173 (2017). https://doi.org/10.1109/SOSE.2017.34
34. Zamagna, R.: Ten Human Resource Blockchain Startups to Watch in 2021 (2020). https://thechainblog.com/future-of-hr-belongs-to-blockchain-technology/
35. Yi, C.S.S., Yung, E., Fong, C., Tripathi, S.: Benefits and use of blockchain technology to human resources management: a critical review. Int. J. Human Resour. Stud. **10**(2), 131140–131140 (2020)
36. Aishwarya, N.: Potential impact of blockchain in HR and people management. J. Emerg. Technol. Innov. Res. **5**(9), 127–130 (2018)
37. Faliagka, E., Tsakalidis, A., Tzimas, G.: An integrated e-recruitment system for automated personality mining and applicant ranking. Internet Res. **22**(5), 551–568 (2012)
38. Brandão, C., Silva, R., dos Santos, J.V.: Online recruitment in Portugal: theories and candidate profiles. J. Bus. Res. **94**, 273–279 (2019)
39. Swaroop, K.R.: E-HRM and how it will reduce the cost in organization. Asia Pac. J. Mark. Manag. Rev. **1**(4), 133–139 (2012)
40. Dulebohn, J.H., Marler, J.H.: E-compensation: the potential to transform practice. In: Greutal, Stone (eds.) The Brave New World of e-HR, pp. 166–189. Jossey-Bass, San Francisco, CA. (2005); Esterhuyse, M., Scholtz, B.: The Intention to Use e-Learning in Corporations
41. Stone, D.L., Dulebohn, J.H.: Emerging issues in theory and research on electronic human resource management (eHRM). Hum. Resour. Manag. Rev. **23**(1), 1–5 (2013)
42. Amara, N.B., Atia, L.: E-training and its role in human resources development. Glob. J. Human Resour. Manag. **4**(1), 1–12 (2016)
43. Mohsin, M., Sulaiman, R.: A study on e-training adoption for higher learning institutions. Int. J. Asian Soc. Sci. **3**(9), 2006–2018 (2013)
44. Aparicio, M., Bacao, F., Oliveira, T.: An e-learning theoretical framework. J. Educ. Technol. Soc. **19**(1), 292–307 (2016)
45. Esterhuyse, M., Scholtz, B.: The intention to use e-learning in corporations. In: CONF-IRM, p. 12 (2016)
46. Jackson, C.B., Quetsch, L.B., Brabson, L.A., Herschell, A.D.: Web-based training methods for behavioral health providers: a systematic review. Adm. Policy Ment. Health 1–24. PMID:29352459 (2018)
47. Agrawal, S.: NITI Aayog Works on Strategy to Leverage Blockchain Technology, 17 March 2018 [Online]. https://economictimes.indiatimes.com/news/. Accessed 12 July 2020
48. Das, P.S.A.A.: Blockchain Technology Report India 2018. Inc42, Delhi (2018)
49. Panda, S.K., Mohammad, G.B., Nandan Mohanty, S., Sahoo, S.: Smart contract-based land registry system to reduce frauds and time delay. Secur. Priv. e172 (2021). https://doi.org/10.1002/spy2.172
50. Panda, S.K., Satapathy, S.C.: Drug traceability and transparency in medical supply chain using blockchain for easing the process and creating trust between stakeholders and consumers. Pers. Ubiquit. Comput. (2021). https://doi.org/10.1007/s00779-021-01588-3
51. Niveditha, V.R., Sekaran, K., Amandeep Singh, K., Panda, S.K.: Effective prediction of bitcoin price using wolf search algorithm and bidirectional LSTM on internet of things data. Int. J. Syst. Syst. Eng. **11**(3–4), 224–236.
52. Sathya, A.R., Panda, S.K., Hanumanthakari, S.: Enabling smart education system using blockchain technology. In: Panda, S.K., Jena, A.K., Swain, S.K., Satapathy, S.C. (eds.) Blockchain Technology: Applications and Challenges. Intelligent Systems Reference Library, vol. 203. Springer, Cham (2021). https://doi.org/10.1007/978-3-030-69395-4_10
53. Lokre, S.S., Naman, V., Priya, S., Panda, S.K.: Gun Tracking system using blockchain technology. In: Panda, S.K., Jena, A.K., Swain, S.K., Satapathy, S.C. (eds.) Blockchain Technology: Applications and Challenges. Intelligent Systems Reference Library, vol. 203. Springer, Cham (2021). https://doi.org/10.1007/978-3-030-69395-4_16

54. Panda, S.K., Daliyet, S.P., Lokre, S.S., Naman, V.: Distributed Ledger Technology in the Construction Industry Using Corda, The New Advanced Society: Artificial Intelligence and Industrial Internet of Things Paradigm. https://doi.org/10.1002/9781119884392.ch2
55. Panda, S.K., Satapathy, S.C.: An investigation into smart contract deployment on Ethereum platform using Web3.js and solidity using blockchain. In: Bhateja, V., Satapathy, S.C., Travieso-González, C.M., Aradhya, V.N.M. (eds.) Data Engineering and Intelligent Computing. Advances in Intelligent Systems and Computing, vol. 1. Springer, Singapore (2021). https://doi.org/10.1007/978-981-16-0171-2_52
56. Panda, S.K., Rao, D.C., Satapathy, S.C.: An investigation into the usability of blockchain technology in internet of things. In: Bhateja, V., Satapathy, S.C., Travieso-González, C.M., Aradhya, V.N.M. (eds.) Data Engineering and Intelligent Computing. Advances in Intelligent Systems and Computing, vol. 1. Springer, Singapore (2021). https://doi.org/10.1007/978-981-16-0171-2_53
57. Varaprasada Rao, K., Panda, S.K.: A design model of copyright protection system based on distributed ledger technology. In: Satapathy, S.C., Lin, J.C.W., Wee, L.K., Bhateja, V., Rajesh, T.M. (eds.) Computer Communication, Networking and IoT. Lecture Notes in Networks and Systems, vol 459. Springer, Singapore (2023). https://doi.org/10.1007/978-981-19-1976-3_17
58. Varaprasada Rao, K., Panda, S.K.: Secure electronic voting (E-voting) system based on blockchain on various platforms. In: Satapathy, S.C., Lin, J.C.W., Wee, L.K., Bhateja, V., Rajesh, T.M. (eds.) Computer Communication, Networking and IoT. Lecture Notes in Networks and Systems, vol. 459. Springer, Singapore (2023). https://doi.org/10.1007/978-981-19-1976-3_18
59. Sadhya, V., Sadhya, H.: Barriers to Adoption of Blockchain Technology (2018). https://aisel.aisnet.org/amcis2018/adoptiondiff/presentations/20. Accessed 28 June 2019
60. Panda, S.K., Dash, S.P., Jena, A.K.: Optimization of block query response using evolutionary algorithm. In: Bhateja, V., Satapathy, S.C., Travieso-González, C.M., Aradhya, V.N.M. (eds.) Data Engineering and Intelligent Computing. Advances in Intelligent Systems and Computing, vol. 1. Springer, Singapore (2021). https://doi.org/10.1007/978-981-16-0171-2_54
61. Nanda, S.K., Panda, S.K., Das, M., Satapathy, S.C.: Automating vehicle insurance process using smart contract and Ethereum. In: Chakravarthy, V.V.S.S.S., Flores-Fuentes, W., Bhateja, V., Biswal, B. (eds.) Advances in Micro-Electronics, Embedded Systems and IoT. Lecture Notes in Electrical Engineering, vol. 838. Springer, Singapore (2022). https://doi.org/10.1007/978-981-16-8550-7_23
62. Gao, W., Hatcher, W.G., Yu, W.: A Survey of Blockchain: Techniques, Applications, and Challenges (2018). https://semanticscholar.org/paper/a-survey-of-blockchain:-techniques,-applications,-gao-hatcher/6d67f30a76b28ce21bd2719. Accessed 28 June 2019
63. Panda, S.K., Elngar, A.A., Balas, V.E., Kayed, M. (eds.): Bitcoin and Blockchain: History and Current Applications, 1st ed. CRC Press (2020). https://doi.org/10.1201/9781003032588
64. Panda, S.K., Jena, A.K., Swain, S.K., Satapathy, S.C. (eds.): Blockchain Technology: Applications and Challenges. Springer, Intelligent Systems Reference Library. https://doi.org/10.1007/978-3-030-69395-4
65. Baker, P.: Blockchain HR Technology: 5 Use Cases Impacting Human Resource (2021). https://www.techtarget.com/searchhrsoftware/tip/Blockchain-HR-technology-5-use-cases-impacting-human-resources. Accessed 28 June 2020
66. Ruel, H., Bondarouk, T., Looise, J.K.: E-HRM: innovation or irritation: an explorative empirical study in five large companies on web based HRM. Manag. Rev. **15**, 364–380 (2004)

Chapter 4
Securing Electronic Health Record System in Cloud Environment Using Blockchain Technology

Dileep Kumar Murala, Sandeep Kumar Panda, and Santosh Kumar Sahoo

Abstract Current online healthcare services Electronic Health Records/Electronic Medical Records (EHR/EMR) play an important role in storing, sharing, and maintaining patients' medical records. Implementing the EHR system in a cloud environment offers many options, including ubiquitous access, elastic computing resources, high-grade fault tolerance, and differentiation with other systems. Cloud storage solves the problem of storing complex medical data in leisure mode, while cloud service providers in transit are treated as non-covered units. Therefore, the cloud service provider has no obligation to ensure privacy and adequate access to the EHR. Consequently, security and privacy issues have become major issues for optimizing EHR systems for cloud environments. To provide security to the cloud environment the people use different cryptographic techniques. Cloud users use Trusted Third Party (TTP) for authentication and auditing, protecting unauthorized users, and authorized confidential data. Trust becomes a major barrier to sharing electronic data in cloud environments. If a third party is compromised, then there is a chance to hack the data because the client does not aware of the location information of storage. So, the trust issue is a challenging task in the cloud environment. To overcome these problems, we have introduced blockchain technology in cloud environments to securely transfer EHR data between authorized entities. Blockchain technology is important in such cases because it provides data-ledger-based features that can be distributed to all companies on the network. This approach is a tamper-proof mechanism because every health transaction information in the blockchain is stored as hash values. This approach provides tamper-proof, trust, confidentiality, integrity, authentication, availability, and easy access control to the cloud user data at the time of data transfer and rest modes.

D. K. Murala (✉) · S. K. Sahoo
Department of Computer Science and Engineering, Faculty of Science and Technology (IcfaiTech), ICFAI Foundation for Higher Education, Hyderabad, Telangana, India
e-mail: drdileepm@ifheindia.org

S. K. Panda
Department of Data Science and Artificial Intelligence, Faculty of Science and Technology (IcfaiTech), ICFAI Foundation for Higher Education, Hyderabad, Telangana, India
e-mail: sandeeppanda@ifheindia.org

S. K. Panda et al. (eds.), *Recent Advances in Blockchain Technology*,
Intelligent Systems Reference Library 237, https://doi.org/10.1007/978-3-031-22835-3_4

Keywords Cloud security · Electronic health record · Blockchain · Cloud privacy · Trust · Tamper-proof mechanism

4.1 Introduction

The twenty-first century saw a huge leap in digital technology which is changing the landscape of the health care system worldwide [1]. The transformation in the medical service industry is a continuous and steady movement from a paper-based health care system to an electronic system. So, there are changes over paper-based to computer-based electronic healthcare records for improvement [2]. Health professionals have to maintain E-Medical Records (EMR) and E-Health Records (EHR). Patients' Personal Health Records (PHR) are controlled by the patient or their relatives. E-Health Data (EHD) is the systematic storage of patients' smart health records in the form of computerized patient records [3]. Advantages of EHRs incorporate the utilization of simple and quick medical information, the capacity to deal with a feasible medical work process, reduce medical mistakes, expand patient security, decrease medical expenses, and improve medical decision making and stronger support [4]. In any case, the progress from traditional health care systems to e-health (electronic health) care presents specific difficulties concerning the privacy and security of medical data [5].

Moreover, the advancement in information and communication technology leads to face various security and privacy-related attacks to the EHD [6]. To provide security and privacy to the EHD, we must follow the three fundamental security goals as Confidentiality, Integrity, and Availability (CIA). Data confidentiality provides only authorized users who should be able to access data. If any unauthorized user wanted to access the data then confidentiality gets compromised. Data integrity provides the E-health data that has not been modified or damaged by any unauthorized users [7]. Data availability ensures that when and where data is required always data will remain accessible even when natural disasters, system failures, and denial-of-service attacks occurred [8]. To provide security, Accountability also plays a crucial role which provides asking the right to the users when something has happened. Privacy plays a very important role in e-health data [9]. To maintain privacy for e-health data access control plays a curtail role. Access control verifies that the user is authorized to access specific information or not [10]. Access control determines the people who access which type of data and up to which level. In EHRs, maintaining security and privacy against different threats and attacks becomes a major problem. There is a real concern about different user's access levels to patients' EHRs data [11]. In EHRs system security could compromises due to information disclosed by unauthorized user accesses and violence's which includes trust, authentication, and availability issues [12]. In EHRs, if the number of administrative users increases, privacy could become a major problem [13].

Nowadays, cloud computing is widely used in the healthcare industry [14]. In the health care system, cloud computing provides various facilities like convenient storage and easy exchange of medical information among various users with less cost [15]. It helps in the collection, storing, and retrieval of health care data, despite the time and space restrictions of various patients, doctors, and healthcare providers. With enhanced efficiency and effectiveness, cloud services offer various advantages like less cost, easy access, effective processing, and dynamic data updating, etc. [16]. Because cloud data can operate on various remote servers and the data can be integrated and managed by many users as a single ecosystem accessed from different regions, it can compromise, leading to confidentiality and security. Its storage on outsider servers normally expands these vulnerabilities. Considering the crowded nature of health data in the open segment, there is a need to make increasingly verify, proficient and viable systems for sharing and getting information among partners [17]. Numerous advancements are now being utilized in cloud situations to give the security and privacy of a smart health care system [18].

The current advanced encryption method like the Attribute-Based Encryption (ABE) method is incapable of solving this problem due to its expensive computation [19]. The majority of approaches like Key-Policy Attribute-Based Encryption (KP-ABE) and Ciphertext-Policy Attribute-Based Encryptions (CP-ABE) are permitted a private key administration place to arbitrarily choose a master key and permit clients to make decoding keys [20]. In the event of a key-manager attack, these solutions cannot prevent data from being attacked. Other unsafe dangers incorporate an introduction to risky harm data, uncovering delicate patient data to other people, sharing and composing of login credentials, or reacting to phishing messages [21]. One answer to defeating these confinements in the present framework is the presentation of a patient-focused electronic health framework. A customized electronic health record framework in which the patient is the ubiquitous consistent supplier for their information (aside from in crises) by all partners through specialists, drug specialists, medical caretakers, and researchers [17, 22, 23]. Blockchain innovation can be utilized as an inherent access control device to help conveyance instruments conveyed in the cloud. This methodology is a carefully designed system because each health exchange data in the blockchain is put away as hash regards. It can guarantee the security, privacy, and accessibility of EHD [18, 24]. The presentation of this innovative development incorporating with ECC cryptographic algorithm gives a safe and effective system for proficient storage, move, and utilization of electronic health records in the cloud environment [25, 26, 27].

4.1.1 Motivation

Cloud users use Trusted Third Party (TTP) for authentication and auditing, protecting unauthorized users, and authorized confidential data. Trust becomes a major barrier to sharing electronic data in cloud environments. If a third party is compromised, then there is a chance to hack the data because the client does not aware of the location information of storage. So, the trust issue is a challenging task in the cloud environment. The approach will be tamper-proof and provide trust, confidentiality, integrity, authentication, availability, and easy access control to the cloud user data at the time of data transfer.

4.1.2 Novelty of Our Work

This approach provides a decentralization storage framework and provides smart contracts-based access control mechanism to improve the security of EHRs data sharing in the cloud environment. To provide security and privacy of medical data, we used ECC cryptographic algorithm for encrypted data. This technique significantly provides system management and provides security. Also, we provide a tampered-proof mechanism for medical data and tracing the source data by using blockchain technology. We implement this approach and generate a comparison study with existing systems and find out network latency for various users count.

4.1.3 Our Contribution

To provide security and privacy of medical data, we proposed an approach for securing EHR systems using blockchain technology in a cloud environment. In this approach, we use the Elliptic Curve Cryptography (ECC) algorithm for generates security between key pairs for public-key encryption by using the mathematics of elliptic curves. This technique significantly provides system management and provides security. Also, we provide a tampered proof mechanism for medical data and tracing the source data by using blockchain technology. We implement this approach and generate a comparison study with existing systems and find out network latency for various users count.

4.2 Background Concepts and Literature Survey

This section will review the previous work done in our research and summarize the vital background knowledge required to understand the paper.

4.2.1 Background Concepts

Blockchain-based EHR system: Blockchain innovation is known for its application in Bitcoin digital currency which is an open record for making and keeping up exchange information and trustworthiness [7]. Blockchain structure is formed as a collection of blocks and their links called chains [8]. Each block encloses serial block number, data to be stored on the block, time-stamp, and hash values of the current block and its previous blocks. These blocks form a chain using a hash chain [16]. Figure 4.1 describes the basic architecture of blockchain technology. The chaining system guarantees the honesty of this protected information structure. Blockchain technology provides security, transparency, agreement, and trustworthy service based on the comparison of past and current blocks hash values [9, 28, 29].

Blockchain technology provides a distributed and fault-tolerant data sharing system in the network [11]. The blockchain system consists of several nodes each node maintained Distributed Ledger Technology (DLT) which is not easy to alter the data [10]. In blockchain for each transaction, it provides the acknowledgment for providing tamper-proof and integrity of data [12]. The acknowledgment consists of the transaction ID. By using transaction-id, we can easily identify the location and those transaction block details. Blockchain provides fault-tolerant data sharing; by using this we can easily eliminate compromised nodes within the network. The blockchain characteristics like tamper-proof, decentralized, fault-tolerant have attracted researchers to develop various applications in a cloud environment [30, 31].

Blockchain innovation is known for its application in Bitcoin digital currency which is an open record for making and keeping up exchange information and trustworthiness [32]. Blockchain consists of a sequence of blocks; each block is useful for storing a list of transaction records like a public ledger. Blockchain provides Peer-to-Peer (P2P) Distributed Ledger Technology (DLT), it allows transparency, unalterable, verifiability, and audit without the involvement of trusted third-party persons. Blockchain provides different characteristics those are decentralization, tamper resistance, immutable, autonomy, fraud-free, secure, user anonymity, auditability, and trust [33]. Integrating blockchain technology into the cloud environment can lead to data provenance [34]. Due to the integration of blockchain technology into the cloud, the cloud centralized network nodes are become a distributed network to record data provenance, fault-tolerant, and secured with a DLT strong cryptographic notion [35, 36].

Blockchain-based access control offers a collection of new security highlights for e-health, with incredible points of interest over customary access control arrangements. The blockchain-based medical data sharing system consists of mainly three roles [37]. Those are the System Manager (SM), the patient, and the hospital. The SM manages the entire system. Whenever a patient visits a doctor, first he/she has to register with the hospital. Then, the hospital will assign a doctor for him/her to make a diagnosis. After completion of treatment, the verifiers perform the verification

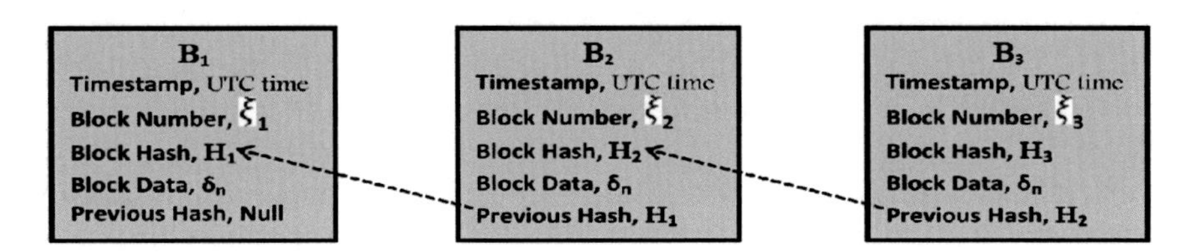

Fig. 4.1 Blockchain technology basic architecture

process on the user's medical results. Finally, the doctor stores the patient's medical data stored in the blockchain. In this system, any doctor can access the patient's historical data with the patient's permission. Blockchain-based EHRs are unique to cloud-based EHRs. In cloud-based EHRs, the CSP takes the responsibility for storing medical data into the cloud, whereas, in blockchain-based EHRs, the server chooses any public parameters for storing different user's medical data in the blockchain network [38, 39]. In cloud-based EHRs, the server generates user's secret keys and distributes them to authorized users, but in blockchain-based EHRs, the server does not have the right to generate users' secret keys.

Elliptic Curve Cryptography (ECC): Neal Koblitz and Victor S. Miller suggested the ECC algorithm in the year 1985 and this was used widely used around 2004 [40]. ECC is a public key cryptography algorithm based on the elliptic curve theory. This theory is useful for generating small, efficient, and faster cryptography keys and requires less computational power for creating keys. With the rapid development of mathematics and computational power, the gap between factoring and multiplying large numbers is reduced, hence the key size will increase very fast [41]. ECC using small length keys compared to RSA but provides the same level of security. These types of cryptography algorithms are suitable for less power consumption devices like mobile phones due to the small-sized key [41, 42].

The working procedure of the ECC algorithm is depending upon the multiplication result of any randomly chosen points and those values on the elliptic curve. Even if we know the multiplication result and exact points used in the curve, it's very difficult to find the values used on the chosen points on the curve to get the multiplication results [40]. ECC algorithm uses discrete logarithms instead of Trapdoor functions. Compare to trapdoor functions, discrete logarithms are significantly difficult to solve, difficult problems generate a very strong security system [41]. ECC algorithm provides a reliable transaction, high security with short keys, and a more efficient processing algorithm.

This theory is useful for generating small, efficient, and faster cryptography keys. Compare to other public-key cryptography algorithms (like Diffie-Hellman, RSA, etc.), the ECC algorithm requires less computational power for creating keys. For generating keys, instead of using large prime numbers, ECC follows the elliptic curve equation [43]. Every public-key cryptography algorithm must have key characteristics for generating keys called Trapdoor Functions. It is very difficult to find good trapdoor functions for providing security in public-key cryptography systems. The working procedure of the ECC algorithm is depending upon the multiplication result of any randomly chosen points and those values on the elliptic curve. Even if we know the multiplication result and exact points used in the curve, it is very difficult to find the values used on the chosen points on the curve to get the multiplication results [44]. ECC algorithm uses discrete logarithms instead of Trapdoor functions. Compare to trapdoor functions, discrete logarithms are significantly difficult to solve and difficult problems generate a very strong security system [45, 46]. ECC works on an elliptic curve mathematical equation its looks like this $y2 = x3 + ax + b$ and the graphical representation of the Elliptic curve is given in Fig. 4.2.

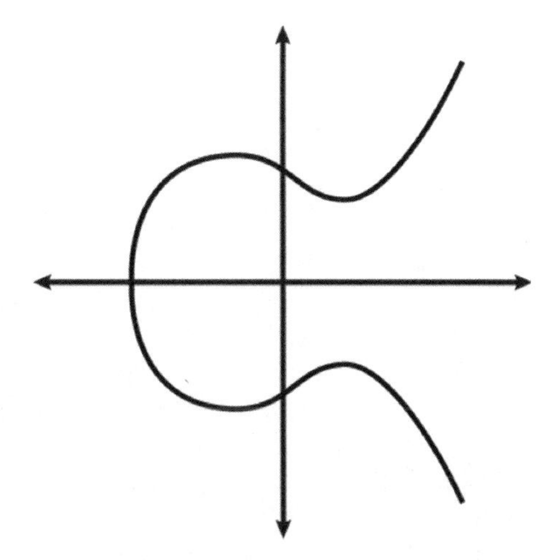

Fig. 4.2 Elliptic curve

ECC using small length keys is compared to RSA but provides the same level of security. These small-size keys are more useful for providing security and consume less power. ECC algorithm provides a reliable transaction, high security with short keys, and a more efficient processing algorithm [43].

Security Issues: There are two major technical issues in cloud computing those are server down issues, and power consumption issues. Server down issue was occurred due to non-availability of resources or denial of service attack. Because of this issue, the authorized users facing various difficulties during access the resources from the cloud server. Another issue related to the cloud was power consumption issues. While accessing cloud services by using physical devices like smartphones, laptops it consumes more power.

Cloud platforms facing various Legal issues those are security violations, privacy violations, data loss, data management issues, electronic discovery, liability for copyright infringement, data breaches, hacking, and cybersecurity issues.

4.2.2 Literature Survey

Farzandipour et al. [2] discussed the transformation in the medical service industry as a continuous and steady movement in the health care system from paper-based records to electronic records [2]. Greenhalgh et al. [4] explained the improvement over those changes. Those computer-based e-records are like Electronic Medical Records (EMR), Electronic Health Records (EHR), Personal Health Records (PHR), and Electronic Health Data (EHD) [4]. EHRs and EMRs are maintained by health professionals, while PHRs are controlled by the patient or their relatives. EHD is a systematic collection of patients' smart health records in the form of computerized patient records [2]. Haas et al. [5] explained the advantages of EHRs incorporating the utilization of simple and quick medical information, the capacity to deal with a feasible medical work process, reduce medical mistakes, expanded patient security, decreased medical expenses, and improved medical decision making and stronger support [5]. The progress from traditional health care systems to e-health (electronic health) care presents specific difficulties concerned with the privacy and security of data [47].

Azaria et al. [14] explained blockchain innovation is known for its application in Bitcoin digital currency which is an open record for making and keeping up exchange information and trustworthiness [14]. Christian Esposito et al. [1] discussed Blockchain structure which is formed as a collection of blocks and its links called chain [1]. Each block encloses serial block number, data to be stored on the block, timestamp, and cryptographic hash values of the current block and its previous blocks. These blocks are formed as a hash chain. Blockchain technology provides a distributed and fault-tolerant data sharing system in the network [6, 48–50]. The blockchain system consists of several nodes each node maintained Distributed Ledger Technology (DLT), it's not easy to alter the data. In the blockchain, for each transaction, it provides the acknowledgment for providing tamper-proof and integrity of data [5]. The acknowledgment consists of the transaction ID. By using transaction id, the location and those transaction block detail can easily be identified. Blockchain provides fault-tolerant data sharing; by using this we can easily eliminate compromised nodes within the network [7]. The blockchain characteristics like tamperproof, decentralized, fault-tolerant have attracted researchers to develop various applications in a cloud environment [8]. Blockchain technology is used in various domains to solve various problems. The author proposed a model by using blockchain technology in the biomedical domain [10, 51, 52]. This approach solves various biomedical-related problems. In paper [9], Jiaxing, Li et al. proposed a framework that is more focused on the patient's private data in the cloud environment. This framework was mainly designed for rural areas where cost plays a massive role. This approach ensures cost-effectiveness. In these frameworks, different medical professionals, patients, and policymakers connected remotely through a cloud. Patients share their data and get medical services from different medical professionals remotely within the cloud. This approach provides a symptom analysis mechanism where collecting data and data delivery are the key points. Wiljer et al. [21] proposed cloud supported e-health systems provide more benefits to patients and medical institutions to efficiently manage their EHRs [21]. Cloud-based EHRs also suffer from various security and privacy challenging issues and threats. To protect patients' privacy from various attacks and threads, the EHRs manager performs encryption on EHRs data before storing it. Eltayieb et al. [40] provided the third-party-based cryptographic key management approach for protecting patients' E-health records in the cloud [40]. In their approach, the third-party-based server provides the secret key to all the patients. As a result, the third-party server can retrieve the patients' EHRs and may perform unauthorized activities on e-health records therefore the patient's data privacy will be affected.

Sun et al. [53] proposed an e-health system to maintain privacy to the patients' data without having a third-party server in his work [53]. In this approach, the patient's e-health records are outsourced by the patients. Before outsourcing data into the cloud, the doctor must provide e-health records to the patients. This approach provides security to the patient's data but, this approach can't be directly adopted in e-Health systems because the communication and computation costs become a heavy burden. Furthermore, at the time of treatment, the doctor provides trusted reports. Sometimes malicious doctors or unauthorized users wanted to tamper with the cloud server outsourcing EHRs data. Detect such malicious activities become a very difficult task. Moreover, the existing approach doesn't consider the timeliness of EHRs. To overcome third-party-related issues, improve security and privacy and reduce cost-related issues for sharing EHRs data in a cloud environment using blockchain has been stated at the theoretical level within the prototype described by Zhang [54]. He proposed an efficient and secure de-duplication called the Healthdep approach scheme for cloud-assisted E-Health systems. It was mainly developed for rural areas where cost is an important issue. His approach ensures cost-effectiveness and focuses on patient's private data in the cloud environment. This approach doesn't suitable for group users in a public cloud environment. Trust and confidentiality-related issues are more. Rifi et al. [18] proposed an E-health system to maintain privacy to the patients' data without having a third-party server [18]. In this paper, he identified the specific approaches and various benefits for improving security and privacy for exchanging medical data in a cloud environment using blockchain technology. The primary limitation of this approach is performance.

4.3 Proposed System Model

In a public cloud environment, securely transferring EHRs data between authorized users becomes a major problem. The existing EHRs systems suffered from security and privacy issues (Data confidentiality, Availability, Decentralized access, Data Authorization, User authentication, Data Integrity, and Data privacy) for securely sharing data between authorized users in a cloud environment. Unauthorized users can easily access or damaged cloud storage data without the data owner's permission. To overcome these problems we proposed securing electronic health record sharing systems in a cloud environment using Blockchain technology. This approach provides a decentralization storage framework and provides smart contracts-based access control mechanism to improve the security of EHRs data sharing in the cloud environment.

4.3.1 System Architecture

EHRs system contains a collection of e-health records; each record contains medical history and personal information about patients. Each record contains the Patient ID and Area Id of the patient. EHR system using Blockchain technology in the cloud architecture is shown in Fig. 4.3. It contains different components that are described as follows:

EHRs Manager: EHR Manager plays an important role in securely sharing data between an authorized user and a cloud server using blockchain technology. The major responsibility of the EHRs manager is to control different user transactions, data storage processes, and provide data access permissions to the authorized user in the blockchain network. Another major role of EHRs managers is to make smart contracts by using strict user policies.

Smart contracts: It defines different operations that allow access control systems. The end users interrelate with the contract system by using the Application Binary Interface (ABI) and contract address. In smart contracts, the system has to identify user's requests, authenticate user requests, and provide the data access permissions for requested users through the messages or transactions. In this approach, smart contracts act as major software, through these different users access all operations

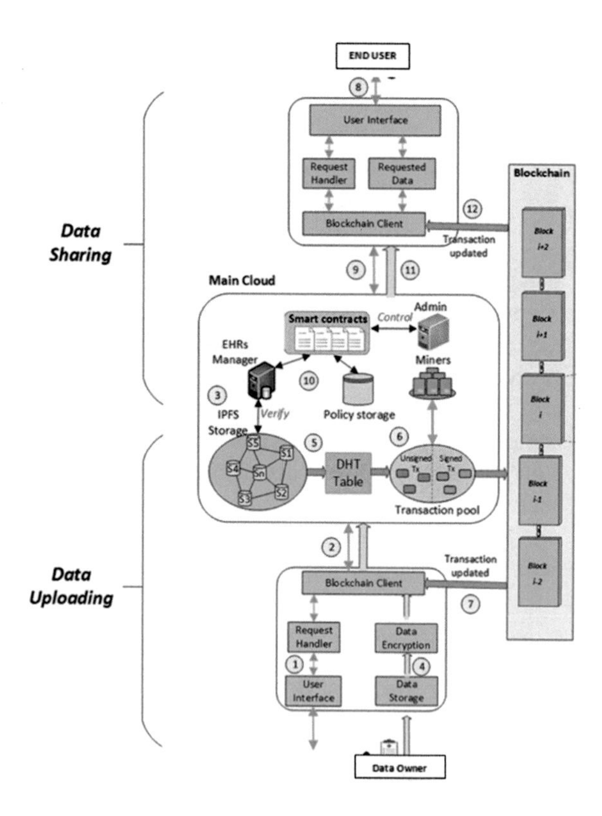

Fig. 4.3 EHR system using blockchain technology in the cloud

in the blockchain network. Smart contracts within Ethereum blockchain provides achieve automated recording of events that are accessible to all participating stakeholders in the network. The efficiency of the blockchain solution is highly dependent on the coding of the smart contract and also the consensus algorithm used to verify and confirm a transaction.

Admin: The admin plays a vital role. Admin performs various operations in a cloud environment like the addition of new users, deletion and change access permissions, or revoke access permissions, and also manages different transactions.

Decentralized storage: In the blockchain network, it is not possible for storing and sharing a large amount of data. To overcome this problem, we use InterPlanitary File System (IPFS). IPFS is useful for storing and sharing data in the blockchain network. By using IPFS, we can reduce single-point failures, increase data retrieval speed, and have high storage throughput. In our work, the major advantages of IPFS are to encrypt and store the medical record in IPFS nodes. The hash values of each medical data are recorded and stored in Distributed Hash Table (DHT). The users access files based on the hash values of their contents. In this approach, we integrate smart contracts to IPFS to enhance user access control and decentralization. We construct the IPFS platform with a network of individual nodes. Each node stores patient's location ID. The system architecture is mainly divided into two phases those are data upload phase and the data sharing phase.

Phase-1 Data Upload: In the data upload phase, the data owner sends the request for creating Blockchain to the Cloud Server (CS). The CS sends the request to the EHR Manager for confirmation. The EHRs manager sends the upload request to the smart contract in the form of a signal for the verification process. The smart contract visits the policy storage and verifies the policy list. If the data owner is an authorized user, then EHRs sends the acknowledgment to the data owner for uploading data into the server. Once the request is accepted, the data owner encrypts the data file with EHRs public key by using Elliptic Curve Cryptography (ECC) asymmetric algorithm. The encrypted file was uploaded to Interplanetary File System (IPFS) server. The IPFS server automatically generates and returns the hash value for the uploaded file to the EHRs manager and stores the same hash value in the Distributed Hash Table (DHT). If any modifications are occurring to the uploaded files the EHRs manager can easily detect those modifications. In addition to that, the transaction pool stores all the uploaded files' metadata information. In this approach, the role of miners is to create all transactions into permanent blocks, verifies those blocks, and sends them to other miners for confirmation. If other miners are also agreed, then the confirmation signatures are appended to the blocks. Finally, all other users in the blockchain network receive those blocks and verify the copy of each block with the blockchain client. In this technique, each transaction can monitor in a distributed manner.

Phase-2 Data Sharing: The end-user sends the request to the cloud server for accessing data. Before accessing data, the end-user must create a blockchain account. It is comprised of a public and a private key of end-users for authentication. The Cloud Server sends the request to the EHR Manager for verification. The EHR manager sends those data requests to a smart contract for the verification process in the form of a signal. The smart contract visits the policy storage and verifies the policy list. After verification smart contract returns the results to the EHRs manager. If the end-user was an unauthorized user, The EHRs manager blocks the requests coming from this user. Otherwise, the EHRs manager analyses the user's requestIDs (Area ID, Patient ID), verify the hash values of those requests in DHT and retrieves the required data from the IPFS storage system. The IPFS system provides encrypted files to the EHRs manager. The EHRs manager decrypts the retrieved files with the help of its private key by using the ECC cryptography algorithm and sends the data to the end-user. Like the data upload process, each transaction creates in the form blocks and inserted into a transaction pool for the mining process. Each block is verified by the different miners and added sequentially to the Blockchain for sharing with other users. The algorithm of EHRs data access protocol is given below.

Algorithm 1.1.: EHR data access protocol

Input: T_i

Output: Outcome
Initialization: (By the EHR Manager)

1: Receive a Request Ti from a cloud User
2: Get the public key of the request
    PUKEY ← Ti.get_Sender _ PublicKey()
3: Send the public key to Admin
    msg.sender ← EHRsManager
Verification (By the smart contract and Administrator)
4: **if** PUKEY is existing in the policy-list
    Policy-list (PUKEY) ← true (T)
5: **end**
6: Decrypt the transaction
    decodeTi ← abiDecoder : decode_Method (Ti)
7: Specify request information
    Loc_Address ← server.get _Data(decodeTi ([DataIndex]))
8: Specify AreaID and PatientID
    AreaID ← Loc_Address (Index[AreaID])
    PatientID ← Loc_Address (Index[PatientID])
EHRs Data Retrieval (By the smart-contract)
9: **while** condition True do
10: **if** policy_List (AreaID) → T then
11: **if** policy_List (PatientID) → T then
12: Result ← retrieve_EHRs (PUKEY; Loc_Address)
13: break;
14: **else**
15: Outcome ← fine (PUKEY; act)
16: break;
17: **else**
18: Outcome ← fine (PUKEY; act)

4.3.2 Use Case Diagram of EHR System

In our proposed methodology, actors are mainly categorized into 3 types those are data owner/health provider, end-users, and cloud service provider (CSP). Figure 4.4 describes the roles and responsibilities of each user associated with the EHR system. The role of the healthcare provider is to register with CSP with valid details. The CSP verifies health care provider details and generates ID and sends it to the Health care provider. The health care provider encrypts the uploaded data and creates a block for each e-health record and sends those data to CSP. The major role of CSP is to authenticate different user's requests and verifies each request. The user views the available data in the cloud and sends the request for accessing those data to the data owner. Then, the data owner view and verify the requests coming from various users, then, send the requested data to the users and produce a key for each transaction. The end-user based on the verification key downloads the requested data and decrypts the file by using a key generated by the data owner and access the data.

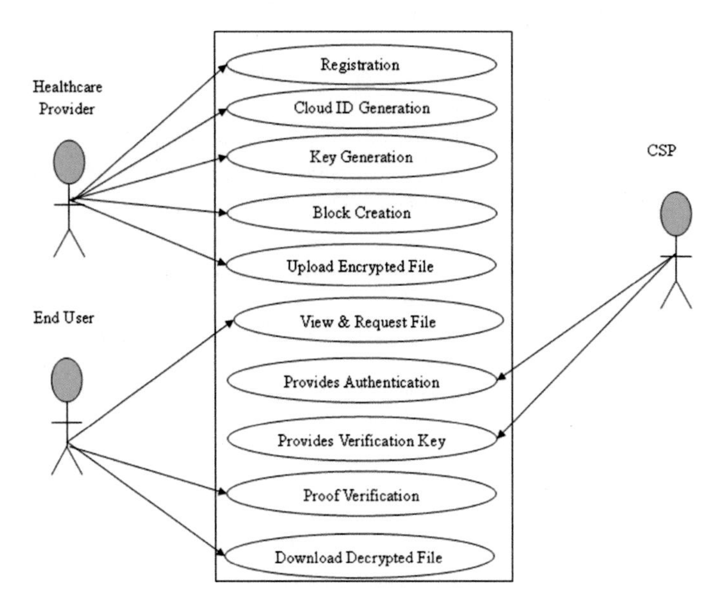

Fig. 4.4 The flow of information and medicine

4.4 Security and Network Performance

4.4.1 How Does Blockchain Provide Data Security?

Blockchain provides data security from unauthorized access and provides a tamper-proof mechanism for securing the data in the blockchain network.

Encryption and Validation: Blockchain performs encryption of user's data to prevent unauthorized modification without using a cryptographic signature. Blockchain provides decentralized data accessing in the network, because of this nature any node in the network verifies the file signature. If any changes occurred, then the signature is rendered invalid.

Securing data in a Decentralized manner: Blockchain technology is decentralized in nature. It does not depend on central control or third-party persons for sharing data between the users in a blockchain network. Blockchain uses various protocols across the network to securely transfer the data between authorized users. Blockchain provides a data ledger facility for storing data honest and accurate manner.

Difficulties in hacking: In Blockchain, data can be store in the form of blocks, each block contain information and link with previous blocks. These blocks of information are transferred through the blockchain network between various users. Blockchain provides a decentralized and distributed ledger across each node in a network. The data in the blockchain is updated and synchronized continuously. If an attacker tries to break the blockchain network and corrupt the data we can easily find out the single point failure in a network and using blockchain it's very difficult to tamper with the data. Blockchain is not only secured, reliable but also transparent which makes it even more appealing to various fields.

The network performance can be analyzed by considering transmission time as a measure. Here, the main aim is to minimize the network latency, decrease file loss rate, and improve the security in our approach. All the notations used for the analysis of performance are described in Table 4.1.

4.4.2 Security and Network Performance

Assume that all the channels are reliable and a group of users is accessing multiple data centers in Blockchain. Let L be the storage of loss file then, L is represented as:

$$\mathrm{L} = \mathrm{LFS} \cdot \mathrm{FLP}. \tag{4.1}$$

As all the file block sizes are the same (32 KB), the latency time of broadcast a file block to the data center is $\frac{bs}{bw_i}$. Here, we ignore the links between the users and the data center. Then the total latency of transmission is $\sum_{i=1}^{m} \frac{bs}{bw_i}$.

Table 4.1 Notations used in this approach

Notations	Meaning
LFS	Lost file size
FLP	File loss probability
F	The fraction of malicious nodes
AT	Attack times (successful)
M	File blocks count
N	P2P network nodes count
P	Probability of attack
A	Total files size
TTTF	Total Transmission Time of File
TTF	The transmission time of File
Bs	Block size
Bw	Bandwidth between the user and the data center
|FBS|	All File Block replicas (in number)
Ni	Node i
U	Users count
BWSi	Ni Links' latency
FSi	Ni file size
R	File replica
K	Probability of producing honest node
Q	Probability of producing attacker next block
Qz	Probability of never catch up z blocks by the attacker

Let N1, N2, ..., Nn represents nodes of this architecture. Then, a node can be chosen randomly having the smallest transmission latency. Therefore, the ith node TTF can be represented as $\frac{FS}{\text{arg}min_k BWS_k}$. Summing all TTF's we get

$$\text{TTTF} = \sum_{k=1}^{n} \frac{FS}{\text{arg}min_k BWS_k}. \tag{4.2}$$

4.4.3 Security Analysis

In our approach, the replicas of each file block are stored randomly nearby nodes in the blockchain network. The probability of getting a file block replica by the attackers is $\frac{r}{n}$ ($r = 1, 2, 3$ ...), but the probability of receiving all m file blocks is $\left(\frac{r}{n}\right)^m$.

If the count size of n and m is high, the probability of receiving all file blocks or replicas by the attackers is considerably tending to 0. The attacker's file block accessing speed is depending on computation power. Assume that more than half of the computational power of the network is accessed by the attacker, so the attacker can create a longer chain for fake trading information, because, all the network users only can distinguish the longest chain in the network. Nakamoto proofed the following formula [47].

$$Q_z = \begin{cases} 1, if K \leq Q \\ \left(\frac{Q}{K}\right)^Z, if K > Q \end{cases} \tag{4.3}$$

Let $K > Q$. If the number of blocks increases then the successful probability of an attack will be exponentially declined. The potential progress of attackers follows the poison distribution for the calculation of the average time spent on the generation of blocks. Then using Eq. (4.3), the probability of attacker catching can be computed by the following formula

$$AT = \begin{cases} \sum_{x=0}^{k} f(x).\left(\frac{Q}{K}\right)^{z-k}, if K \leq z \\ \sum_{x=0}^{\infty} f(x), if K > z \end{cases} \tag{4.4}$$

where $f(k) = \frac{\lambda_2^k.e^{-\lambda_2}}{k!}$ the value of $\lambda_2 = z.\frac{Q}{K}$.

Then the Eq. (4.4) can be rewritten as follows to avoid the infinite trail of distribution.

$$\mathrm{AT} = 1 - \sum_{k=0}^{z} \mathrm{f(k)}.\left(1 - \left(\frac{\mathrm{Q}}{\mathrm{K}}\right)^{\mathrm{z-k}}\right). \tag{4.5}$$

So using Eqs. (4.1) and (4.5), the function of file loss in the architecture will be calculated as follows.

$$L = \left(\frac{r}{n}\right)^m .AT.\frac{bs}{bw}. \tag{4.6}$$

This proposed approach is evaluated various security and privacy performance metrics like (Data confidentiality, Availability, Decentralized access, Data Authorization, User authentication, Data Integrity, and Data privacy) shown in Table 4.2 with this comparison analysis, our approach provides more security and privacy against various attacks and threats.

Network latency: It's defined as the total time taken for sending and processing the request between the sender and receiver. Figure 4.5 represents the average time taken for processing a different number of requests in a cloud environment. Compare to other existing approaches our approach proves better results in terms of network latency performance.

Table 4.2 Performance metrics comparison of our system with other works

Feature	Zhang [54]	BaDs' [55]	HealthTap [56]	Rifi [18]	Our system
Data confidentiality	No	No	No	Yes	Yes
Availability	No	No	Yes	Yes	Yes
Decentralized access	No	Yes	Yes	Yes	Yes
Data Authorization	Yes	No	Yes	No	Yes
User authentication	Yes	Yes	Yes	No	Yes
Data Integrity	Yes	Yes	Yes	Yes	Yes
Data privacy	Yes	Yes	Yes	Yes	Yes

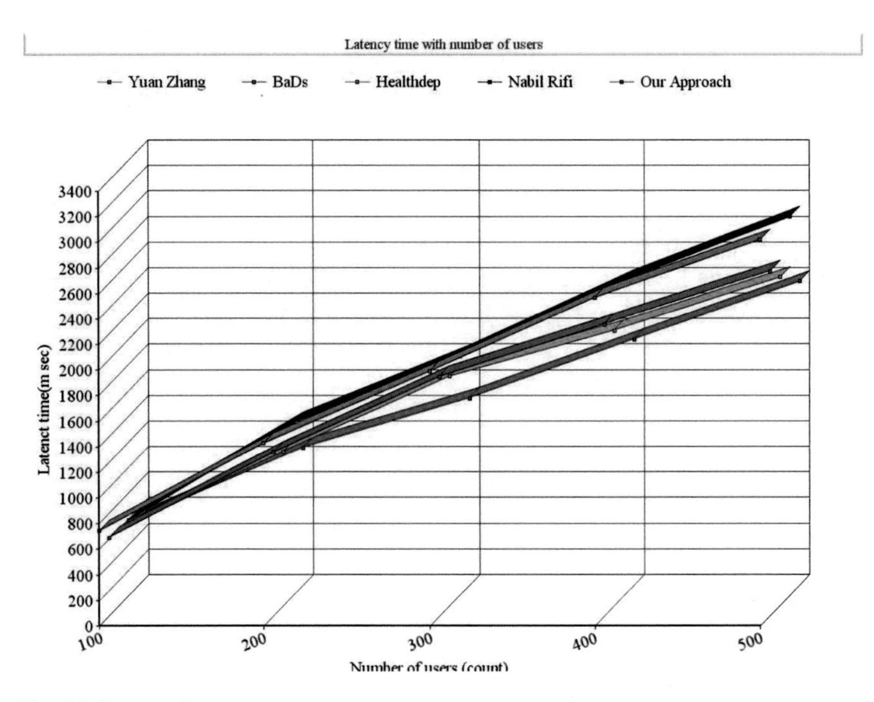

Fig. 4.5 Latency time

4.5 Implementation

The proposed approach was developed by using the Ethereum blockchain platform. Ethereum is a public blockchain platform any user can access. In this approach we used Solidity language for writing smart contracts; to perform compilation and testing we are using Remix IDE. Remix IDE is an online web-based development environment for implement the code and tests the code for smart contracts. The proposed approach was developed by using the Ethereum blockchain platform. Ethereum is a public blockchain platform any user can access. In Ethereum blockchain information is distributed among the various participating nodes can be done by automating processes without requiring manual input from the user and utilizing different features of the Ethereum blockchain such as web3js, JSON-RPC, and Infura.

In this approach we used Solidity language for writing smart contracts; to perform compilation and testing we are using Remix IDE. Remix IDE is an online web-based development environment for implement the code and tests the code for smart contracts. In this approach, for detecting various vulnerabilities in the code at different severity levels we use the SmartCheck tool. Oyente tool was also used to explore smart contract security.

User login: The user can log in with their username and password. If this is a new user, the user must register before logging in. This is a process that has been documented in the cloud. To use cloud documents, every health care provider must register. In this process, your basic information, email, contacts, etc. will be collected and stored in the cloud. Cloud ID is automatically generated for a specific user at the time of registration.

Healthcare Provider: The role of health care providers is to upload and download patients recodes. Before uploading data into the cloud, the healthcare provider encrypted data and crates blocks for each transaction.

Information Selection and Uploading: In this function of the cloud, to upload and maintenance of the data set the PHR is used by the health provider (Fig. 4.6).

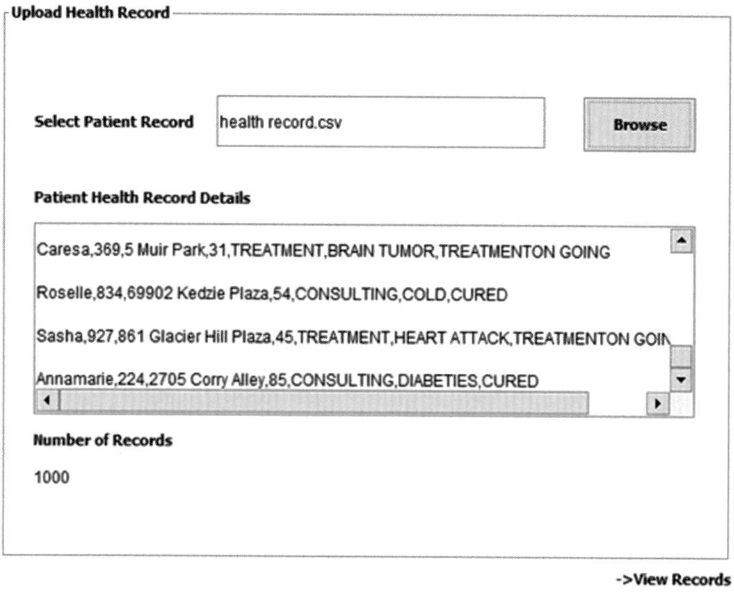

Fig. 4.6 Information selection and uploading

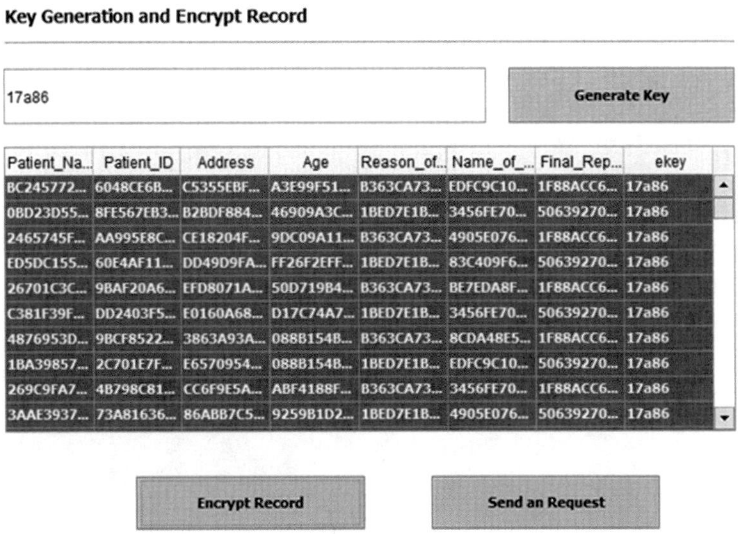

Key Generation and Encrypt Record

17a86

Generate Key

Patient_Na...	Patient_ID	Address	Age	Reason_of...	Name_of_...	Final_Rep...	ekey
BC245772...	6048CE6B...	C5355EBF...	A3E99F51...	B363CA73...	EDFC9C10...	1F88ACC6...	17a86
0BD23D55...	8FE567EB3...	B2BDF884...	46909A3C...	1BED7E1B...	3456FE70...	50639270...	17a86
2465745F...	AA995E8C...	CE18204F...	9DC09A11...	B363CA73...	4905E076...	1F88ACC6...	17a86
ED5DC155...	60E4AF11...	DD49D9FA...	FF26F2EFF...	1BED7E1B...	83C409F6...	50639270...	17a86
26701C3C...	9BAF20A6...	EFD8071A...	50D719B4...	B363CA73...	BE7EDA8F...	1F88ACC6...	17a86
C381F39F...	DD2403F5...	E0160A68...	D17C74A7...	1BED7E1B...	3456FE70...	50639270...	17a86
4876953D...	9BCF8522...	3863A93A...	088B154B...	B363CA73...	8CDA48E5...	1F88ACC6...	17a86
1BA39857...	2C701E7F...	E6570954...	088B154B...	1BED7E1B...	EDFC9C10...	50639270...	17a86
269C9FA7...	4B798C81...	CC6F9E5A...	ABF4188F...	B363CA73...	3456FE70...	1F88ACC6...	17a86
3AAE3937...	73A81636...	86ABB7C5...	9259B1D2...	1BED7E1B...	4905E076...	50639270...	17a86

Encrypt Record

Send an Request

Fig. 4.7 Key generation and encrypted patient records

Key Generation and Encrypted Patient Records: Before uploading data into the cloud the healthcare provider based on a key is generated by the cloud service provider and performs encryption operation on patients' data (Fig. 4.7).

View patient records: After uploading data into the cloud, the healthcare provider can view the patient's records. The data owner can easily fetch the required records by using patient-id (Fig. 4.8).

Send a request to a cloud server: After performing encryption the healthcare provider sends the request to the CSP for creating blocks and stores them into the cloud. After successfully send the request to the CSP the healthcare provider can view the request status with verification key whether the CSP accepts, reject, and is pending the status (Fig. 4.9).

Block Creation: After getting acceptance from CSP the healthcare provider creates and build blocks and view the blocks information uploading data into the cloud server (Fig. 4.10).

View Blocks: Figure 4.11 describe how many blocks and timestamp are created, we can also see those blocks. Each block stores the information whereas timestamp is a small data stored in each block as a unique serial and whose main function is to determine the exact moment in which the block has been mined and validated by the blockchain network.

Upload Acknowledgement: After successfully uploading data into the cloud server, the CSP sends the acknowledgment to the health care provider (Fig. 4.12).

EHR Manager: The EHR manager maintains all the patient records and views the different user's requests and also checks whether the user is authorized one or not. If the requested user is authorized user provides permission to the user to access the data (Fig. 4.13).

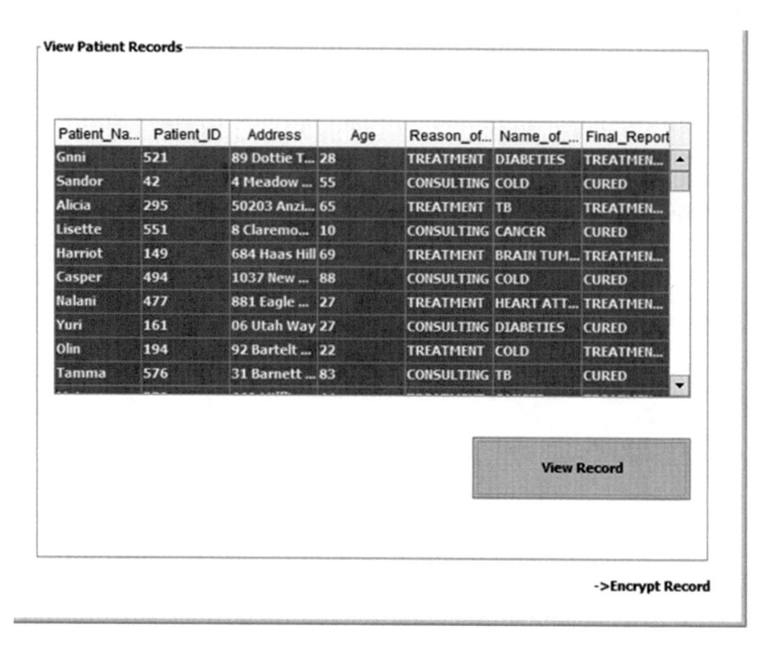

Patient_Na...	Patient_ID	Address	Age	Reason_of...	Name_of_...	Final_Report
Gnni	521	89 Dottie T...	28	TREATMENT	DIABETIES	TREATMEN...
Sandor	42	4 Meadow ...	55	CONSULTING	COLD	CURED
Alicia	295	50203 Anzi...	65	TREATMENT	TB	TREATMEN...
Lisette	551	8 Claremo...	10	CONSULTING	CANCER	CURED
Harriot	149	684 Haas Hill	69	TREATMENT	BRAIN TUM...	TREATMEN...
Casper	494	1037 New ...	88	CONSULTING	COLD	CURED
Nalani	477	881 Eagle ...	27	TREATMENT	HEART ATT...	TREATMEN...
Yuri	161	06 Utah Way	27	CONSULTING	DIABETIES	CURED
Olin	194	92 Bartelt ...	22	TREATMENT	COLD	TREATMEN...
Tamma	576	31 Barnett ...	83	CONSULTING	TB	CURED

Fig. 4.8 View patient records

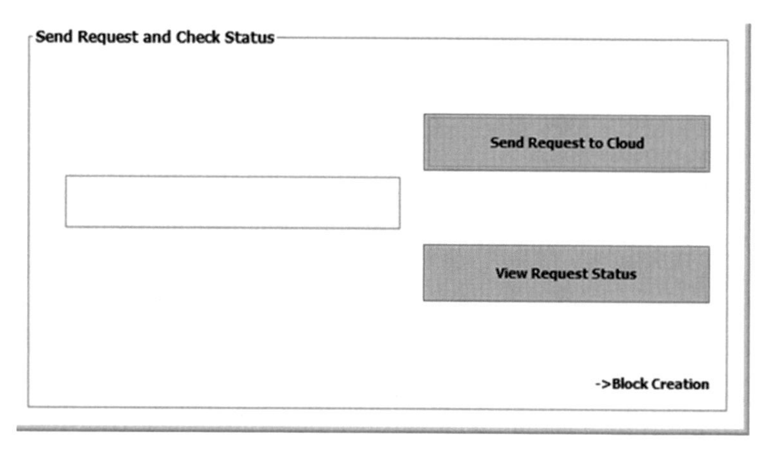

Fig. 4.9 Send a request to a cloud server

Verify user: Before grant permission, to the requested user the healthcare provider verify if the requested user has authorized user or not. If the user is valid user grants permission otherwise reject the user's request (Fig. 4.14).

The search required data: The healthcare provider search required data by using the patient name and retrieves data from the cloud server (Fig. 4.15).

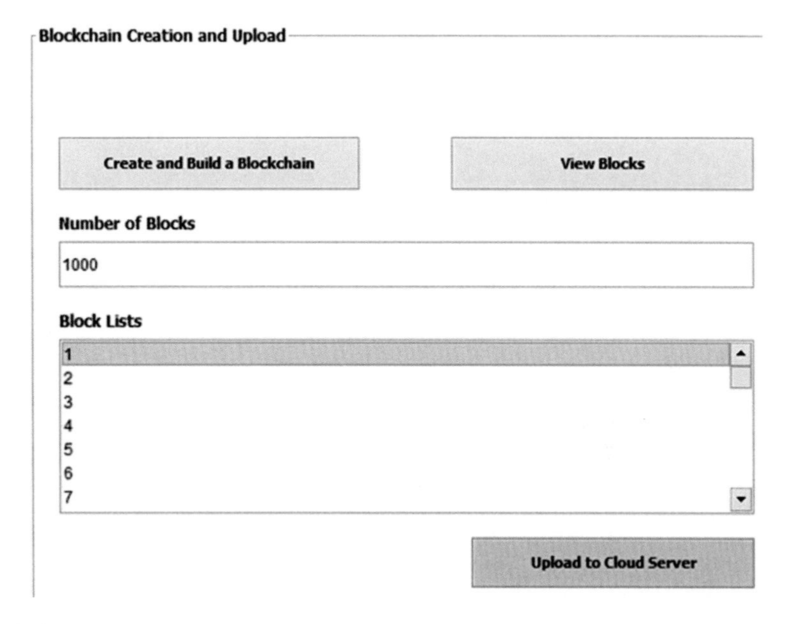

Fig. 4.10 Block creation

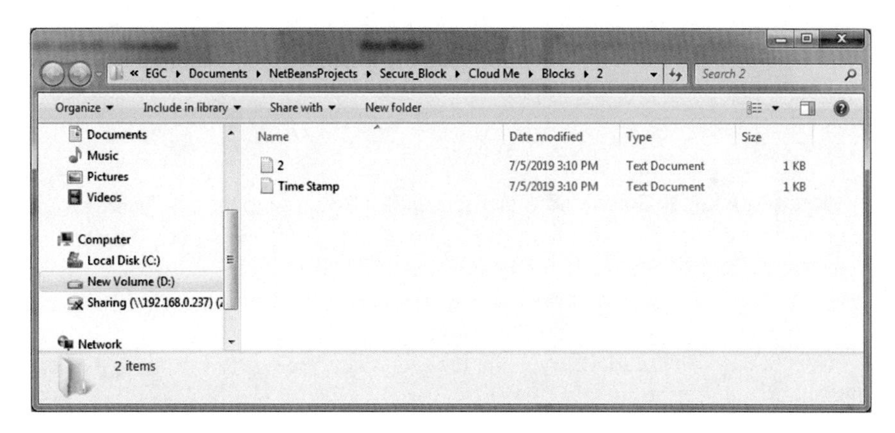

Fig. 4.11 View blocks

Fig. 4.12 Upload acknowledgements

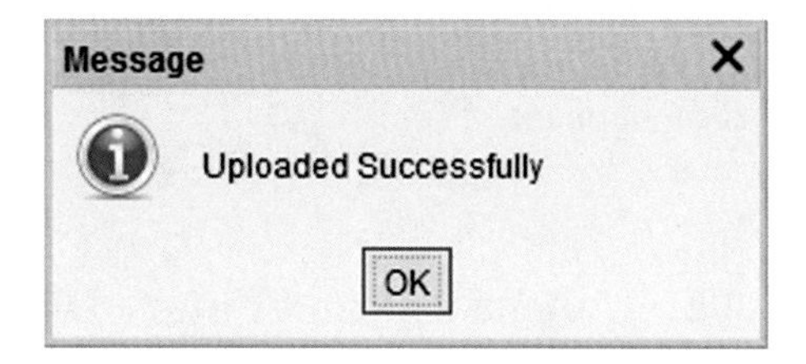

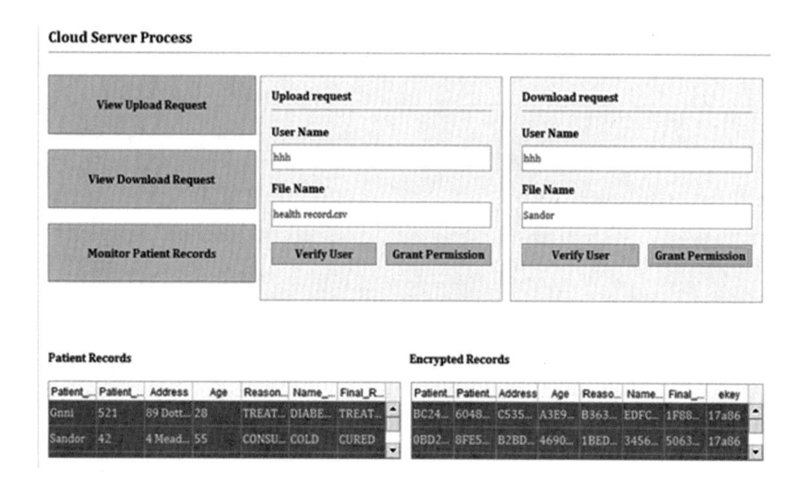

Fig. 4.13 EHR manage

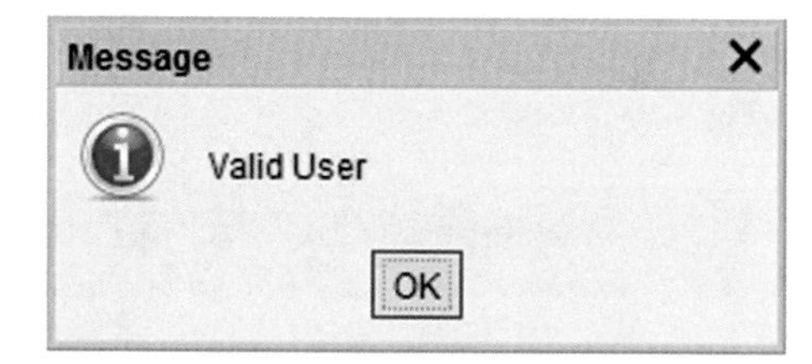

Fig. 4.14 Verify user

Request a key to download: The user sends the request to the healthcare provider to generate the key to download the required data from the server. The end-user verifies the request status. If the healthcare provider provides a key then the end-user receives the key and verifies the secret key. If the key matches then the user downloads the required file from the server (Fig. 4.16).

Secret Key Verification: By using the secret key, the server verifies is the user is an authorized user or not (Fig. 4.17).

Decrypt and Download the file: After successfully downloading data from the server, the user decrypts the downloaded file with the key provided by the healthcare provider and accesses content from the download records (Fig. 4.18).

View Downloads: After decrypting the file, the file contents in the file are in the form of plaintext (Fig. 4.19).

4.6 Conclusions and Future Direction

In cloud computing, security and privacy are two key factors to protect the electronic health record data and resources that are available. To provide security to the

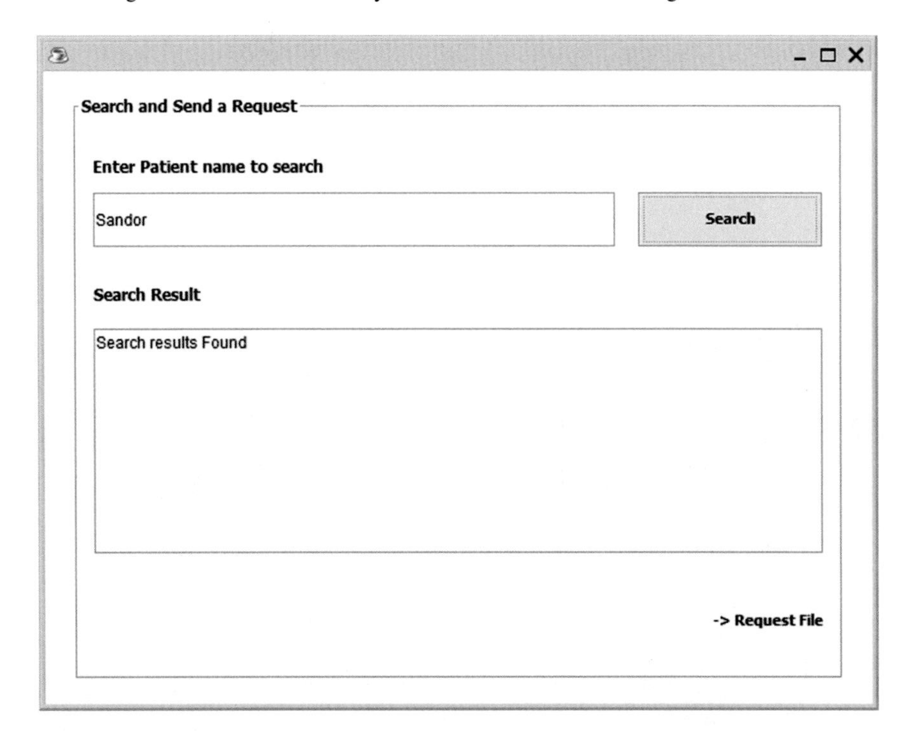

Fig. 4.15 Search required data

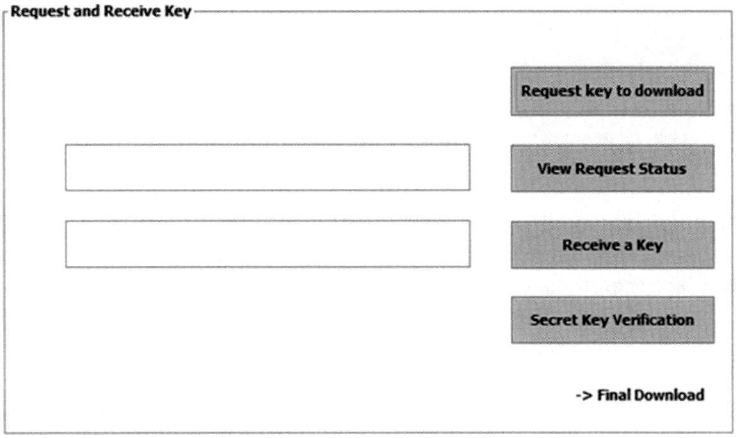

Request and Receive Key
Request key to download
View Request Status
Receive a Key
Secret Key Verification
-> Final Download

Fig. 4.16 Request a key to download

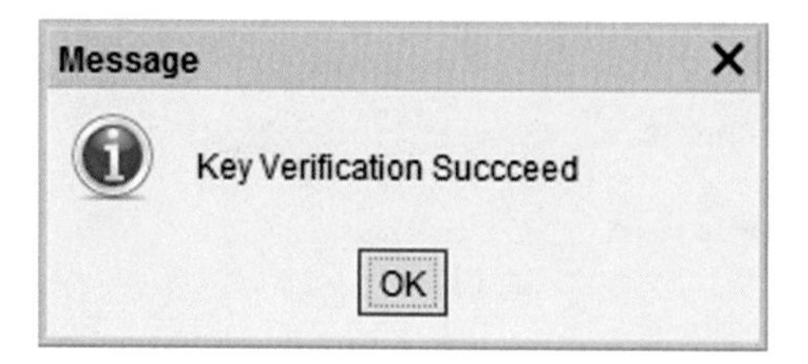

Fig. 4.17 Secret key verification

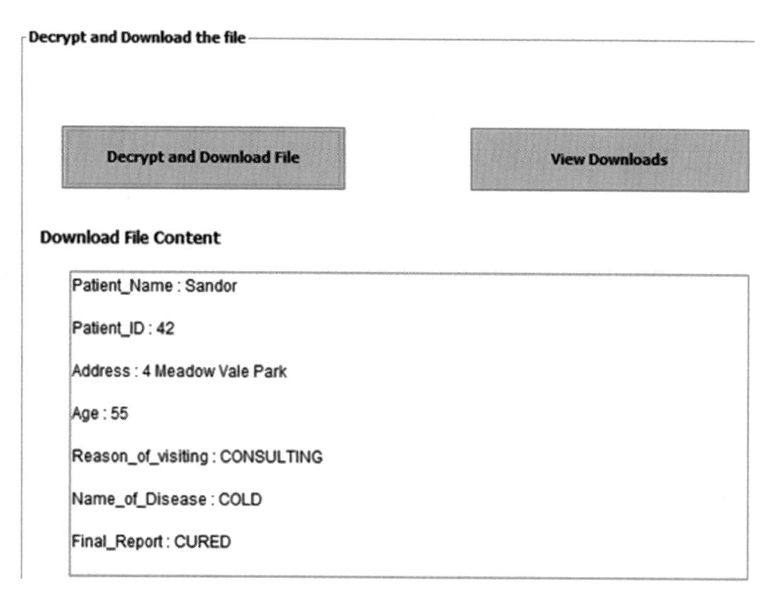

Fig. 4.18 Decrypt and download the file

electronic health record data we introduced blockchain technology in cloud environments to securely transfer EHR data between authorized entities. The existing EHRs systems face security and privacy issues (Data confidentiality, Availability, Decentralized access, Data Authorization, User authentication, Data Integrity, and Data privacy) for securely sharing data between authorized users in a cloud environment. Unauthorized users can easily access or damaged cloud storage data without the data owner's permission. To overcome these problems we proposed securing electronic health record sharing systems in a cloud environment using Blockchain technology. This approach is a tamper-proof mechanism because every health transaction in the blockchain is stored as hash values. Blockchain technology provides a data ledger-based feature that can be distributed to the entire registered user in the network. In this approach, we evaluated various performance metrics and comparison analysis shows the advantages of our system with other existing systems. Our approach can easily identify and prevent unauthorized access controls and achieve privacy and

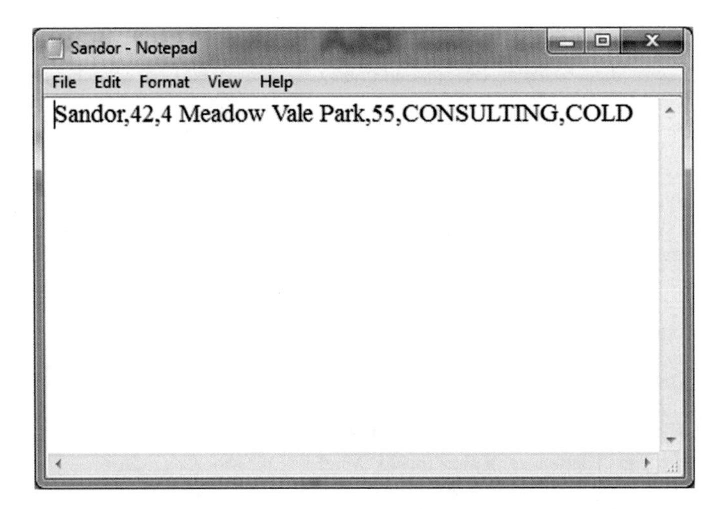

Fig. 4.19 View downloads

network security. Based on the advantages of the proposed system, we trust that our approach provides better security to share the medical data in the cloud. The execution outcome indicates that our approach can permit various clients to share EHRs data of the cloud ineffective and trustworthy way.

Some possible Extensions of work that will be included in the future work using blockchain may include the following features: (1) Cloud servers can store information from various clients. To grant a degree of security dependent on the import of the information, the information can be classified. (2) Some performance metrics like Mean-time-to-detect, the volume of data transfer, computation overhead, communication overhead, and frequency of third-party access will be considered. (3) Implement this approach in various mobile cloud platforms and other technologies like (Edge computing, Fog computing) to provide more security, and easy access.

References

1. Esposito, C., De Santis, A.: Blockchain: A Panacea for Conference on Theory and Practice of Electronic Governance, pp. 402–405. ACM (2012)
2. Farzandipour, M., Sadoughi, F., Ahmadi, M., Karimi: Security requirements and solutions in electronic health records: lessons learned from a comparative study. J. Med. Syst. **34**(4), 629–642 (2010)
3. Fernández-Alemán, J.L., Señor, I.C., Lozoya, P.Á.O., Toval, A.: Security and privacy in electronic health records: a systematic literature review. J. Biomed. Inf. **46**(3), 541–562 (2013)
4. Greenhalgh, T., Hinder, S., Stramer, K., Bratan, T., Russell, J.: Adoption, non-adoption, and abandonment of a personal electronic health record: a case study of HealthSpace. BMJ **2010**, 1–11 (2010)
5. Haas, S., Wohlgemuth, S., Echizen, I., Sonehara, N., Müller, N.: Aspects of privacy for electronic health records. Int. J. Med. Inf. **80**(2), 26–31 (2011)

6. Mouratidisa, H., Islam, S., Kalloniatis, C., Gritzalis, S.: A framework to support the selection of cloud providers based on security and privacy requirements. J. Syst. Softw. 1–18 (2013)
7. Hei, X., Du, X., Lin, S., Lee, I.: Pipac: patient infusion pattern-based access control scheme for wireless insulin pump system. In: Proceedings of IEEE INFOCOM, pp. 3030–3038 (2013)
8. Kang, J., Yu., Rong, Huang, X., Wu, M., Maharjan, S., Xie, S., Zhang, Y.: Blockchain for secure and efficient data sharing in vehicular edge computing and networks. IEEE Internet Things J. **6**(3), 4660–4670 (2018)
9. Li, J., Liu, Z., Chen, L., Chen, P., Wu, J.: Blockchain-based security architecture for distributed cloud storage. In: IEEE International Conference on Ubiquitous Computing and Communications (ISPA/IUCC), pp. 408–411 (2017)
10. Gai, K., Guo, J., Zhu, L., Yu, S.: Blockchain meets cloud computing: a survey. IEEE Commun. Surv. Tutor. **22**(3), 2009–2030 (2020)
11. Bendiab, K.: WiP: a novel blockchain-based trust model for cloud identity management. In: 16th IEEE International Conference on Dependable, Autonomic and Secure Computing (DASC 2018) (2018)
12. Kuo, T.T., Kim, H.E., Ohno-Machado, L.: Blockchain distributed ledger technologies for biomedical and health care applications. J. Am. Med. Inform. Assoc. **24**(6), 1211–1220 (2017)
13. Patra, M.R., Das, R.K., Padhy, R.P.: Crhis: cloud-based rural healthcare information system. In: Proceedings of the 6th International Conference on Theory and Practice of Electronic, pp. 402–405 (2012)
14. Azaria, A., Ekblaw, A., Vieira, T., Lippman, A.: Medrec: Using blockchain for medical data access and permission management. In: 2nd International Conference on Open and Big Data (OBD), pp. 25–30 (2016)
15. Branovic: A workload characterization of elliptic curve cryptography methods in embedded environments. ACM SIGARCH Comput. Archit. News **32**(3), 27–34 (2004)
16. Chase, M.: Multi-authority attribute-based encryption. In: TCC. LNCS, vol. 4392, pp. 515–534. Springer (2007)
17. Jacobs, M.: A Proposed Blockchain Reference Architecture (2016). https://www.linkedin.com/pulse/proposed-blockchain-reference-architecture-mike-jacobs
18. Rifi, N., Rachkidi, E., Agoulmine, N., Taher, N.C.: Towards using blockchain technology for eHealth data access management. In: Fourth International Conference on Advances in Biomedical Engineering (ICABME), pp. 1–4 (2017)
19. Kayna, B., Kaynak, S., Uygun, Ö.: Cloud manufacturing architecture based on public blockchain technology. IEEE Access **8**, 2163–2177 (2020)
20. Chakravarthy, M.H.: Hybrid Elliptical Curve Cryptography Based Secured Cloud Using Mobile Ant Agent (2016). shodhganga.inflibnet.ac.in
21. Wiljer, D., Urowitz, S., Apatu, E., DeLenardo, C., Eysenbach, G., Harth, T., et al.: Patient accessible electronic health records: exploring recommendations for successful implementation strategies. J. Med. Internet Res. **10**(4), 34 (2008)
22. Rothstein, M.A.: Health privacy in the electronic age. J. Leg. Med. **28**(4), 487–501 (2007)
23. Xie, S, Zheng, Z., Chen, W., et al.: Blockchain for cloud exchange: a survey. Comput. Electr. Eng. 1–12 (2020)
24. Swan, M.: Blockchain: Blueprint for a New Economy. O'Reilly Media, Inc., 1005 Gravenstein Highway North, Sebastopol, CA 95472 (2015)
25. Koblitz, N.: Elliptic curve cryptosystems. Math. Comput. **48**(177), 203–209 (2018)
26. Lu, Y.: The blockchain: state-of-the-art and research challenges. J. Ind. Inf. Integr. **15**, 80–90 (2019)
27. Ying, Z., Wei, L., Li, Q., Liu, X., Cui, J.: A lightweight policy preserving EHRs sharing scheme in the cloud. IEEE Access **6**, 53698–53708 (2018)
28. Panda, S.K., Mohammad, G.B., Nandan Mohanty, S., Sahoo, S.: Smart contract-based land registry system to reduce frauds and time delay. Secur. Priv. e172 (2021). https://doi.org/10.1002/spy2.172
29. Panda, S.K., Satapathy, S.C.: Drug traceability and transparency in medical supply chain using blockchain for easing the process and creating trust between stakeholders and consumers. Pers. Ubiquit. Comput. (2021). https://doi.org/10.1007/s00779-021-01588-3

30. Liu, L.S., Shih, P.C., Hayes, G.R.: Barriers to the adoption and use of personal health record systems. In: Proceedings of iConference, pp. 363–370 (2011)
31. Sathya, A.R., Panda, S.K., Hanumanthakari, S.: Enabling smart education system using blockchain technology. In: Panda, S.K., Jena, A.K., Swain, S.K., Satapathy, S.C. (eds.) Blockchain Technology: Applications and Challenges. Intelligent Systems Reference Library, vol. 203. Springer, Cham (2021). https://doi.org/10.1007/978-3-030-69395-4_10
32. Novo, O.: Blockchain meets IoT: an architecture for scalable access management in IoT. IEEE Internet Things J. **5**, 1184–1195 (2018)
33. Shaikh, P., Kaul, V.: Enhanced security algorithm using hybrid encryption and ECC. IOSR J. Comput. Eng. **16**(3), 80–85 (2014)
34. Lemieux, V.L.: Trusting records: is blockchain technology the answer. Rec. Manag. J. **26**(2), 110–139 (2016)
35. Lokre, S.S., Naman, V., Priya, S., Panda, S.K.: Gun tracking system using blockchain technology. In: Panda, S.K., Jena, A.K., Swain, S.K., Satapathy, S.C. (eds.) Blockchain Technology: Applications and Challenges. Intelligent Systems Reference Library, vol. 203. Springer, Cham (2021). https://doi.org/10.1007/978-3-030-69395-4_16
36. Panda, S.K., Daliyet, S.P., Lokre, S.S., Naman, V.: Distributed Ledger Technology in the Construction Industry Using Corda, The New Advanced Society: Artificial Intelligence and Industrial Internet of Things Paradigm. https://doi.org/10.1002/9781119884392.ch2
37. Niveditha, V.R., Sekaran, K., Amandeep Singh, K.., Panda, S.K.: Effective prediction of bitcoin price using wolf search algorithm and bidirectional LSTM on internet of things data. Int. J. Syst. Syst. Eng. **11**(3–4), 224–236
38. Qi, X., Asamoah, K.O., Sifah, E., Gao, J.: MeDShare: trust-less medical data sharing among cloud service providers via blockchain. IEEE Access **5**, 14757–14767 (2017)
39. Panda, S.K., Satapathy, S.C.: An investigation into smart contract deployment on Ethereum platform using Web3.js and solidity using blockchain. In: Bhateja, V., Satapathy, S.C., Travieso-González, C.M., Aradhya, V.N.M. (eds.) Data Engineering and Intelligent Computing. Advances in Intelligent Systems and Computing, vol. 1. Springer, Singapore (2021). https://doi.org/10.1007/978-981-16-0171-2_52
40. Eltayieb, N., Elhabob, R., Hassan, A., Li, F.: A blockchain-based attribute-based signcryption scheme to secure data sharing in the cloud. J. Syst. Architect. **102**, 1–11 (2019)
41. Dariua, R.: 4 Features of Blockchain Technology (2016). https://due.com/blog/4-features-blockchain-technology
42. Panda, S.K., Rao, D.C., Satapathy, S.C.: An investigation into the usability of blockchain technology in internet of things. In: Bhateja, V., Satapathy, S.C., Travieso-González, C.M., Aradhya, V.N.M. (eds.) Data Engineering and Intelligent Computing. Advances in Intelligent Systems and Computing, vol. 1. Springer, Singapore (2021). https://doi.org/10.1007/978-981-16-0171-2_53
43. Bethencourt, J., Sahai, A., Waters, B.: Ciphertext-policy attribute-based encryption. In: IEEE Symposium, pp. 321–334 (2007)
44. Casola, V., Castiglione, A., Choo, K.-K.R., Esposito, C.: Healthcare-related data in the cloud: challenges and opportunities. IEEE Cloud Comput. **3**(6), 10–14 (2016)
45. Yujiao, H.W.: Secure cloud-based EHR system using attribute-based cryptosystem and blockchain. J. Med. System. **42**(152), 1–9 (2018)
46. Panda, S.K., Dash, S.P., Jena, A.K.: Optimization of block query response using evolutionary algorithm. In: Bhateja, V., Satapathy, S.C., Travieso-González, C.M., Aradhya, V.N.M. (eds.) Data Engineering and Intelligent Computing. Advances in Intelligent Systems and Computing, vol. 1. Springer, Singapore (2021). https://doi.org/10.1007/978-981-16-0171-2_54
47. Frank, R.: Blockchain and distributed ledger technologies. ACM Comput. Surv. **52**(6) (2017)
48. Nanda, S.K., Panda, S.K., Das, M., Satapathy, S.C.: Automating vehicle insurance process using smart contract and Ethereum. In: Chakravarthy, V.V.S.S.S., Flores-Fuentes, W., Bhateja, V., Biswal, B. (eds.) Advances in Micro-Electronics, Embedded Systems and IoT. Lecture Notes in Electrical Engineering, vol. 838. Springer, Singapore (2022). https://doi.org/10.1007/978-981-16-8550-7_23

49. Varaprasada Rao, K., Panda, S.K.: A design model of copyright protection system based on distributed ledger technology. In: Satapathy, S.C., Lin, J.C.W., Wee, L.K., Bhateja, V., Rajesh, T.M. (eds) Computer Communication, Networking and IoT. Lecture Notes in Networks and Systems, vol. 459. Springer, Singapore (2023). https://doi.org/10.1007/978-981-19-1976-3_17
50. Varaprasada Rao, K., Panda, S.K.: Secure electronic voting (E-voting) system based on blockchain on various platforms. In: Satapathy, S.C., Lin, J.C.W., Wee, L.K., Bhateja, V., Rajesh, T.M. (eds) Computer Communication, Networking and IoT. Lecture Notes in Networks and Systems, vol. 459. Springer, Singapore (2023). https://doi.org/10.1007/978-981-19-1976-3_18
51. Panda, S.K., Elngar, A.A., Balas, V.E., Kayed, M. (eds.): Bitcoin and Blockchain: History and Current Applications, 1st ed. CRC Press (2020). https://doi.org/10.1201/9781003032588
52. Panda, S.K., Jena, A.K., Swain, S.K., Satapathy, S.C. (eds.): Blockchain Technology: Applications and Challenges. Springer, Intelligent Systems Reference Library. https://doi.org/10.1007/978-3-030-69395-4
53. Sun, J., Zhu, X., Zhang, C., Fang, Y.: Hcpp: cryptography-based secure EHR system for patient privacy and emergency healthcare. In: Proceedings of IEEE ICDCS, pp. 373–382 (2011)
54. Zhang, Y.: Blockchain-based public integrity verification for cloud storage against procrastinating auditors. IEEE Trans. Cloud Comput. 1–15 (2019)
55. Zhang, Y., He, D., Choo, K.-K.: BaDS: blockchain-based architecture for data sharing with ABS and CP-ABE in IoT. Wirel. Commun. Mob. Comput. **2018**, 1–9 (2018)
56. Zhang, Y., Xu, C., Li, H., Yang, K., Zhou, J., Lin, X.: Healthdep: an efficient and secure deduplication scheme for cloud-assisted eHealth systems. IEEE Trans. Ind. Inf. **14**(9), 4101–4112 (2018)

Chapter 5
Blockchain: The Foundation of Trust in Metaverse

A. R. Sathya

Abstract Metaverse is a computer-generated virtual environment which is a combination of real and digital world facilitated by web technologies, Virtual reality and Internet. Despite considerable benefits and attention, a logical question in how to protect its consumers' digital content and data in the metaverse. In this regard, one of the potential solutions could be blockchain due to its unique features like decentralization, immutability, and transparency. This chapter intends to provide a preliminary information of metaverse and blockchain and how these two technologies can be related to each other. The impact of blockchain technologies on metaverse is also discussed in brief.

Keywords Metaverse · Web technology · Blockchain technology

5.1 Introduction

Over the past few decades, the digitizing of services has emerged as a paradigm for increasing productivity in the domains of business, education, entertainment and any other system that may be combined with online access. As service access reaches its maximum levels of efficiency, performance, and quality the emphasis has switched to the consumer experience. As a result, there is a growing need for greater customer service that is more involved, and service providers are eager to raise their current standards. In recent days metaverse has been getting much attention and is considered as an evolution of digitalization and social interaction. It is termed as next generation web application that to provide 3D virtual environments where humans can meet and communicate with other people or the applications and play in unimaginable ways through digital avatars [1]. Basically, avatars are the digital representations of users, and they are legally equal to their real-world counterparts in the metaverse. The metaverse smoothly merges the real and virtual worlds, enabling avatars to engage in a wide range of activities like creative displays, recreation, networking, and trading.

A. R. Sathya (✉)
Department of Data Science & Artificial Intelligence, Faculty of Science & Technology, ICFAI Foundation for Higher Education, Hyderabad 501203, Telangana, India
e-mail: sathya.renu@gmail.com

S. K. Panda et al. (eds.), *Recent Advances in Blockchain Technology*,
Intelligent Systems Reference Library 237, https://doi.org/10.1007/978-3-031-22835-3_5

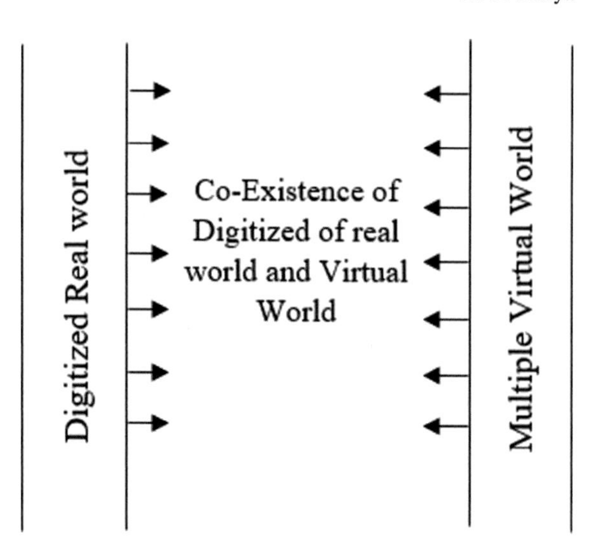

Fig. 5.1 Metaverse developmental stages

It is the evolution of internet application and is combination of many emerging technologies like Internet of Things (IoT), Virtual/Augmented Reality, Artificial Intelligence, Blockchain technology and Cloud Computing. Though the actual concept goes 30 years back, metaverse gained popularity recently after the Facebook's investment on Metaverse development. Many other giants in the industry like Microsoft, Roblox too have showed interest in developing metaverse framework [2]. This clearly indicates that metaverse has the potential to change the future of internet applications.

In [3] the concept of twin technology is proposed to achieve metaverse. Figure 5.1 shows the how metaverse development can be achieved through various sequential stages. First stage is the representations of our actual physical real environments as digital environment as a digital world which will be updated periodically to reflect the changes to their digital counterparts. Second is creating multiple copies of the digital world as virtual world where humans create their own avatars and are considered as digital natives of that digital world. There may be limited connectivity among these digital worlds but eventually they may get merged as one at the final stage i.e. co-existence of digitized real word and virtual world.

5.2 Limitations of 2D Learning Environments

Online education is growing more and more popular, particularly in higher education. This pattern was intensified by the COVID-19 pandemic, which has disrupted all activities based on attendance at all educational institutes [4]. Since beginning, asynchronous and synchronous systems have dominated online education. Asynchronous tools are flexible and some of the standard asynchronous learning tools are Blackboard, Moodle and some social network. Synchronous systems allow for simultaneous online meetings of educators and students in a digital, virtual setting. Synchronous tools include Webex, Zoom, Teams etc.

Both types dependent on software based on two-dimensional digital environments that span in-plane digital frames with width and height but no depth. Application based on 2D web environments has known limitations and constraints. Regular use of synchronous online platforms results in conditions like Zoom Fatigue [5]. Similarly asynchronous learning leads to Emotional isolation, a bad emotion for participation motivation [6]. Though collaborative applications like wikis and blogs can improve the user engagement to an extent but fails to address the users emotional stress and natural communication. Few of the limitations of 2D learning environment are listed below.

1. In 2D environment users will have limited or low self-perception. Through a picture or a live webcam stream of a headshot, they are depicted as disembodied creatures without any personalization options.
2. Videoconferencing sessions are seen as individual video calls to participate in rather than as virtual spaces for group meetings. Extended meetings often result in participants slouching and becoming disengaged.
3. 2D platforms provide few opportunities for participant involvement. Students only have the opportunity to participate passively in learning activities only if teachers start instructs them.
4. Through smileys and emoticons, users have relatively few possibilities for expressing their emotions.

3D, immersive spatial environments can be used to overcome all of these limitations. Some of the immersive technologies are shown in Fig. 5.2.

5.3 Core Attributes of Metaverse

Science fiction is where the Metaverse's most well-known concepts originated. Metaverse is popularly depicted as a representation of real life that is based in a virtual one. Thus, the following are the fundamental characteristics of the Metaverse:

1. Being Synchronous and live 2. Being persistent 3. Full active economy 4. An experience 5. Varied contributors 6. Greater interoperability (Fig. 5.3).

- **Live and Synchronous**: While pre-planned and independent events will happen, it will be a live and real-time experience continually happening for everyone exactly like it does in real-life.
- **Persistent**: Keeps continuing indefinitely without a pause, stop or a reset.
- **Economy value**: People and business would be able to sell, buy, own, create, invest or compensated for a wide range of activities that result in value that is acknowledged by others.

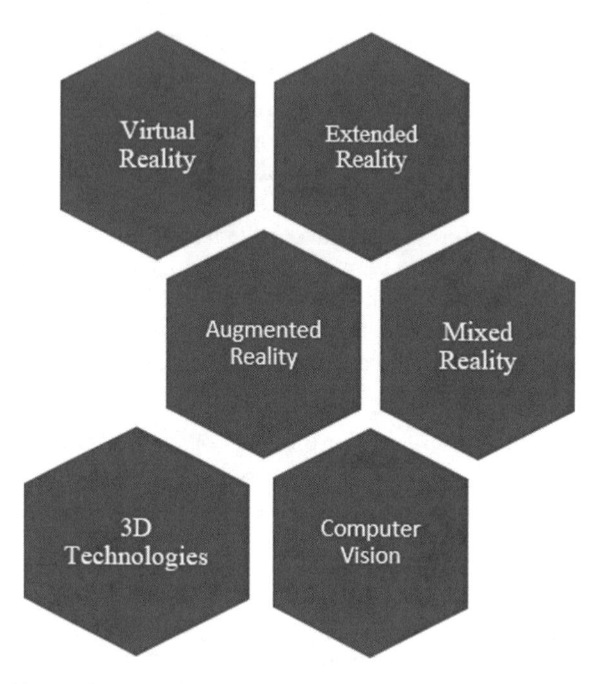

Fig. 5.2 3D and immersive technologies

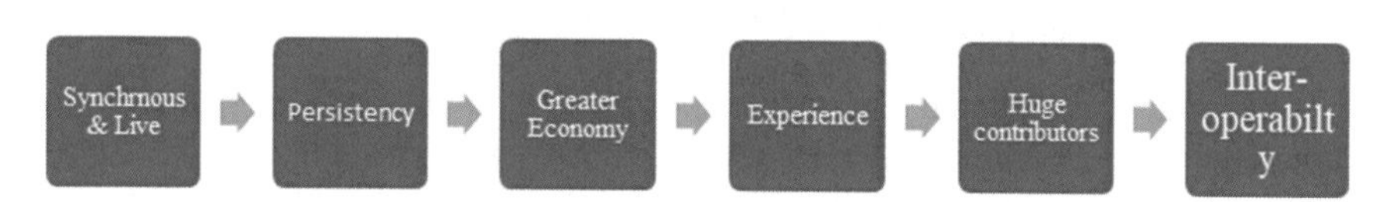

Fig. 5.3 Attributes of metaverse

- **Experience**: It spans both real and virtual, open and closed platforms, private and public networks.
- **Contributors**: It is packed with content and experiences created by many contributors like individuals, informally organized or commercially focused enterprises.
- **Interoperability**: The data, digital assets, information, and other interoperability between the experiences should be remarkable. Items created for one application should be able to be transported to the other.

5.4 Characteristics of Metaverse

Metaverse is a brand-new web application that combines a number of cutting-edge technologies and exhibits cross-technology traits. The Metaverse exhibits sociality

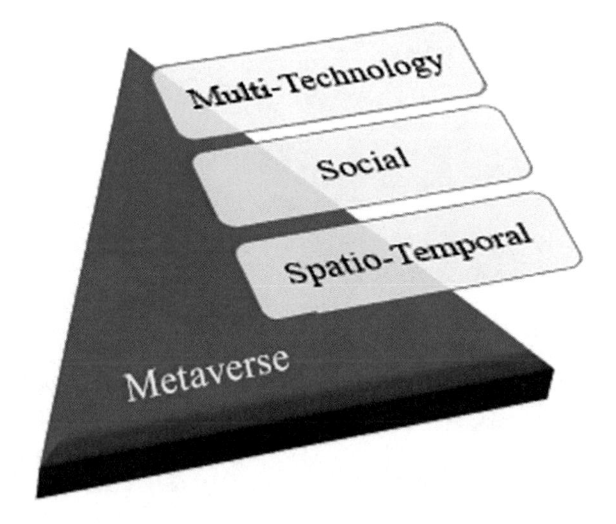

Fig. 5.4 Metaverse characteristics

as a novel social form. It also exhibits spatial and temporal characters as it is highly related to the real world. Figure 5.4 shows the major metaverse attributes.

Cross Technology: Metaverse is integration of variety of technologies. The live experience from the virtual and augmented reality [7] and the copy of real world into a digital world by twin technology [8], economic system using blockchain [9], digital avatars in the virtual world using Artificial Intelligence [10], IoT and many more cutting-edge technologies are combined in the world of metaverse[11–16].

Social: By its definition, metaverse looks like a new social structure. It includes social, legal, cultural and economic systems that are closely connected to reality but having their own individual traits.

Spatio-Temporal: Metaverse exhibits a hyper spatio-temporal quality. It is a mirror image of the real world offering an open, free and engaging experience with no time and space constraints (Fig. 5.5).

5.5 How Blockchain and Metaverse Related?

It is said that people can socialize in metaverse. In that case what would be mode of transaction in the virtual world and what would be the virtual currency. Under such circumstances Blockchain becomes essential to develop an environment where one can create and monetize digital assets in metaverse.

The metaverse can be seen as a chain of complete and fully integrated economic system in terms of creation and usage of digital assets. The process of creation, distribution and consumption of digital items in the digital world is the basic element of the digital economy. The most tedious task in metaverse is developing such an economy system. This is due to the fact that traditional economists are unfamiliar with the development and utilization of digital items that can be traded in the virtual

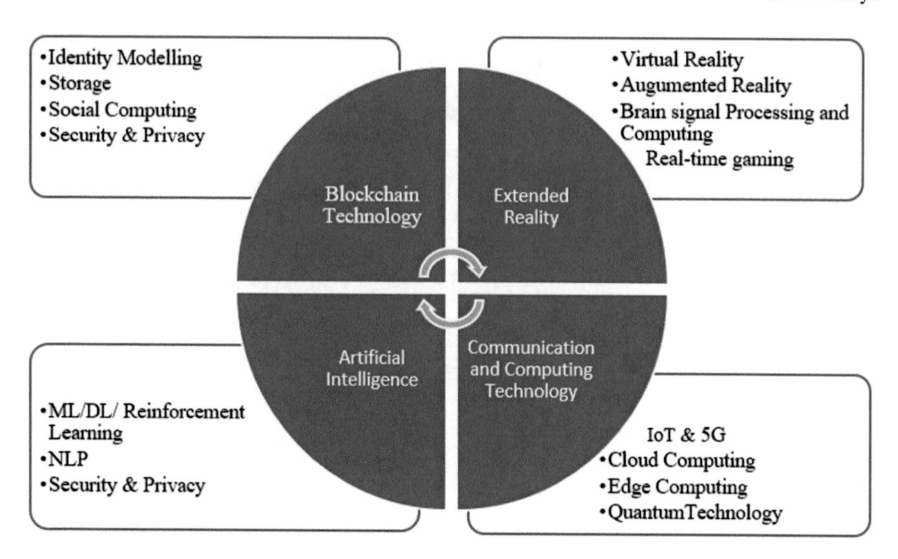

Fig. 5.5 Shows the technology convergence in metaverse

world [17–19]. Furthermore, the centralized economic model of the physical world cannot perform well in a public, equitable, and self-organized digital space due to the large transactions involved. Given the connection between the physical and digital worlds, the metaverse should make it feasible for currency to circulate, removing barriers to life, services, training, working, etc. Hence, to trade the digital assets of the avatars efficiently in the virtual world, it is essential to have an economic system in a decentralized way in metaverse.

It is known that blockchain is a decentralized application which is widely being adopted in various applications. Because it can connect disparate minor sectors and give a solid economic system, blockchain is regarded as one of the metaverse's core infrastructures and helps supply the metaverse with laws that are transparent, open, effective, and dependable. Blockchain is anticipated to deliver a range of opportunities to the metaverse and spark a fresh wave of technological advancement and industrial change. The blockchain architecture includes Data layer, Network layer, Consensus layer, Incentive layer, Contract layer, and Application layer. The relationship between blockchain and metaverse over the architectural layers is shown in Fig. 5.6.

Data and Network Layer: The data layers support the process of data transmission and verification. This would in turn supports in various data transmission and verification mechanisms of metaverse economic model.

Consensus Layer: The consensus mechanisms of blockchain can help metaverse in solving credit problem of metaverse transactions.

Contract and Application Layer: The smart contract of blockchain provides a trustable environment offering a transparent execution of transactions as per the ruled defined.

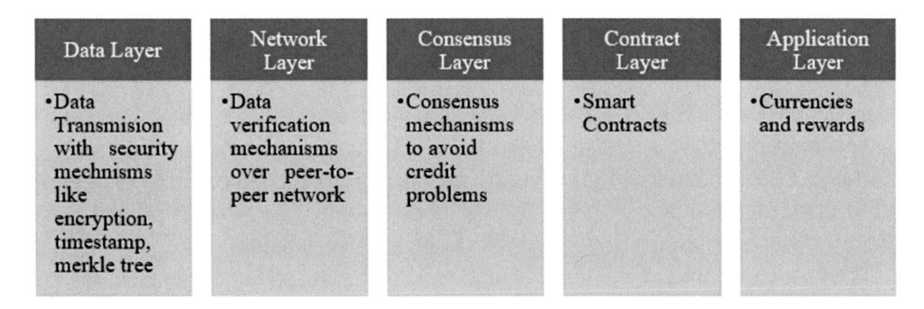

Fig. 5.6 Metaverse in blockchain

It will be challenging to determine the value of the assets exchanged in metaverse without blockchain technology. Particularly when such virtual components interact economically with the real-world economy. Therefore, it would be appropriate to investigate blockchain solutions in the metaverse.

From a technical standpoint, blockchain-based approaches for the metaverse should include data collection, storage, sharing, interoperability, and privacy are discussed below [20–23]. From the economic perspective, since blockchain core functionalities are digital assets, it can very well maintain the economic operations of metaverse. Blockchain can be viewed as the core of metaverse.

5.6 Role of Blockchain in Metaverse

Data collection and Quality: Metaverse receives data from different applications. The quality of the data determines the creation of objects in metaverse. Blockchain validates all the transactions completely thereby improving the quality of data in metaverse.

Data Sharing: The major advantage of the metaverse is its integration with augmented reality on both digital and physical objects. The continuous exchange of AR and VR data, determines the success of metaverse. Hence, effective data sharing mechanisms results in creation of novel, cutting-edge applications that help with problem-solving in the real world. In such scenarios blockchain's encoding systems helps in effective and secure data sharing.

Data Interoperability: Stakeholders in the metaverse may use different applications, access assets in many virtual worlds, and hold those assets. As these applications are developed in different environments, interoperability of data among these applications are limited. Using blockchain's cross-chain protocol it is possible to exchange data between blockchains in different virtual worlds. Therefore, User and data migration will be simplified.

Data Integrity: The data in metaverse has to be consistent and accurate otherwise the stakeholders will lose trust in metaverse. The immutability of blockchain assures the integrity of data in metaverse.

5.7 Blockchain Use Cases in Metaverse

As mentioned in Fig. 5.4 blockchain and metaverse are closely related where blockchain becomes a vital platform to develop and transact digital assets in the virtual world. Access to a digital environment devoid of the intervention of a centralized institution is made possible by the blockchain applications in the metaverse. Some of the popular applications of blockchain is given below.

Cryptocurrencies: Like how the real world is built on Fiat currencies, the need for cryptocurrencies in metaverse is inevitable. In this regard, blockchain systems already have integrated a number of processes, including creation, recording, and trading for cryptocurrencies which are very much essential in metaverse.

NFTs: The primary application of blockchain in metaverse is non-Fungible tokens. As we create a replica of real world as a digital world it is necessary to have copy of real-world possessions in digital space too and NFTs will greatly support in proof of ownership [24]. Using NFTs it is possible to create the digitized version of real-world assets.

Metaverse Transactions: Transactions in metaverse is just not token transferring but includes purchasing, renting and acquisition of items like in real world. However, transaction in metaverse comes with its own challenges which can be overcome by integrating blockchain and IIoT [25–28] (Fig. 5.7).

Online Video Interaction: During this pandemic the world has been kept alive with the advancements of telecommunication. But online interaction cannot replace

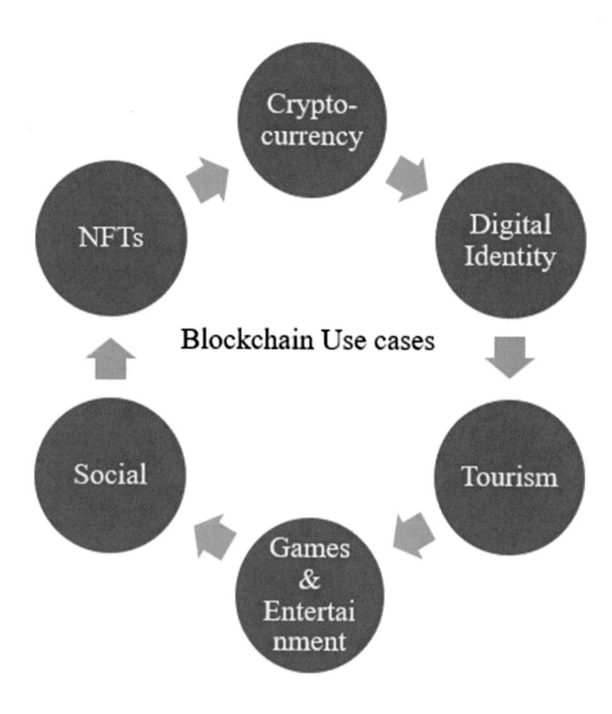

Fig. 5.7 Possible blockchain use cases in metaverse

the face-to-face communication as it has delayed or less interaction, non-cooperation and misinterpretation of comments or feedbacks. In Metaverse, avatars can be used to interact and communicate with others which would help in improvised communication.

Decentralized Market: Decentralized Finance (DeFi), the advanced blockchain technology can can uphold the business and market in a decentralized way in metaverse. Several researchers are ongoing to discuss the representation of DeFi in metaverse markets [29, 30].

Real-Estate: Real estate property dealings can be dealt in metaverse. Similar to real world users can, for instance, gather and resell residences to the public in the metaverse, as well as plan art galleries, music concerts, video game tournaments, and other events.

Peer-to-Peer transactions: Users in the metaverse would be able to conduct direct transactions without the need for intermediaries due to peer-to-peer transactions driven by cryptocurrency.

Digital Identity: As users will be having avatars in metaverse, it is necessary to have identity verification mechanism to ensure the identity of users and avoid false identity in virtual environment. Blockchain based identity verification mechanisms can be used in such scenarios.

5.8 Anticipated Application Areas of Metaverse

Using digital twin technology metaverse creates a mirror image of real world which is also a critical tool for developing smart city. Using digital technology, it is possible to capture data like humans, automobiles, objects and space which might help in creating a recognizable, maintainable, and digital twin city in space. It can enhance urban management and services, increase resource use efficiency, and raise citizen quality of life [31, 32].

Smart City: It is well known that metaverse imitates the real world to virtual world with the help of twin technology which is essential for smart city development. Quality of life and resource utilization can be improved with these technologies.

Games and Entertainment: The live immersive experiences in games and entertainment sector can improve the user experience and fun. The advancement of interactive technology has significantly increased the player's sense of immersion in the game, which can improve its usability, playability, and enjoyment.

Virtual Workspace and Meetings: The conventional remote workspace model's shortcomings can be improvised in metaverse, which can also enhance its features and expand the possibilities for virtual workplace.

Digital Tourism: The limitations of time, space and other factors can easily be overcome by the development of interactive technology, twin technology and allows the users to experience live and unique experience in exploring beautiful locations.

Medical: Metaverse creates a virtual and relaxing environment which can be helpful for mental conditions where the users are allowed to interact with virtual imaginary characters.

Education: The development of Metaverse can support serious gaming, early childhood education, and children's education. Immersive, simulating of realistic scenes to promote the understanding of learning content and can avoid the difficulty real-time experiment.

Social and Economic: User can experience different social interactions as they can communicate with anyone anytime in the virtual space.

5.9 Blockchain Based Solutions for Metaverse Challenges

In Sect. 5.2 we have seen the role of blockchain in metaverse. However, there are certain challenges which can also be resolved using blockchain [33, 34].

- **Data Acquisition**: Metaverse deals with huge amount of unstructured and transactional data, the integrity of the services and reliability is determined by the quality of data. As the amount of data will be increasing day by day, the data acquisition systems will be overloaded. Hence, duplicate and erroneous data may accumulate, impairing the data's quality. This can be resolved by using blockchain as the transaction records are validated and are traced, thereby making it easier for different applications to get verified and authenticate transactional data. The cryptographic security mechanisms of blockchain protects the data from any attack. Therefore, the metaverse data acquisition systems based on blockchain are considered reliable.
- **Data Interoperability**: The transfer of a user's digital assets, such as avatars and NFTs, from one digital environment to another is prohibited. This lack of openness forms a challenging task. The ability to effectively manage interactions between virtual worlds, which is a severe drawback of the traditional process, is necessary for the metaverse's interoperability. Therefore, a cross-chain protocol that enables the exchange of assets like NFTs, payments and between virtual worlds would be the ideal answer. Therefore, companies can use cross-blockchain technology to eliminate middlemen in the metaverse.
- **Data Privacy**: Personal details can be used by attackers to deceive other users. Thus, privacy of personal details is essential. This leads to the use of blockchain where privacy of the data is guaranteed by the cryptographic techniques. Additionally, by implementing zero-knowledge proof on the blockchain, people can easily access identifying information that is crucial to the metaverse while also keeping their privacy and control over their belongings.

5.10 Conclusion

On a final note, the technology giants like Facebook, Google, Apple has huge plans for metaverse, due to the availability of sophisticated computing devices and intelligent wearables, our digitized future will be more interactive, more living, more embodied, and more multimedia. There are still numerous challenges to be overcome before metaverse is fully integrated into our day today life. Though it is anticipated that metaverse can be the future, the fact is unknown for sure. But the Metaverse seems to provide a wide range of opportunities, and Blockchain is undoubtedly a key technology for the Metaverse's fundamental elements, such as digital ownership proof, fund transfers, governance, data access, and interoperability.

References

1. Lee, L.-H., Braud, T., Zhou, P., Wang, L., Xu, D., Lin, Z., Kumar, A., Bermejo, C., Hui, P.: All one needs to know about metaverse: a complete survey on technological singularity, virtual ecosystem, and research agenda (2021). arXiv:2110.05352
2. Fennimore, J.: Roblox: 5 Fast Facts You Need to Know. https://heavy.com/games/2017/07/roblox-youtube-free-download-corporation-baszucki-cassel-nerfmodder/. Accessed 16 Dec 2021
3. Xi, N., Chen, J., Gama, F., Riar, M., Hamari, J.: The challenges of entering the metaverse: an experiment on the effect of extended reality on workload. Inf. Syst. Front. 1–22 (2022)
4. Boltz, L.O., Yadav, A., Dillman, B., Robertson, C.: Transitioning to remote learning: lessons from supporting K-12 teachers through a MOOC. Br. J. Edu. Technol. **52**(4), 1377–1393 (2021)
5. Bailenson, J.N.: Nonverbal overload: a theoretical argument for the causes of Zoom fatigue. Technol. Mind Behav. **2**, 61 (2021)
6. Anderson, T., Vargas, P.R.: A critical look at educational technology from a distance education perspective. Digit. Educ. Rev. (37), 208–229 (2020)
7. Koutitas, G., Smith, S., Lawrence, G.: Performance evaluation of AR/VR training technologies for EMS first responders. Virtual Real. **25**(1), 83–94 (2021)
8. Tao, F., Zhang, H., Liu, A., Nee, A.Y.: Digital twin in industry: state-of-the-art. IEEE Trans. Industr. Inf. **15**(4), 2405–2415 (2018)
9. Jeon, H., Youn, H., Ko, S., Kim, T.: Blockchain and AI meet in the metaverse. In: Advances in the Convergence of Blockchain and Artificial Intelligence, p. 73 (2022)
10. Dinh, T.N., Thai, M.T.: AI and blockchain: a disruptive integration. Computer **51**(9), 48–53 (2018)
11. Pillai, B., Biswas, K., Muthukkumarasamy, V.: Cross-chain interoperability among blockchain-based systems using transactions. Knowl. Eng. Rev. **35**, E23 (2020). https://doi.org/10.1017/S0269888920000314
12. Panda, S.K., Mohammad, G.B., Nandan Mohanty, S., Sahoo, S.: Smart contract-based land registry system to reduce frauds and time delay. Secur. Priv. e172 (2021). https://doi.org/10.1002/spy2.172
13. Panda, S.K., Satapathy, S.C.: Drug traceability and transparency in medical supply chain using blockchain for easing the process and creating trust between stakeholders and consumers. Pers. Ubiquit. Comput. (2021). https://doi.org/10.1007/s00779-021-01588-3
14. Niveditha, V.R., Sekaran, K., Amandeep Singh, K., Panda, S.K.: Effective prediction of bitcoin price using wolf search algorithm and bidirectional LSTM on internet of things data. Int. J. Syst. Syst. Eng. **11**(3–4), 224–236

15. Varaprasada Rao, K., Panda, S.K.: A design model of copyright protection system based on distributed ledger technology. In: Satapathy, S.C., Lin, J.C.W., Wee, L.K., Bhateja, V., Rajesh, T.M. (eds.) Computer Communication, Networking and IoT. Lecture Notes in Networks and Systems, vol. 459. Springer, Singapore (2023). https://doi.org/10.1007/978-981-19-1976-3_17
16. Varaprasada Rao, K., Panda, S.K.: Secure electronic voting (E-voting) system based on blockchain on various platforms. In: Satapathy, S.C., Lin, J.C.W., Wee, L.K., Bhateja, V., Rajesh, T.M. (eds.) Computer Communication, Networking and IoT. Lecture Notes in Networks and Systems, vol. 459. Springer, Singapore (2023). https://doi.org/10.1007/978-981-19-1976-3_18
17. LaDuke, W.: Traditional ecological knowledge and environmental futures. Colo. J. Int. Envtl. L. Pol'y **5**, 127 (1994)
18. Sathya, A.R., Panda, S.K., Hanumanthakari, S.: Enabling smart education system using blockchain technology. In: Panda, S.K., Jena, A.K., Swain, S.K., Satapathy, S.C. (eds.) Blockchain Technology: Applications and Challenges. Intelligent Systems Reference Library, vol. 203. Springer, Cham (2021). https://doi.org/10.1007/978-3-030-69395-4_10
19. Lokre, S.S., Naman, V., Priya, S., Panda, S.K.: Gun tracking system using blockchain technology. In: Panda, S.K., Jena, A.K., Swain, S.K., Satapathy, S.C. (eds.) Blockchain Technology: Applications and Challenges. Intelligent Systems Reference Library, vol. 203. Springer, Cham (2021). https://doi.org/10.1007/978-3-030-69395-4_16
20. Gadekallu, T.R., Huynh-The, T., Wang, W., Yenduri, G., Ranaweera, P., Pham, Q.-V., da Costa, D.B., Liyanage, M.: Blockchain for the Metaverse: A Review (2022). arXiv:2203.09738
21. Panda, S.K., Daliyet, S.P., Lokre, S.S., Naman, V.: Distributed Ledger Technology in the Construction Industry Using Corda, The New Advanced Society: Artificial Intelligence and Industrial Internet of Things Paradigm. https://doi.org/10.1002/9781119884392.ch2
22. Panda, S.K., Satapathy, S.C.: An investigation into smart contract deployment on Ethereum platform using Web3.js and solidity using blockchain. In: Bhateja, V., Satapathy, S.C., Travieso-González, C.M., Aradhya, V.N.M. (eds.) Data Engineering and Intelligent Computing. Advances in Intelligent Systems and Computing, vol. 1. Springer, Singapore (2021). https://doi.org/10.1007/978-981-16-0171-2_52
23. Panda, S.K., Rao, D.C., Satapathy, S.C.: An investigation into the usability of blockchain technology in internet of things. In: Bhateja, V., Satapathy, S.C., Travieso-González, C.M., Aradhya, V.N.M. (eds.) Data Engineering and Intelligent Computing. Advances in Intelligent Systems and Computing, vol. 1. Springer, Singapore (2021). https://doi.org/10.1007/978-981-16-0171-2_53
24. Ethereum: Non-fungible tokens (nft). https://ethereum.org/en/nft/#gatsby-focus-wrapper. Accessed 16 Dec 2021
25. Zhang, M., Cheng, Y., Deng, X., Wang, B., Xie, J., Yang, Y., Zhang, J.: Accelerating transactions relay in blockchain networks via reputation. In: IEEE/ACM 29th International Symposium on Quality of Service (IWQOS), pp. 1–10. IEEE (2021)
26. Wang, G., Shi, Z., Nixon, M., Han, S.: Chainsplitter: towards blockchain-based industrial IoT architecture for supporting hierarchical storage. In: IEEE International Conference on Blockchain (ICBC), pp. 166–175. IEEE (2019)
27. Panda, S.K., Dash, S.P., Jena, A.K.: Optimization of block query response using evolutionary algorithm. In: Bhateja, V., Satapathy, S.C., Travieso-González, C.M., Aradhya, V.N.M. (eds.) Data Engineering and Intelligent Computing. Advances in Intelligent Systems and Computing, vol. 1. Springer, Singapore (2021). https://doi.org/10.1007/978-981-16-0171-2_54
28. Nanda, S.K., Panda, S.K., Das, M., Satapathy, S.C.: Automating vehicle insurance process using smart contract and Ethereum. In: Chakravarthy, V.V.S.S.S., Flores-Fuentes, W., Bhateja, V., Biswal, B. (eds.) Advances in Micro-Electronics, Embedded Systems and IoT. Lecture Notes in Electrical Engineering, vol. 838. Springer, Singapore (2022). https://doi.org/10.1007/978-981-16-8550-7_23
29. Dai, C.: Dex: a dapp for the decentralized marketplace. In: Blockchain and Crypt Currency, pp. 95–106. Springer, Singapore (2020)

30. Panda, S.K., Elngar, A.A., Balas, V.E., Kayed, M. (eds.): Bitcoin and Blockchain: History and Current Applications, 1st ed. CRC Press (2020). https://doi.org/10.1201/9781003032588
31. Madine, M., Salah, K., Jayaraman, R., Al-Hammadi, Y., Arshad, J., Yaqoob, I.: appxchain: Application-level interoperability for blockchain networks. IEEE Access **9**, 87 777–87 791 (2021)
32. Panda, S.K., Jena, A.K., Swain, S.K., Satapathy, S.C. (eds.): Blockchain Technology: Applications and Challenges. Springer, Intelligent Systems Reference Library. https://doi.org/10.1007/978-3-030-69395-4
33. Yang, Q., Zhao, Y., Huang, H., Xiong, Z., Kang, J., Zheng, Z.: Fusing blockchain and AI with metaverse: a survey. IEEE Open J. Comput. Soc. (2022)
34. Ning, H., Wang, H., Lin, Y., Wang, W., Dhelim, S., Farha, F., Ding, J., Daneshmand, M.: A Survey on Metaverse: the State-of-the-Art, Technologies, Applications, and Challenges (2021). arXiv:2111.09673

Chapter 6
Survey on Blockchain Technology and Security Facilities in Online Education

Rohini Jha

Abstract Securing digital records is the crucial paradigm for the intelligent environment. Hence, several authentication mechanisms were implemented to improve the trustworthy of the online application in data sharing. In addition, blockchain models have been implemented that include the cryptographic function to offer security for the online education academic system. However, in some cases, the security efficiency against unknown attacks has degraded the digital application environment and security range. As a result, in order to determine the most appropriate security architecture for the online education environment, the benefits and cons must be thoroughly examined. Considering this, the present chapter has aimed to design the literature survey studies incorporating online education and blockchain security. In addition, the performance of the blockchain has been validated with different online education features are user verification, certificate sharing, and online teaching facilities. Finally, the performance has been measured based on the execution time, and the merits and limitations have been described separately.

Keywords Blockchain · Online Education · Cryptographic function · Security range · Certificate verification · E-learning

6.1 Introduction

Blockchain technology has recently been an emerging technology in various fields [1]. Nowadays, it is widely used in different growing sectors like information technology (IT), medical [2], banking, and education. Generally, in a connected network, Blockchain defines a design that reserves transactional information known as a block, and the databases in which the data are stored is called chain [3]. Popularly, Blockchain is also mentioned as a digital ledger. It is a form of data management that is mainly used in maintaining the records of the transaction [4]. It is one of the reliable techniques such that the failure of one block does not affect the working of the

R. Jha (✉)
Birla Institute of Technology, Ranchi, India
e-mail: rohinijha@bitmesra.ac.in

S. K. Panda et al. (eds.), *Recent Advances in Blockchain Technology*,
Intelligent Systems Reference Library 237, https://doi.org/10.1007/978-3-031-22835-3_6

entire network application [5]. Also, it does provide high security, decentralization, and transparency in the network [6]. Because of its advantages, it is used in different fields like business, health, finance, and education [7]. Researchers worldwide are developing different techniques using blockchain features in different fields [8], as it can store past information and cannot be removed or altered by anyone [9]. In recent times, scholars have been exploring the features of Blockchain in the educational field [10]. In the educational field, it is mainly used in storing students' data securely [11]. Also, it provides the service of controlling the details of students by themselves [12]. Digital-based education is one of the platforms created recently using the application of blockchain technology [13]. E-learning is one such platform, which is trusted world widely for grading systems and for higher education [14]. As the Blockchain provides high security, it is widely used in various fields. In recent times, the application of Blockchain in Education is proliferating [15]. By using Blockchain, the student's details are digitalized and secured. Also, nowadays, it is used in education to validate, share, and issue certificates online [16]

Moreover, in some educational institutions, Blockchain is used to store the payment transaction records of students (e-transcripts) and certificates of students [17]. The blockchain trends are described in Fig. 6.1. Currently, due to COVID-19, most educational institutions prefer the online mode of learning through the internet [18]. In such cases, blockchain technology plays a crucial role, as it can store the information of students' securer manner [19]. Also, the digital certificates and the learning data are stored in a highly secure manner with the help of Blockchain [20]. This chapter provides a survey on how blockchain technology is helpful in online Education [21]. Thus, this chapter contributes to the different existing blockchain technologies used in the educational sector [38], their benefits, and their challenges. Moreover, this survey presents the merits and demerits of each technique. [39] While the institutions, university education learners, and annual graduates continue to rise, the need for easily verifiable degree credentials creates new business prospects [40]. This work shows two financial projections in which the product cost is divided between the student and the company as the agency's primary stakeholders [41]. In addition, the students want proof of certification [42] that is inexpensive and simple to verify, while companies require a reliable and speedy confirmation of degrees while recruiting [43]. Furthermore, as many students graduate each year, the challenge of forged diplomas is significant [44]. In India, it is simple to get digital documents [45]. Companies that hire vast numbers of recent graduates spend a ton of cash to verify applicants' academic credentials and records [46, 47]. An Authorization that employs blockchains can solve this issue. In addition, the Blockchain is decentralized, a digital ledger managed collaboratively by multiple computers known as nodes [34]. A user cannot modify the information in the database without the consensus of all those who preserve the records [48]. Educational files for people details have included education practices and student personal details [31]. These documents are vital for a person's future Education and profession [49]. The fundamental characteristic of archives is their primordial nature, which allows them to reconstruct the original historical circumstances [50]; hence, these documents are crucial to students, academic facilities, and possible employers [51]. The advent of technology has led

Blockchain in Industry trends	Filament	Samsung (Business machine)	Slock it	Tactical Analysis
	R3 Corda	Private Blockchain Strategies	Linux Foundation	Etheriun association
Blockchain trends in software applications	Paxos	Distribute Computing	Fault tolerant Byzantine	Practical Byzantine, Bitcoin etherium casper

Fig. 6.1 Blockchain trends

to the digitization of records. Unlike traditional paper-based record keeping, digital data is kept on a file system with significant variability [52]; consequently, they may be readily updated during storing, transfer, and sharing operations.

The profession of medicine is entering an era fraught with both new potential and new difficulties because of the advent of big data. Recently, an increasing number of infectious illnesses, such as COVID-19, characterized by powerful infectious and devastating capabilities, have captured individuals' attention. So, the research has indicated that the dissemination of medical information to a certain degree that the rapid spread of illnesses among huge populations [66]. In order to conduct the research study based on the medical diagnosis system, many research organizations require a substantial quantity of accurate samples. Moreover, the methods from the field of AI are being applied to diagnostics in the treatment of refractory disorders like hyperactivity, glaucoma, and Parkinson's disease [67]. At last, the exchange of medical information is the critical importance to the development of medical system advancement, in addition to delivering clinical information that may be used for diagnostic and natural sciences.

Furthermore, electronic medical records (EMR) can also act as a potential foundation for judgment when resolving legal conflicts arising from medical care provisions [68]. Also, EMRs had evolved into the most comprehensive database of patients throughout the entirety of the necessary medical process; yet, the EMR data centers that are now in use are not without flaws [69]. EMRs are often kept in a personal system, which presents a challenge because patient information may get dispersed among several different medical centers if the data moves from one health care facility to the next as a result of the natural progression of their lives. Patients lose simple access to their medical records following a diagnosis because the information is entered into EMRs at communities and health institutions. This is the case even when the individual is the legal owner of the information.

6.2 Blockchain Based Trust Scheme for E-learning

The distribution of student certificates has become a vital comprehension for students. Therefore, Mishra et al. [21] have created a tamper-proof authentication procedure in the blockchain context. In this instance, the created model was supported by several frameworks, including authentic, unchangeable, etc. Ultimately, prototype implementations were utilized for the results of the proposed method. Therefore, it has achieved a high level of security, although the transaction procedure is lengthier.

Chen et al. [22] introduced a blockchain-based application to the school sector to reduce the digital certificate change. An inventive application was coupled with a cryptocurrency trust mechanism to safeguard stored data. Consequently, crucial indicators are added to the current paradigm, and a higher confidentiality rate is achieved. However, software architecture has required additional time.

The education documents are a more critical asset; thus, they were digitized and saved in intelligent apps. However, harmful occurrences frequently impair smart programmers. Li and Han [23] proposed a safe storage system that protects digital documents and a reliable data transmission mechanism to address this problem. In addition, the developed technique employs a crypto technique to protect information. The period it takes to complete the process is minimal. However, the approach has resulted in a meager rate of confidentiality.

A blockchain-based system for managing academic transaction records was described by Lizcano et al. [24]. In addition, depending on the university education characteristics of the pupils, connect multiple were delivered securely using blockchain technology. Finally, multiple student inquiries are effectively addressed with a high percentage of confidentiality. However, it lacked security if some harmful events were supplied. The Blockchain in the educational section is described in Fig. 6.2.

Today, online education has become a popular method; Guo et al. [25] had built an electronic verification management program to preserve the storage's private range. The crypto architecture with a validation layer was revised in this suggested framework to safeguard the data sets. Moreover, if the client failed the test, the program or data in question was locked. Consequently, the suggested technique includes a monitoring framework for hazardous occurrences. Consequently, it has increased the confidentiality rating of the data stored. However, the design is intricate.

Yun et al. [26] presented a novel Deep Q Networks (DQN) Fracture-based Bitcoin technique for developing blockchain strategies sharing systems. Furthermore, the DQN method has effectively identified the harmful risk. Deep reassurance learning is used to retrain and predict the system's more exemplary parameters. Hence it optimizes the public Blockchain's current security spectrum and throughput. However, the initial countermeasure is necessary for its effectiveness. Hence, the discussed summary is described in Table 6.1.

Chi et al. [27] designed a Hyperledger-based Fabric data sharing architecture for the blockchain technology's crucial capabilities and safety functionality. The group detection approach has split the organizations to facilitate data exchange. The

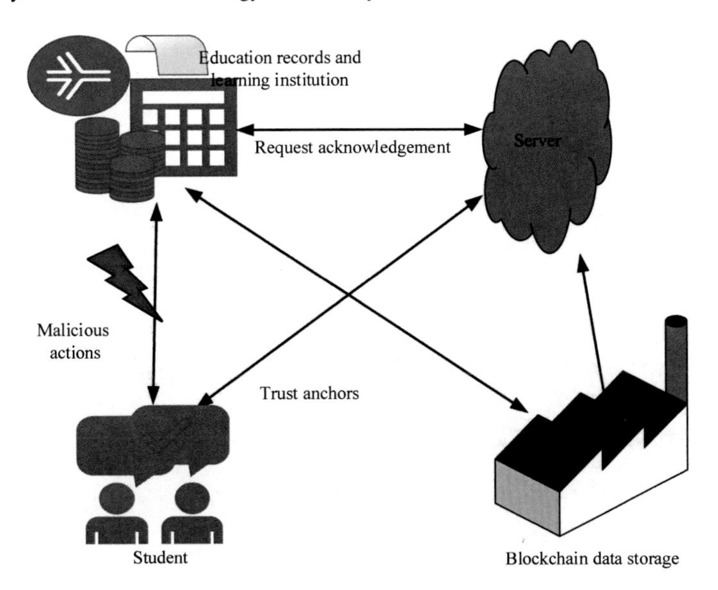

Fig. 6.2 Blockchain educational system

repercussions are calculated on the basis of degree-of cooperation, calculation time, and immediate cost. Nonetheless, the principal concerns do not apply to all cases. Cao et al. [28] propose a novel Two Arch2 algorithmic technique for blockchain stability in Network-based industrialization. The stability has been improved, and the bitcoin price has been drastically decreased. However, fewer parameters are optimized by this Arch 2 model.

The majority of the support for decentralized interaction comes from scalable systems that include distributed trust and security [68, 79–81]. Moreover, the Decentralization has ensured that the way of maintaining information is not centralized, that it is not exported, and that it is not under the jurisdiction of any other organization but rather the original business [69, 82–84]. Any alteration to the data needs to be validated by all of the persons affected before it can be saved. Hence, it is the key merits of the assessment of Blockchain, which guarantees that the scheme is dispersed and repetitive, making it difficult for exchanges to be falsified [70], rescinded, or duplicated. The rise of virtual currencies is the main application of the blockchain framework. The use of blockchain technology has been expanded into additional domains, such as smart contracts, healthcare, quality assurance, credit encryption, procurement, and medication, amongst others. In addition, blockchain technology provides a defense mechanism against social engineering. Since there have been relatively few studies conducted in this area, implementing the blockchain paradigm in educational settings is also a developing field of research [71]. As a result, the most likely direction that e-learning will head shortly is toward virtual learning and online education [72]. Since the use of the blockchain framework is highly effective in safeguarding different datasets than the need for other techniques,

Table 6.1 Merits and demerits of discussed models

Author	Methods	Merits	Demerits
Mishra et al. [21]	Tamper-proof	985 of confidential rate has been recorded, blockchain strategy has been implemented to hide the data from the third parties	The recorded transaction is very high
Chen et al. [22]	Academic blockchain application	Trust mechanism has been implemented for securing the data broadcasting function	High computation complexity
Li and Han [23]	Safe storage system	The crypto models are implemented to hide the Certificate sharing process	In some cases retrieving data is difficult
Lizcano et al. [24]	Students detail records transaction	Here, the trust model is implemented in communication Channel for the authenticity verification	Designing this verification model is very complex
Guo et al. [25]	Electronic verification management	It has reduced the execution time and resource usage during the certificate sharing	But the presence of malicious behavior has damaged the verification model
Yun et al. [26]	DQN	It has identified the harmful actions exactly	But the modification is often required based on application specific
Chi et al. [27]	Hyper ledger	Here the blockchain has been designed in the Hyper ledger by by incorporating the verification module. It helps to authenticate the Present users in the education system	High computation cost
Cao et al. [28]	Two Arch2 algorithm	It has minimized the resource usage	However, it is complex in design

the implementation of personal device policies in any institution has been facilitated by its utilization [73, 85–87]. Furthermore, blockchain technology has enabled the storage of educational records in a trustworthy decentralized system, the provision of trustworthy cryptographic keys, the recognition of information source sharing through the use of payment systems [74], and the protection of learning resources through the data encryption. Also, data integrity, Immutability, and dependability are further advantages that may be gained from employing blockchain technology.

6.3 Blockchain Facilities in Online Education

The Blockchain Intelligent Receives Instructions is achieved through the applications as a result of the Blockchain's distinctive technology development in the format of the massive online applications. Moreover, the education values have afforded the info technology infrastructure innovation forced to proceed the classroom instruction required for students all over the world to acquire digital educational materials, suggest the best classes dependent on the quality of necessities and standings showcased, and use digital games to run and execute organizational tasks for Blockchain-related online studies [29]. Each achievement and output of online Education has resulted in better academic courses for the student studies. The principal benefits of the blockchain network over traditional relational databases are its dispersion, decentralization, and trustworthy encryption. Moreover, in an online learning scenario, the adaptability of the package online teaching programs must also be considered to ensure dispersed data storage. In addition, the college courses are essentially a real-time process, and the learning status of each student is similarly dynamic. Furthermore, capturing the multimedia data as earlier is equally vital to decentralization and reliable encryption. Due to its maximum bandwidth, the Blockchain-based architecture with the digital blockchain platform has been deployed [30, 89, 90].

The potential for Blockchain to upset traditional corporate procedures is becoming evident. There have been described four sorts of blockchain measures: Efficiency play, record keeper, blockchain disruptor, and technology platform market. Following the institution-centric strategy, various applications have emerged inside the education sector. Student-centric alternatives validate obtained credentials, whereas institution-centric alternatives primarily assist institutional operations. Moreover, several use instances address various parts of the business domain, including simplifying the verification, virtual importance of higher education travel documents, forever securing issuance certificates [54], validating the certification program [55], recognizing credits automatically, etc. [56]. In addition, the preliminary examination and assessment of proposed projects reveal that most prioritize the learner while simplifying the document [31, 91, 92]. Due to its decentralized structure, blockchain technology offers benefits such as authenticity, anonymity, trustworthiness, and institutional and time autonomy. In addition, Education is the current domain where distributed ledger technology has the potential to be applied and where obstacles are being addressed.

University education is a complicated system with several obstacles that may be mitigated by applying this project [53]. This article provides a summary of existing systems, their actual condition, and the trajectory of their growth, as well as specific suggestions and enhancements for current methods that would improve them and, subsequently, the process of knowledge. It also suggests degrees of blockchain incorporation in a secondary learning system that, when coupled with information gathering, can extend and broaden the quality of study to the advantage of institutions, businesses, and learners [32].

Utilizing blockchains, a conceptual framework for e-learning platforms is suggested. In particular, the term learning group should be utilized as a way to illustrate the actual application situations. The unique qualities of the blockchain technology, such as virtual money, smart contracts, etc., can solve the three constraints as mentioned above of existing e-learning platforms. For instance, if the student meets the objective or performs the work, the coding of the shared ledger will pay the student. Multiple potential implementations of the suggested conceptual paradigm for utilizing digital currency are presented in this article [33]. The summary of the discussed literature is detailed in Fig. 6.3.

6.4 Certificate Sharing Using Blockchain

Applications of blockchains are proliferating quickly in the present day, providing the potential to address weak points in the area of online learning, like the complexity of e-learning evaluation, the absence of a truly united e-learning main exam, and the neediness of online educational certificates. Moreover, the blockchain platform for e-learning training and evaluation, which consists of an altogether fresh network framework based on the range of formal and informal blockchain systems, and also four principal cryptographic protocol strategies for the implementation of e-learning evaluation and credit transfer, digital certificate disbursal and data encryption, digitally signed confirmation, and e-learning discount coupon allocation [75]. It has been proved that the suggested approach is a good contender for establishing fairer, better, and more accessible online learning environments [33]. In the Blockchain-based security concept in online Education, initially, the student's data has been converted into the digital format after that blockchain concept is executed as the crypto models. Hence, Qin Liu et al. [34] have designed the non-tampering features in the blockchain application for the verification process to identify the actual users to prove the originality score. Hence, the designed model has gained a high-security range. However, this model is inefficient in front of the harmful unknows malicious events. The critical public infrastructure (PKi) was extensively used to create, manage, distribute, store, and revoke digital certificates crucial in establishing secure connections.

Ever since the beginning of the millennium, online learning has entered a period in which it is experiencing fast expansion due to the rapid advancement of computers involving the internet. Moreover, E-learning, also recognized as online classes or digital training, is a browser teaching method that utilizes information systems and Cloud computing to facilitate the dissemination of content and quick learning. Other names for distance classes include online Education and online learning. Online instruction, which utilizes the Web as its channel, overcomes the limitations imposed

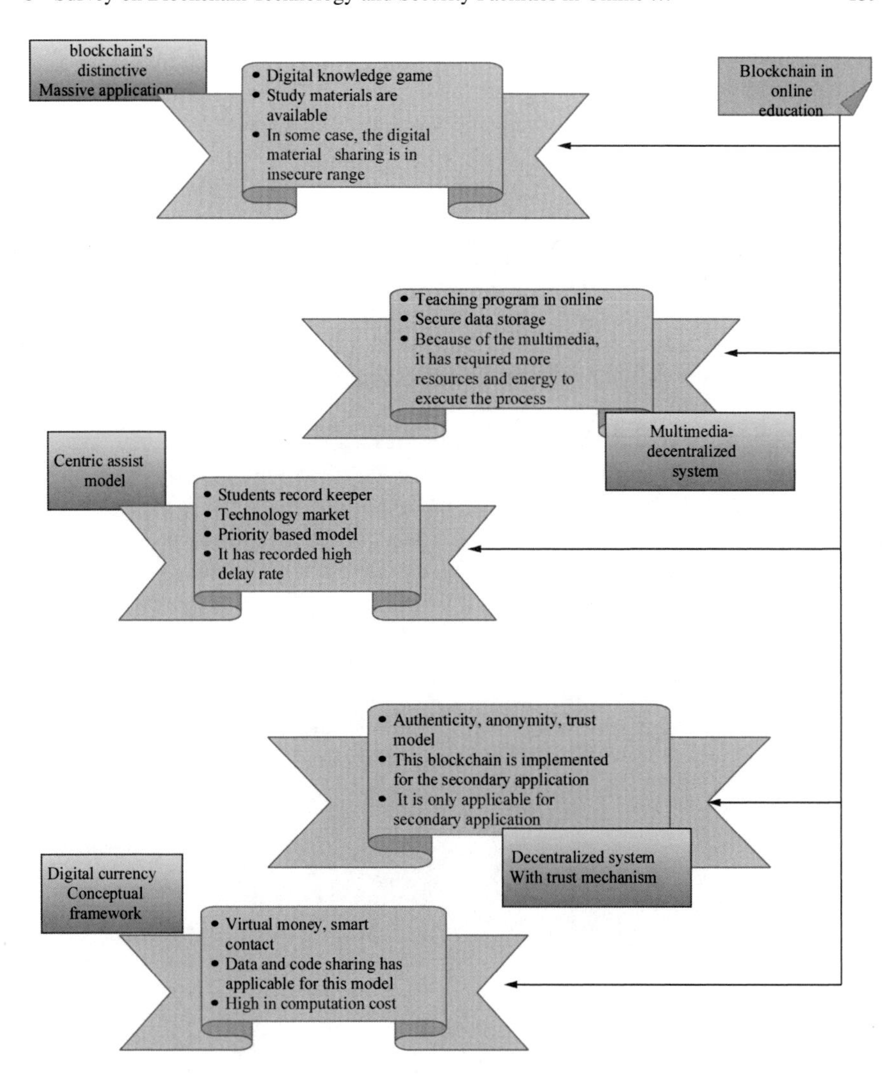

Fig. 6.3 Blockchain facilities and drawbacks for online Education

by location, atmosphere, time, and instructors by providing students with access to high-quality learning activities at any moment and in all locations [76].

A PKi platform authenticates individuals with their associated public-key and establishes the basis for the integrity of public-key cryptographic protocols in public-key authentication and encryption. Owing to the inclusion of a centralized certification authority, the classic PKi systems are susceptible to security flaws such as solitary and other unknown attacks [35]. The framework of the student record in Blockchain is described in Fig. 6.4. In addition, the classic centralized PKi network and the safety risks present potential methods for addressing issues by developing blockchains.

Two frameworks are utilizing Blockchain that is trustworthy and bulletin board. However, it is costly in cost. Certificate fraud is the educational industry's most significant problem [57]. To avoid certificate fabrication and create a safe environment for document verification, software for an authorized model is developed utilizing Blockchain-based contracts [58]. Moreover, the preservation of Blockchain prohibits data manipulation and facilitates the implementation of achievement testing. The mechanism produces a fresh block for each blockchain client. This Blockchain is duplicated among all the servers in the peer-to-peer connection, which increases the system's stability. When a client is authenticated, the system electronically provides the user with both their credentials and a Fast Response number. This facilitates the exchange of credentials among employers. Also, the technique called alpha-blending has been utilized to embed the watermark in the digital certificated to maximize the security range [36]—this aids in preventing various attacks during various stages of credential distribution. The Summary of Certificate Sharing is detailed in Table 6.2.

Traditional document certifications and digital certificates have challenges with preservation and maintenance. In addition, the problems in the inconvenient verification, low dependability, anti-tampering, and anti-counterfeiting [37]. This article suggested a scheme for constructing a decentralized certificate scheme with new decentralized and cryptographic protocols, where a collection of blockchain credential systems designed to provide blockchain credential services for entrepreneurial competitions among university students has evolved [59]. This solution uses smart contracts to implement specific certificate administration, issuance, verification, and revoke operations. Signatory information, certification templates, and certification

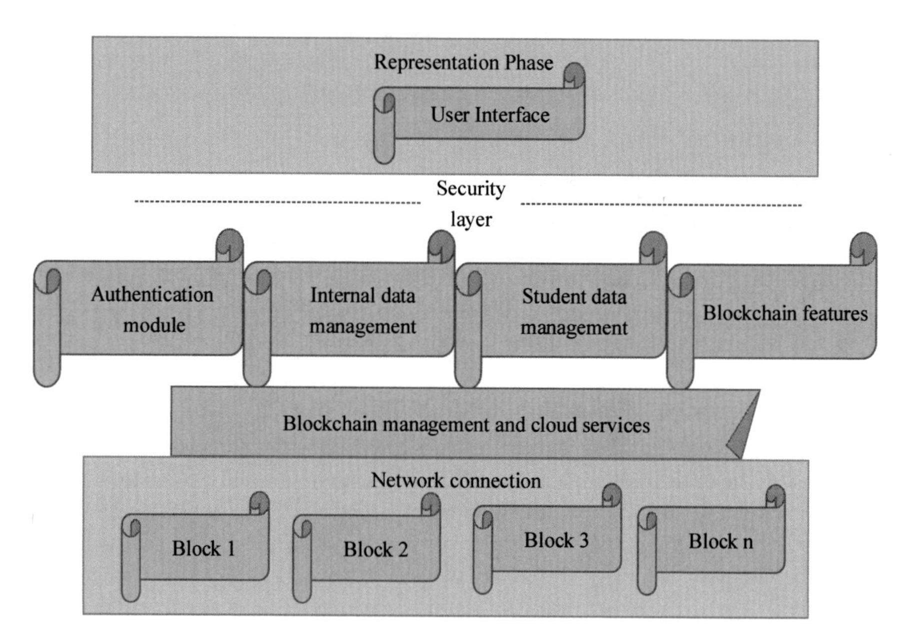

Fig. 6.4 Students records in Blockchain

Table 6.2 Summary of Certificate Sharing

Authors	Mechanism	Merits	Drawbacks
Li et al. [33]	Cryptographic protocol	Credit transfer, digital certificate is applicable by this designed protocol	Complex in design
Liu et al. [34]	Non-tampering features	High security range has been recorded	But detecting the unknown attack is too difficult
Li et al. [35]	Public-key-infrastructure	Here, the centralized certificate verification model has been implemented to improve the originality range of the stored digital certificate	Its economy is too high
Poorni et al. [36]	Alpha-blending	It has prevented different types of known and unknown attacks	But, the design of this model has required more resources
Xie et al. [37]	Decentralized cryptographic model	Signatory information has been implemented as the data integrity model	High computation time has been recorded

knowledge are maintained in a verification contract to structure the digital education data, allowing for more simple queries and validations.

The use of blockchains is an attractive option for addressing the challenges presented by online Education, which include inadequate certification, a shortage of acknowledgment, and poor information protection. Currently, the most common applications for this technology may be found in banking, the Network, and the IoT. In addition, the financial sector technologies have included digital money, currency transfers, currency exchanges, and payment services. There is no need for a person to get involved in executing smart contracts, which may be used for everything from shares and commodities to loans [77]. Moreover, the context of the IoT and blockchain technology has made it possible for objects to interact independently and detect faults. This technique has also seen some preliminary applications in the educational world thus far. For example, Mike Sharple postulated the Blockchain might be used to realize cloud processing of educational material, which would comprise the so-called information money [78]. Several academics have proposed using blockchain technology to verify credit cards, encrypt the data, and distribute data storage. In addition, the Blockchain scheme and Mozilla's web system were utilized in constructing an online educational certificate system by the digital media lab. Furthermore, the technology behind the cryptocurrency has been integrated into the process of product design within the industrial sector. The Sony-Global Educational blockchain technology infrastructure might publicly exchange the record data and learning material securely without giving any information to the intelligent system. Therefore achieving the education fairness and enabling the data broadcasting services to be digitalized.

A certificate authority is a verification document utilized to identify the online users on a particular web. A digital card is an electronic characteristic of a certification body that contains proof of ownership of the key pair and a key that could be used to confirm the authenticity of the element conversing over the network users. The user must extract the signature using the given public key to validate the certificate. After this, the output was changed with the hashing operation performed on the clear text. In the event that the results remain unchanged, it will be possible to demonstrate that the certification has still not been falsified or interfered with in any way.

It is possible to make extensive help from electronic documents in e-commerce and electronics-related activities. The certificate sharing process in Blockchain is described in Fig. 6.5. Moreover, these credentials can be utilized in the protection mechanisms of digital payments, online subscription services, online shopping, and connecting directly to secure websites, among other applications. Digital certificates can have their validity temporarily revoked before their expiry date in certain conditions. The credential subscriber is required to promptly remove the certificate application from the authorization center whenever there is a transition in the user's personal identification information or whenever the subscriber's secret key is lost, leaked, or presumed to be made public.

Moreover, at the same time, the authorization process is forced to place the credential released to the public digital certificates list at the appropriate time. Hence, the

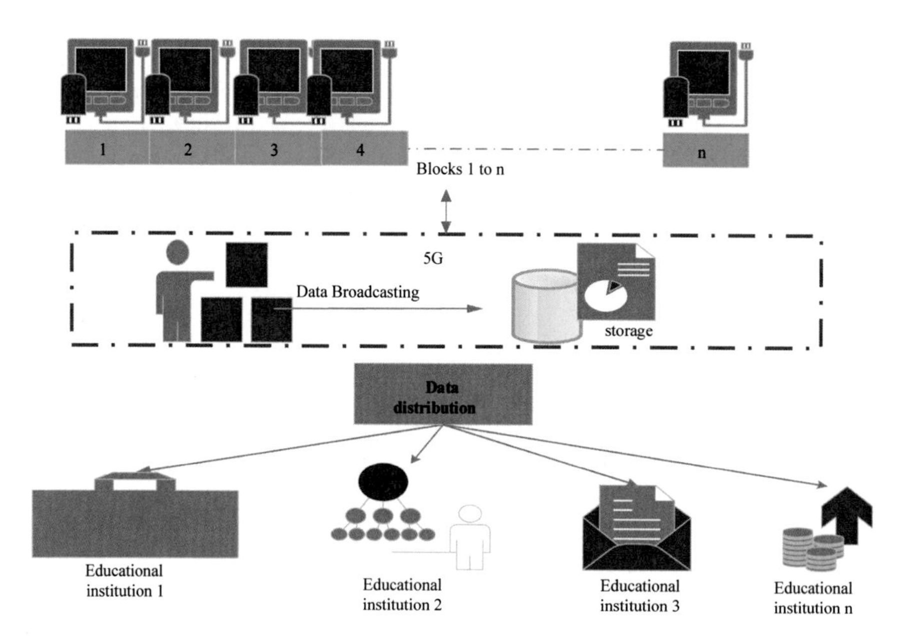

Fig. 6.5 Certificate sharing using Blockchain

certificates are no longer valid when revoked, which marks the expiry time of the certificate's lifetime. Furthermore, additional users who have already gotten the secret key and have used it to sign various electronic files or protocols can duplicate the digital certificate if the secret key is inadvertently disclosed to a third party. In the Unpredictable outcomes, destroying valuable property might be the outcome of this course of action. The ongoing usage of conventional certificate authorities has a much more severe effect if these events occur. When checking the legitimacy of a certification, it is necessary to examine the specific server to determine whether the certificate has been previously invalidated.

6.5 Blockchain Security in Medicine

The present COVID-19 pandemic crisis has a significant impact on daily lives. An efficient contact tracking system must be created to prevent the spread of infectious illnesses. Unfortunately, present methods have major privacy problems that compromise the identification and position of users and customers. Although some confidentiality methods have been presented, there are several centralization-related problems [61]. To address these concerns, a Privacy-Preserving Tracing (PPT) system has been introduced in Blockchain-based Healthcare applications. In PPT, the 5G-integrated system connects the underlying architecture, allowing anybody with a smartphone or wearable gadget linked to the 5G network to do location checks to see that they have had potential touch with a specific identification without breaching their security. A reputable medical facility can successfully identify individuals and their respective close relationships. The proposed PPT approach offers online privacy, transparency, dependability, and identification with high computing & communications efficiency and reduced latency, as determined by a comprehensive security study.

Recently, there has been an increasing interest in employing drones to provide medical services. Moreover, such applications often entail issues about the safety of flight, the transfer of data, and the protection of individual privacy; hence, issues about the reliability of connectivity and the enhancement of data protection should be tackled first. Hence, the 5G communication system and the technology behind Blockchain are now considered potential solutions to these issues. This piece begins with classifying the many medical uses of drones, and then it moves on to discuss the primary difficulties associated with these possibilities [62]. Then, the fundamentals of Blockchain and 5G technology investigate the inherent qualities of both technologies to ensure reliable communication and increased data protection. An illustrated case is provided and then analyzed within the context of disaster aid. The analysis findings show that there is a significant opportunity to use Blockchain and 5G technology to further the medical uses of drones.

The exchange of patient health records has a significant positive impact on the study of infection treatment and epidemics. In recent years, eHealth systems based on blockchain technology have gained many results in the transmission and administration of electronic records. Nevertheless, there are still certain obstacles to overcome.

Solutions built on permission blockchains have a large bandwidth and are scalable, but they are susceptible to rollback threats and can result in privacy leaks. Furthermore, the Design of the public Blockchain was more open and safe, but at the expense of sustainability, and there are no benefits for health organizations to participate in the networks. In addition, the information retrieval process in Blockchain ledger electronic health record systems is ineffective due to the fundamental makeup of blockchains. So, the Data exchange and privacy-preserving (DEPP) blockchain [63] was designed with micro blocks and critical blocks of the patients to use the stored medical details. Hence, the data can retrieve more quickly. In addition, a reputational system is being developed to encourage participation in DEPP on the part of medical facilities. DEPP can provide medical data exchange for individuals in a fashion that is respectful of their privacy by using typically involved techniques. Then several metrics were measured to evaluate the performance of the DEPP.

Most hospitals only keep copies of their client's information on-site and have no backup. This creates a substantial risk of either the loss of data or file corruption. Even while many medical institutions are moving their digital data to the web, cloud storage that comes with its own set of security risks. During the current COVID-19 epidemic, several different health care facilities have been the targets of denial attacks and ransomware assaults [64]. As a result of these assaults, numerous emergency responders were rendered inoperable, leaving thousands of people without access to medical treatment. Another issue with these old database procedures is that companies frequently lose or mix up the data about the individual, which, should go without saying, results in significant consequences. A large number of academics are currently focusing their efforts on Blockchain technologies in order to enhance the preservation of medical information. Hence, this comprehensive analysis has described the secure data storage facilities in the medical institution. It assesses the present technologies and their design, making future Blockchain technologies much more accessible.

At present, medical certifications are quite significant for many people because they wish to avail themselves of health benefits such as taxable income, legal procedures, insurance applications, and many other things. The merits and demerits of Blockchain in the medical application are described in Fig. 6.6. Before the development of computers, medical certifications were only present in the form of physical copies. Producing, distributing, and maintaining these certifications continues to be a substantial challenge. The digitalization of medical reports and papers gives rise to possible safety concerns. For example, there is a possibility that the confidentiality of healthcare documentation may be compromised if certifications were forged. In addition, persons were always needed to present themselves at the medical clinics providing the credentials and wait to receive them. At the moment, the architecture of any modern healthcare links IoT gadgets and development tools that interact with the Infosystems. This is the case in all healthcare industries. The combination of blockchains with the IoT can substantially impact the healthcare sector by enhancing operational easiness, security, and openness while creating new commercial opportunities [65]. In light of this, a method that makes use of blockchain technology and safeguards patients' right to privacy was implemented in IoT-based medical systems.

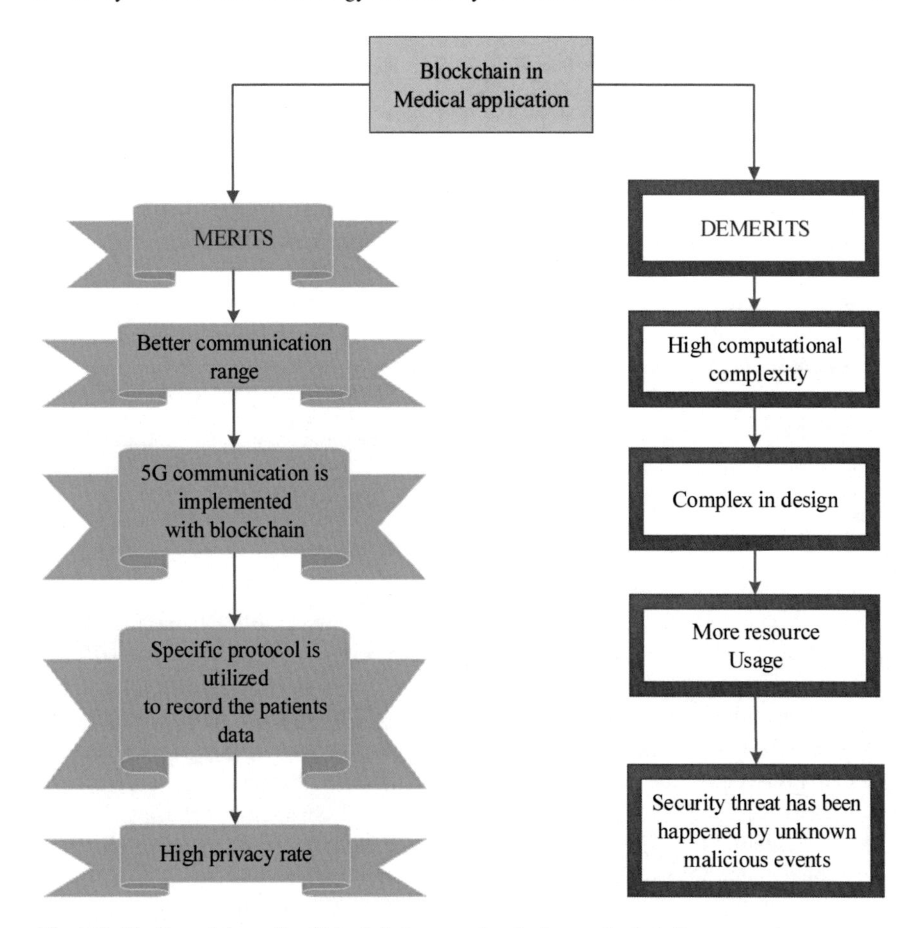

Fig. 6.6 Merits and demerits of blockchain execution in the medical studies

Moreover, the suggested scheme creates an interface that allows users and medical clinics to communicate with one another in order to produce and manage health records. In addition, the suggested system provides safety by stipulating the ground rules through a shared ledger. The findings and discussion indicated that the suggested plan is more effective than the already used methods.

There is a significant amount of promise for an AI-based blockchain based on the Electronic Health Records (EHR) designed to maintain dependable, safe, and resilient storage solutions for EHR. This electronic medical record would also make it easier for researchers, clinicians, and governmental authorities to obtain data as a means of study, carry out analytics, and assist well-studied choices. Depending on the medical reports and the Computed tomography results, an Artificial Neural system is used to categorize the patients as possibly COVID-19 affirmative or possibly COVID-19 negative [66]. This classification is done for each patient individually. Blockchain uses the Inter-Planetary protocol to record the patient data who may or may not have

COVID-19. After validation and verification of entities, only authorized parties have been able to gain access to EHR. For the purpose of determining whether or not the patient may be infected with COVID-19.

Hence, the performance has been evaluated for several different AI-based methods using measures such as learning curve and accuracy. In addition, the 6G considerably reduces the network delay and dependability difficulties, making transmitting data in the real-time environment easier. Furthermore, in the midst epidemic, the volume of data being created is very significant. As a result, we decided to use the Inter-Planetary protocol since it provides a solution that is both expensive and satisfies all of the demanding standards [67]. In the end, the evaluation was conducted for MedBlock's architecture performance in security, network, and memory and found that it was superior to other systems. The summary of discussed literature is described in Table 6.3.

6.6 Survey Analysis

Typically, centralized storage and administration modes are utilized, making systems susceptible to various assaults. In addition, the records of various educational stages are maintained on separate storage servers within the academic system. The storage servers are often built to permit accessibility only by current staff, without interchange. A server outage might as quickly lose data or leaking. Organizations often implement security measures to secure personal details that prohibit record accessibility. However, organizations lack safe and efficient channels for sharing data. For example, learners may have challenges when transferring from one university to the next until maintaining the completion of their prior studies.

To check the working function of the discussed blockchain model, some security models were taken are TPS [21], Multi authority (MA) Encryption [23], single Authority (SA) encryption, and Education Secure Recorded (ESR)model. Hence, the model TPS has recorded the lowest execution duration that, is described in Fig. 6.7.

The average forecasts of all algorithms are essential since they can reduce neural network function's reliance on a few distinct data characteristics. The compared schemes are mini-blockchain (MBC), Least significant bit (LSB); consequently, dropout is designed to randomly eliminate hidden and transparent units and sensory characteristics during the learning of convolutional neural systems [60]. Monte Carlo (MC) dropout models and Deep Learning Without dropout (DLWoD) are taken to check the performance of the designed scalable Blockchain. Hence, the validated execution cost is described in Fig. 6.8. In this manner, co-adaptation and overfitting are avoided, and the channel's generalization ability and achievement tests are enhanced.

Table 6.3 Summary of discussed literatures

Zhang et al. [61]	PPT	Here, the 5G cellular connection is linked with the privacy module for the better communication. Hence, better communication has been established	However, security threat is occur against the harmful attack
Chen et al. [62]	Blockchain in drone assist medical system	Here, the Blockchain was introduced for the drone assist medical system. It has ensured the trust worthy communication	But, it has recorded high computational complexity
Zou et al. [63]	DEPP blockchain	Here, several blocks were designed to store the data in a systematic manner	It has required more resources to execute the process
Kumar et al. [64]	Comprehensive analysis in block chain	For the future enhancement of the blockchain technologies, the demerits of the present Blockchain has been studied	None
Namasudra et al. [65]	Blockchain in IoT	It has managed the medical records that stored in the blockchain environment	High energy consumption was recorded
Mistry et al. [66]	AI-blockchain with EHR	Here, the Inter-Planetary protocol is utilized to record the patient's data within short duration	Design complexity has recorded

6.7 Conclusion

This survey has been conducted to measure the functioning of the Blockchain in the online education environment. Here, the facility of Blockchain in the education system has been analyzed in different sectors. Finally, the graphical representation has validated the execution cost for designing the particular blockchains in the education system. Besides, the intelligent models are also implemented along with the Blockchain to afford the most acceptable security range and prediction performance. This has shown the availability and flexibility of the blockchain features based on application range. In addition, to maximize the blockchain security capacity, the

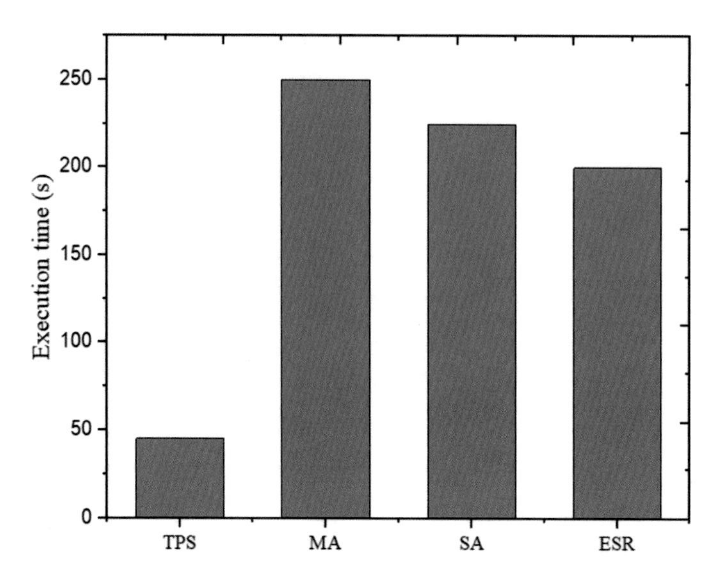

Fig. 6.7 Execution time validation

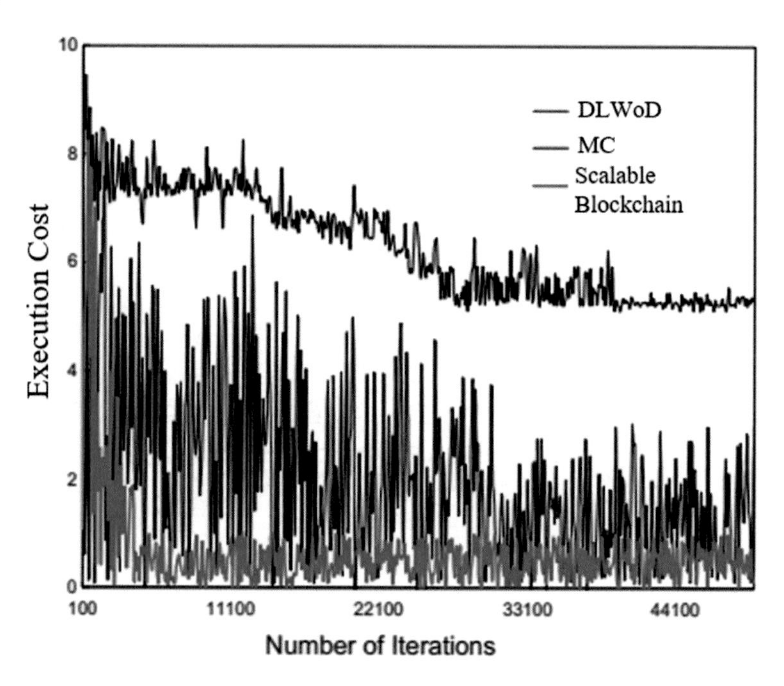

Fig. 6.8 Transaction duration

mathematical model is incorporated into the crypto models, and then the security range and the execution duration have been noted. Moreover, the Blockchain is implemented for the extensive data application; hence its efficiency is validated by the run duration. Hence, the execution time parameter is considered the chief metric.

References

1. Goumidi, H., Aliouat, Z., Harous, S.: Vehicular cloud computing security: a survey. Arab. J. Sci. Eng. **45**, 2473–2499 (2020). https://doi.org/10.1007/s13369-019-04094-0
2. Rangwani, D., Om, H.: A secure user authentication protocol based on ECC for cloud computing environment. Arab. J. Sci. Eng. **46**, 3865–3888 (2021). https://doi.org/10.1007/s13369-020-05276-x
3. Alzahrani, B.A.: Secure and efficient cloud-based IoT authenticated key agreement scheme for e-health wireless sensor networks. Arab. J. Sci. Eng. **46**, 3017–3032 (2021). https://doi.org/10.1007/s13369-020-04905-9
4. Darwish, M.A., Yafi, E., Al Ghamdi, M.A., Almasri, A.: Decentralizing privacy implementation at cloud storage using blockchain-based hybrid algorithm. Arab. J. Sci. Eng. **45**, 3369–3378 (2020). https://doi.org/10.1007/s13369-020-04394-w
5. Almagrabi, A.O.: A3C: Access appropriate analogous computing for cloud-assisted edge users. Arab. J. Sci. Eng. (2021). https://doi.org/10.1007/s13369-021-05611-w
6. Luo, R.C., Chung, L.Y., Lien, C.H.: A novel symmetric cryptography based on the hybrid haar wavelets encoder and chaotic masking scheme. IEEE Trans. Ind. Electron. **49**(4), 933–944 (2002). https://doi.org/10.1109/TIE.2002.801252
7. Stallings, W.: Cryptography and network security, 4/E. Pearson Education India (2006)
8. Sultan, N.A.: Reaching for the "cloud": How SMEs can manage. Int. J. Inf. Manag. **31**(3), 272–278 (2011). https://doi.org/10.1016/j.ijinfomgt.2010.08.001
9. Che, J., Duan, Y., Zhang, T., Fan, J.: Study on the security models and strategies of cloud computing. Procedia Eng. **23**, 586–593 (2011). https://doi.org/10.1016/j.proeng.2011.11.2551
10. Jensen, M., Schwenk, J., Gruschka, N., Iacono, L.L.: On technical security issues in cloud computing. In: 2009 IEEE International Conference on Cloud Computing, pp. 109–116 (2009). doi:https://doi.org/10.1109/CLOUD.2009.60
11. Banerjee, M., Lee, J., Choo, K.K.R.: A blockchain future for internet of things security: A position paper. Digit. Commun. Netw. **4**(3), 149–160 (2018). https://doi.org/10.1016/j.dcan.2017.10.006
12. Ouaddah, A., Elkalam, A.A., Ouahman, A.A.: Towards a novel privacy-preserving access control model based on blockchain technology in IoT. In: Rocha, Á., Serrhini, M., Felgueiras, C. (eds.) Europe and MENA cooperation advances in information and communication technologies. Adv. Intell. Syst. Comput. vol. 520. Springer, Cham (2017). https://doi.org/10.1007/978-3-319-46568-5_53
13. Takabi, H., Joshi, J.B.D., Ahn, G.: Security and privacy challenges in cloud computing environments. IEEE Secur. Priv. **8**(6), 24–31 (2010). https://doi.org/10.1109/MSP.2010.186
14. Kwak, K.S., Ullah, S., Ullah, N.: An overview of IEEE 802.15.6 standard. 2010 3rd International Symposium on Applied Sciences in Biomedical and Communication Technologies (ISABEL 2010), pp. 1–6 (2010). https://doi.org/10.1109/ISABEL.2010.5702867
15. Goyal, V., Pandey, O., Sahai, A., Waters, B.: Attribute-based encryption for fine-grained access control of encrypted data. In: Proceedings of the 13th ACM conference on Computer and communications security, pp. 89–98. Acm (2006). https://doi.org/10.1145/1180405.1180418
16. Gennaro, R., Lysyanskaya, A., Malkin, T., Micali, S., Rabin, T.: Algorithmic tamper-proof (atp) security: theoretical foundations for security against hardware tampering. In: Naor, M. (eds.) Theory of cryptography. TCC 2004. Lect. Notes Comput. Sci., vol. 2951. Springer, Berlin, Heidelberg (2004). https://doi.org/10.1007/978-3-540-24638-1_15

17. Gilad-Bachrach, R., Dowlin, N., Laine, K., Lauter, K., Naehrig, M., Wernsing, J.: Cryptonets: Applying neural networks to encrypted data with high throughput and accuracy. Proceedings of The 33rd International Conference on Machine Learning, vol. 48, pp. 201–210. PMLR (2016)
18. Chatterjee, A., Kaushal, M., Sengupta, I.: Accelerating sorting of fully homomorphic encrypted data. In: Paul, G., Vaudenay, S. (eds.) Progress in Cryptology—INDOCRYPT 2013. Lecture Notes in Computer Science, vol. 8250. Springer, Cham (2013). https://doi.org/10.1007/978-3-319-03515-4_17
19. Li, L., El-Latif, A.A.A., Niu, X.: Elliptic curve ElGamal based homomorphic image encryption scheme for sharing secret images. Signal Process. **92**(4), 1069–1078 (2012). https://doi.org/10.1016/j.sigpro.2011.10.020
20. Lin, X., Sun, X., Ho, P.H., Shen, X.: GSIS: A secure and privacy-preserving protocol for vehicular communications. IEEE Trans. Veh. Technol. **56**(6), 3442–3456 (2007). https://doi.org/10.1109/TVT.2007.906878
21. Mishra, R.A., Kalla, A., Braeken, A., Liyanage, M.: Privacy protected blockchain based architecture and implementation for sharing of students' credentials. Inf. Process. Manag. **58**(3), 102512 (2021). https://doi.org/10.1016/j.ipm.2021.102512
22. Chen, G., Xu, B., Lu, M., Chen, N.S.: Exploring blockchain technology and its potential applications for education. Smart Learn. Environ. **5**(1), 1–10 (2018). https://doi.org/10.1186/s40561-017-0050-x
23. Li, H., Han, D.: EduRSS: A blockchain-based educational records secure storage and sharing scheme. IEEE Access **7**, 179273–179289 (2019). https://doi.org/10.1109/ACCESS.2019.2956157
24. Lizcano, D., Lara, J.A., White, B., Aljawarneh, S.: Blockchain-based approach to create a model of trust in open and ubiquitous higher education. J. Comput. High. Educ. **32**(1), 109–134 (2020). https://doi.org/10.1007/s12528-019-09209-y
25. Guo, J., Li, C., Zhang, G., Sun, Y., Bie, R.: Blockchain-enabled digital rights management for multimedia resources of online Education. Multimed. Tools. Appl. 1–21 (2019). https://doi.org/10.1007/s11042-019-08059-1
26. Yun, J., Goh, Y., Chung, J.M.: DQN based optimization framework for secure sharded blockchain systems. IEEE Internet Things J. **8**(2), 708–722 (2020). https://doi.org/10.1109/JIOT.2020.3006896
27. Chi, J., Li, Y., Huang, J., Liu, J., Jin, Y., Chen, C., Qiu, T.: A secure and efficient data sharing scheme based on Blockchain in industrial internet of things. J. Netw. Comput. Appl. **167**, 102710 (2020). https://doi.org/10.1016/j.jnca.2020.102710
28. Cao, B., Wang, X., Zhang, W., Song, H., Lv, Z.: A many-objective optimization model of industrial internet of things based on private Blockchain. IEEE Netw. **34**(5), 78–83 (2020). https://doi.org/10.1109/MNET.011.1900536
29. Oganda, F.P., Lutfiani, N., Aini, Q., Rahardja, U., Faturahman, A.: Blockchain Education Smart Courses of Massive Online Open Course Using Business Model Canvas. In: 2020 2nd International Conference on Cybernetics and Intelligent System (ICORIS), IEEE (2020). doi:https://doi.org/10.1109/ICORIS50180.2020.9320789
30. Guo, J., Li, C., Zhang, G., Sun, Y., Bie, R.: Blockchain-enabled digital rights management for multimedia resources of online education. Multimed. Tools. Appl. **79**(15), 9735–9755 (2020). https://doi.org/10.1007/s11042-019-08059-1
31. Kamišalić, A., Turkanović, M., Mrdović, S., Heričko, M.: A preliminary review of blockchain-based solutions in higher Education. International workshop on learning technology for Education in cloud, Springer, Cham (2019). doi:https://doi.org/10.1007/978-3-030-20798-4_11
32. Juričić, V., Radošević, M., Fuzul, E.: Creating student's profile using blockchain technology. In: 2019 42nd International Convention on Information and Communication Technology, Electronics and Microelectronics (MIPRO), IEEE (2019). doi:https://doi.org/10.23919/MIPRO.2019.8756687
33. Li, C., Guo, J., Zhang, G., Wang, Y., Sun, Y., Bie, R.: A blockchain system for E-learning assessment and certification. In: 2019 IEEE International Conference on Smart Internet of Things (SmartIoT), IEEE (2019). doi: https://doi.org/10.1109/SmartIoT.2019.00040

34. Liu, Q., Guan, Q., Yang, X., Zhu, H., Green, G., Yin, S.: Education-industry cooperative system based on Blockchain. In: 2018 1st IEEE International Conference on Hot Information-Centric Networking (HotICN), IEEE (2018). doi: https://doi.org/10.1109/HOTICN.2018.8606036
35. Li, Y., Yu, Y., Lou, C., Guizani, N., Wang, L.: Decentralized public key infrastructures atop Blockchain. IEEE Netw. **34**(6), 133–139 (2020). https://doi.org/10.1109/MNET.011.2000085
36. Poorni, R., Lakshmanan, M., Bhuvaneswari, S.: DIGICERT: a secured digital certificate application using Blockchain through smart contracts. In: 2019 International Conference on Communication and Electronics Systems (ICCES), IEEE (2019). doi: https://doi.org/10.1109/ICCES45898.2019.9002576
37. Xie, R., Wang, Y., Tan, M., Zhu, W., Yang, Z., Wu, J., Jeon, G.: Ethereum-blockchain-based technology of decentralized smart contract certificate system. IEEE Internet Things Mag. **3**(2), 44–50 (2020). https://doi.org/10.1109/IOTM.0001.1900094
38. Sharma, S., Batth, R.S.: Blockchain technology for higher education sytem: A mirror review. In: 2020 International Conference on Intelligent Engineering and Management (ICIEM), IEEE (2020). doi: https://doi.org/10.1109/ICIEM48762.2020.9160274
39. Liu, L., Han, M., Zhou, Y., Parizi, R.M., Korayem, M.: Blockchain-based certification for Education, employment, and skill with incentive mechanism. Blockchain cybersecurity, trust and privacy, pp. 269–290. Springer, Cham (2020). https://doi.org/10.1007/978-3-030-38181-3_14
40. Johnson, K., Krueger, B.S.: Who supports using cryptocurrencies and why public education about blockchain technology matters?. Blockchain and the public sector, pp. 127–149. Springer, Cham (2021). https://doi.org/10.1007/978-3-030-55746-1_6
41. Santos, H., Batista, J., Marques, R.P.: Digital transformation in higher Education: the use of communication technologies by students. Procedia Comput. Sci. **164**, 123–130 (2019). https://doi.org/10.1016/j.procs.2019.12.163
42. Sun, X., Zou, J., Li, L., Luo, M.: A blockchain-based online language learning system. Telecommun. Syst. **76**(2), 155–166 (2021). https://doi.org/10.1007/s11235-020-00699-1
43. Priya, N., Ponnavaikko, M., Aantonny, R.: An efficient system framework for managing identity in educational system based on blockchain technology. In: 2020 International Conference on Emerging Trends in Information Technology and Engineering (ic-ETITE), IEEE (2020). doi: https://doi.org/10.1109/ic-ETITE47903.2020.469
44. Rahardja, U., Hidayanto, A.N., Hariguna, T., Aini, Q.: Design framework on tertiary education system in Indonesia using blockchain technology. In: 2019 7th International Conference on Cyber and IT Service Management (CITSM), vol. 7. IEEE (2019). doi: https://doi.org/10.1109/CITSM47753.2019.8965380
45. Turkanović, M., Hölbl, M., Košič, K., Heričko, M., Kamišalić, A.: EduCTX: A blockchain-based higher education credit platform. IEEE access **6**, 5112–5127 (2018). https://doi.org/10.1109/ACCESS.2018.2789929
46. Palma, L.M., Vigil, M.A.G., Pereira, F.L., Martina, J.E.: Blockchain and smart contracts for higher education registry in Brazil. Int. J. Netw. Manag. **29**(3), e2061 (2019). https://doi.org/10.1002/nem.2061
47. Yumna, H., Khan, M.M., Ikram, M., Ilyas, S.: Use of Blockchain in Education: A systematic literature review. In: Asian Conference on Intelligent Information and Database Systems, pp. 191–202. Springer, Cham (2019). https://doi.org/10.1007/978-3-030-14802-7_17
48. Efanov, D., Roschin, P.: The all-pervasiveness of the blockchain technology. Procedia. Comput. Sci. **123**, 116–121 (2018). https://doi.org/10.1016/j.procs.2018.01.019
49. Al Harthy, K., Al Shuhaimi, F., Al Ismaily, K.K.J.: The upcoming Blockchain adoption in Higher-education: requirements and process. 2019 4th MEC international conference on big data and smart city (ICBDSC), IEEE (2019). doi: https://doi.org/10.1109/ICBDSC.2019.8645599
50. Vidal, F., Gouveia, F., Soares, C.: Analysis of blockchain technology for higher Education. In: 2019 International Conference on Cyber-Enabled Distributed Computing and Knowledge Discovery (CyberC), IEEE (2019). doi: https://doi.org/10.1109/CyberC.2019.00015

51. Williams, P.: Does competency-based Education with blockchain signal a new mission for universities? J. High. Educ. Policy Manag. **41**(1), 104–117 (2019). https://doi.org/10.1080/1360080X.2018.1520491
52. Ahel, O., Lingenau, K.: Opportunities and challenges of digitalization to improve access to Education for sustainable development in higher Education. Universities as Living Labs for Sustainable Development, pp. 341–356. Springer, Cham (2020). doi: https://doi.org/10.1007/978-3-030-15604-6_21
53. Sousa, M.J., Rocha, A.: Digital learning: Developing skills for digital transformation of organizations. Future Gener. Comput. Syst. **91**, 327–334 (2019). https://doi.org/10.1016/j.future.2018.08.048
54. Jackson, N.C.: Managing for competency with innovation change in higher Education: Examining the pitfalls and pivots of digital transformation. Bus. Horiz. **62**(6), 761–772 (2019). https://doi.org/10.1016/j.bushor.2019.08.002
55. Xiao, J.: Digital transformation in higher Education: critiquing the five-year development plans (2016–2020) of 75 Chinese universities. Distance Educ. **40**(4), 515–533 (2019). https://doi.org/10.1080/01587919.2019.1680272
56. Kutnjak, A., Pihiri, I., Furjan, M.T.: Digital Transformation Case Studies Across Industries–Literature Review. In: 2019 42nd International Convention on Information and Communication Technology, Electronics and Microelectronics (MIPRO), IEEE (2019). doi: https://doi.org/10.23919/MIPRO.2019.8756911
57. Azarenko, N.Y., Mikheenko, O.V., Chepikova, E.M., Kazakov, O.D.: Formation of innovative mechanism of staff training in the conditions of digital transformation of economy. In: 2018 IEEE International Conference "Quality Management, Transport and Information Security, Information Technologies" (IT&QM&IS), IEEE (2018). doi: https://doi.org/10.1109/ITMQIS.2018.8525021
58. Wang, Y., Luo, F., Dong, Z., Tong, Z., Qiao, Y.: Distributed meter data aggregation framework based on Blockchain and homomorphic encryption. IET Cyber-Phys. Syst.: Theory Appl. **4**(1), 30–37 (2019). https://doi.org/10.1049/iet-cps.2018.5054
59. Wang, J., Peng, F., Tian, H., Chen, W., Lu, J.: Public auditing of log integrity for cloud storage systems via Blockchain. In: Li, J., Liu, Z., Peng, H. (eds,) Security and Privacy in New Computing Environments. SPNCE 2019. Lecture Notes of the Institute for Computer Sciences, Social Informatics and Telecommunications Engineering, vol. 284. Springer, Cham (2019). https://doi.org/10.1007/978-3-030-21373-2_29
60. Chu, C.H.: Task offloading based on deep learning for Blockchain in mobile edge computing. Wirel. Netw. **27**(1), 117–127 (2021). https://doi.org/10.1007/s11276-020-02444-7
61. Zhang, C., Xu, C., Sharif, K., Zhu, L.: Privacy-preserving contact tracing in 5G-integrated and blockchain-based medical applications. Comput. Stand. Interfaces **77**, 103520 (2021). https://doi.org/10.1016/j.csi.2021.103520
62. Chen, J., Wang, W., Zhou, Y., Ahmed, S.H., Wei, W.: Exploiting 5G and Blockchain for medical applications of drones. IEEE Netw. **35**(1), 30–36 (2021). https://doi.org/10.1109/MNET.011.2000144
63. Zou, R., Lv, X., Zhao, J.: SPChain: Blockchain-based medical data sharing and privacy-preserving eHealth system. Inf. Process. Manag. **58**(4), 102604 (2021). https://doi.org/10.1016/j.ipm.2021.102604
64. Kumar, S., Bharti, A.K., Amin, R.: Decentralized secure storage of medical records using Blockchain and IPFS: A comparative analysis with future directions. Security and Privacy **4**(5), e162 (2021). https://doi.org/10.1002/spy2.162
65. Namasudra, S., Sharma, P., Crespo, R.G., Shanmuganathan, V.: Blockchain-based medical certificate generation and verification for IoT-based healthcare systems. IEEE Consum. Electron. Mag. (2022). https://doi.org/10.1109/MCE.2021.3140048
66. Mistry, C., Thakker, U., Gupta, R., Obaidat, M.S., Tanwar, S., Kumar, N., Rodrigues, J.J.P.C.: MedBlock: An AI-enabled and blockchain-driven medical healthcare system for COVID-19. In: ICC 2021-IEEE International Conference on Communications, IEEE (2021). doi: https://doi.org/10.1109/ICC42927.2021.9500397

67. Jeong, W.Y., Choi, M.: Design of recruitment management platform using digital certificate on Blockchain. J. Inf. Process. Syst. **15**(3), 707–716 (2019)
68. Chowdhury, M.J.M., Colman, A., Kabir, M.A., Han, J., Sarda, P.: Blockchain as a notarization service for data sharing with personal data store. In: 2018 17th IEEE International Conference on Trust, Security and Privacy in Computing and Communications/12th IEEE International Conference on Big Data Science and Engineering (TrustCom/BigDataSE), IEEE (2018). doi: https://doi.org/10.1109/TrustCom/BigDataSE.2018.00183
69. Sun, H., Wang, X., Wang, X.: Application of blockchain technology in online Education. Int. J. Emerg. Technol. Learn. 13(10) (2018)
70. Sunarya, P.A., Khoirunisa, A., Nursaputri, P.: Blockchain family deed certificate for privacy and data security. In: 2020 Fifth International Conference on Informatics and Computing (ICIC), IEEE (2020). doi: https://doi.org/10.1109/ICIC50835.2020.9288528
71. Alshahrani, M., Beloff, N., White, M.: Revolutionising higher Education by adopting Blockchain technology in the certification process. In: 2020 International Conference on Innovation and Intelligence for Informatics, Computing and Technologies (3ICT), IEEE (2020). doi: https://doi.org/10.1109/3ICT51146.2020.9311970
72. Bhaskar, P., Tiwari, C.K., Joshi, A.: Blockchain in education management: present and future applications. Interact. Technol. Smart Educ. **18**(1), 1–17 (2020). https://doi.org/10.1108/ITSE-07-2020-0102
73. Bele, R.S., Mehare, J.P.: A review on digital degree certificate using blockchain technology. IJCRT **9**(2), 2320–2882 (2021)
74. Cheng, J.C., Lee, N.Y., Chi, C., Chen, Y.H.: Blockchain and smart contract for digital certificate. In: 2018 IEEE international conference on applied system invention (ICASI), IEEE (2018). doi: https://doi.org/10.1109/ICASI.2018.8394455
75. Kumutha, K., Jayalakshmi, S.: Blockchain Technology and Academic Certificate Authenticity—A Review. Expert Clouds and Applications, vol. 209, pp. 321–334. Springer, Singapore (2022). doi: https://doi.org/10.1007/978-981-16-2126-0_28
76. Huynh, T.T., Huynh, T.T., Pham, D.K.: Issuing and verifying digital certificates with Blockchain. In: 2018 International Conference on Advanced Technologies for Communications (ATC), IEEE (2018). doi: https://doi.org/10.1109/ATC.2018.8587428
77. Gopal, N., Prakash, V.V.: Survey on Blockchain based digital certificate system. Int. J. Eng. Res. Technol. (IRJET) 5(11) (2018)
78. Capece, G., Levialdi Ghiron, N., Pasquale, F.: Blockchain technology: redefining trust for digital certificates. Sustainability **12**(21), 8952 (2020). https://doi.org/10.3390/su12218952
79. Panda SK, Mohammad GB, Nandan Mohanty S, Sahoo S. Smart contract-based land registry system to reduce frauds and time delay. Security and Privacy. (2021); e172. https://doi.org/10.1002/spy2.172
80. Panda, S.K., Satapathy, S.C.: Drug traceability and transparency in medical supply chain using blockchain for easing the process and creating trust between stakeholders and consumers. Pers. Ubiquit. Comput. (2021). https://doi.org/10.1007/s00779-021-01588-3
81. V.R. Niveditha, Karthik Sekaran, K. Amandeep Singh, Sandeep Kumar Panda, Effective prediction of bitcoin price using wolf search algorithm and bidirectional LSTM on internet of things data, Int. J. Syst. Syst. Eng. **11**(3–4), pp. 224–236
82. Sathya A.R., Panda S.K., Hanumanthakari S. (2021) Enabling Smart Education System Using Blockchain Technology. In: Panda S.K., Jena A.K., Swain S.K., Satapathy S.C. (eds.) Blockchain technology: applications and challenges. Intell. Syst. Ref. Libr. vol 203. Springer, Cham. doi:https://doi.org/10.1007/978-3-030-69395-4_10
83. Lokre S.S., Naman V., Priya S., Panda S.K.: Gun tracking system using blockchain technology. In: Panda S.K., Jena A.K., Swain S.K., Satapathy S.C. (eds.) Blockchain technology: applications and challenges. Intell. Syst. Ref. Libr., vol 203. Springer, Cham. (2021) doi:https://doi.org/10.1007/978-3-030-69395-4_16
84. Sandeep Kumar Panda,Shanmukhi Priya Daliyet,Shagun S. Lokre,Vihas Naman, Distributed Ledger Technology in the Construction Industry Using Corda, The New Advanced Society: Artificial Intelligence and Industrial Internet of Things Paradigm, doi:https://doi.org/10.1002/9781119884392.ch2

85. Panda S.K., Satapathy S.C.: An Investigation into Smart Contract Deployment on Ethereum Platform Using Web3.js and Solidity Using Blockchain. In: Bhateja V., Satapathy S.C., Travieso-González C.M., Aradhya V.N.M. (eds.) Data Eng. Intell. Computing. Adv. Intell. Syst. Comput., vol 1. Springer, Singapore. (2021) doi:https://doi.org/10.1007/978-981-16-0171-2_52
86. Panda S.K., Rao D.C., Satapathy S.C. An Investigation into the Usability of Blockchain Technology in Internet of Things. In: Bhateja V., Satapathy S.C., Travieso-González C.M., Aradhya V.N.M. (eds.) Data engineering and intelligent computing. Adv. Intell. Syst. Comput. vol 1. Springer, Singapore. (2021) doi:https://doi.org/10.1007/978-981-16-0171-2_53
87. Panda S.K., Dash S.P., Jena A.K. Optimization of Block Query Response Using Evolutionary Algorithm. In: Bhateja V., Satapathy S.C., Travieso-González C.M., Aradhya V.N.M. (eds.) Data engineering and intelligent computing. Adv. Intell. Syst. Comput., vol 1. Springer, Singapore. (2021) doi:https://doi.org/10.1007/978-981-16-0171-2_54
88. Nanda, S.K., Panda, S.K., Das, M., Satapathy, S.C.: Automating vehicle insurance process using smart contract and ethereum. In: Chakravarthy, V.V.S.S.S., Flores-Fuentes, W., Bhateja, V., Biswal, B. (eds.) Advances in micro-electronics, embedded systems and IoT. Lect. Notes Electr. Eng, vol 838. Springer, Singapore. (2022). doi:https://doi.org/10.1007/978-981-16-8550-7_23
89. Panda, S.K., Elngar, A.A., Balas, V.E., & Kayed, M. (Eds.).: Bitcoin and blockchain: history and current applications (1st ed.). CRC Press. (2020). doi:https://doi.org/10.1201/9781003032588
90. Blockchain Technology: Applications and Challenges, Editors: Panda, S.K., Jena, A.K., Swain, S.K., Satapathy, S.C. (Eds.), Springer, Intelligent Systems Reference Library. https://doi.org/10.1007/978-3-030-69395-4
91. Varaprasada Rao, K., Panda, S.K.: A design model of copyright protection system based on distributed ledger technology. In: Satapathy, S.C., Lin, J.CW., Wee, L.K., Bhateja, V., Rajesh, T.M. (eds.) Computer communication, networking and IoT. Lect. Notes Netw. Syst, vol 459. Springer, Singapore. (2023). doi:https://doi.org/10.1007/978-981-19-1976-3_17
92. Varaprasada Rao, K., Panda, S.K.: Secure electronic voting (E-voting) system based on blockchain on various platforms. In: Satapathy, S.C., Lin, J.CW., Wee, L.K., Bhateja, V., Rajesh, T.M. (eds.) Computer communication, networking and IoT. Lect. Notes Netw. Syst. vol 459. Springer, Singapore. (2023). doi:https://doi.org/10.1007/978-981-19-1976-3_18

Chapter 7
A Govern Chain—Integration of Government Function with Blockchain Technology

Sanjay Fuloria

Abstract Blockchain is a chain of blocks connected using encryption. Blockchains provide secure, indelible, and transparent transactions. There are decentralized applications of blockchain. Decentralized blockchains provide relief from single points of failure. Since the controls are also decentralized, the system is not dependent on any authority. There is a growing list of applications where the government is using blockchain technology in its interactions with the citizens. These applications are called Govern Chains. This chapter discusses the use of blockchains in public procurement, land registration, and electronic voting.

Keywords Blockchain · Govern chain · Transparency · Security · Land registrations · E-voting · Decentralized · Encryption · Applications

7.1 Introduction

Blockchain is a secure ledger that records all transactions up to this point. It is a large and growing file made up of blocks (or blocks of transactions) that are connected (chained) together using encryption. These include a coded version of the previous block, the timing of the transaction, and the headline of the block. There are other fields like the software version number, details of the previous block, the objective of the process, and a cryptographic number called the 'number used once' (nonce). The time in seconds is also mentioned. This time is measured from January 1st, 1970. No hacker can enter this blockchain database and modify the facts. This feature makes blockchain technology very useful for the government. The governments across the world are moving towards e-governance. In a few years from now, the citizens across the world will have e-governments to deal with. The term e-government refers to different technologies that can be used in the public sector as a conceptual framework for experimenting with them [1]. There has always been an effort to automate and streamline the functioning of the government [3–5]. This will enhance the level of

S. Fuloria (✉)
Professor, Operations Management and IT at IBS, Hyderabad, India
e-mail: sanjay.fuloria@ibsindia.org

S. K. Panda et al. (eds.), *Recent Advances in Blockchain Technology*,
Intelligent Systems Reference Library 237, https://doi.org/10.1007/978-3-031-22835-3_7

public service delivery. The beneficiaries will include entrepreneurs, citizens, and the various agencies of the government.

Assume a BTC blockchain (one of the cryptocurrencies) records various transactions. In other words, blockchain is a financial ledger that records all past money transactions (between community members) back to the beginning. It is a trustworthy register (or ledger) that stores all transactions up to this point. It is a large and increasing file. So, a BTC blockchain (like Bitcoin), currently includes huge amounts of data (in Giga Bytes) and records all bitcoin transactions between persons. that John Smith has 20 BTCs, for example. Two major aspects contribute to BTC's unique properties: 1. Decentralization 2. The technique for verifying transactions, intimately linked to the blockchain structure It is achieved by having multiple computers (nodes) around the world hold the whole blockchain (which should be like all other nodes' blockchain files). For example, unlike banking systems that are centralized where each bank retains clients' data (including the data of transactions) on its own computers (safely backed up), blockchain nodes are run by volunteers all over the world. There is no central authority or data repository. People with node computers are independent, except that their computers host blockchain nodes. Nodes are always connected via a protocol, forming a worldwide peer-to-peer network. This resembles any peer to peer (P2P) network. Every node has a blockchain file copy with it. Some of them (nodes or miners) can validate recent global transactions (in BTCs). There are also computers that merely verify transactions. They are termed miners despite not being nodes. Consider the following: If A wants to send B 100 EUR (in BTCs), A's computer (or smartphone) first informs the network's miners that A want to send B 100 EUR. This is an invalidated transaction, and it is not yet recorded into the register. A miner should first check if A has enough money (at least 100 EUR plus some for the provision). Now the miner checks this by going over the entire blockchain, seeing all of A's transactions, and determining how much money is available. Because there is no such thing as an account state, the money one has is always calculated again, based on all previous transactions. This is a great feature that makes the blockchain more secure. Manipulating another person's account requires changing past transactions, which is impossible.

7.2 Applications that Are Decentralized

Such applications are called DApps (Decentralized Applications). They combine on-chain logic of the application with traditional application engineering. The user and machine interfaces are provided by embedded applications, web applications, mobile or server-side applications. Such decentralized arrangements lead to a better execution of smart contracts.

DL networks' technologies, organization, and governance differ. These factors boost DL network trust, performance, and scalability. Off-chain applications and smart contracts use software engineering techniques. Decentralization and immutable ledgers require adaptations.

DL ecosystems often include distributed and support services in addition to key network services. These services enable better storage by making it distributed. The decentralized resources can be named in a user-friendly manner. The sources of data that are not on the chain can also be accommodated. Smart contracts are enabled as well.

DL technologies and services support decentralized application architects and developers. To productize DApps, they need business-grade development. They also require a supportive ecosystem and other services that can be provided by third parties. This would lead to a wider use of such technologies and add-on services.

How do we obtain decentralized trust? This can be obtained by the distributed property of ledger and their control that is decentralized. If a database is installed on a single computer, the owner/operator of the computer has all the access rights. The owner/operator can let another person access the database, but he/she reserves the right to override any of the new person's actions. Even in a distributed system, the control remains with the owner/operator. A distributed system does provide some benefits like redundancy, removing single point of failure, scalability, and resilience to failures. On the other hand, if we want decentralization of control, we need another approach. In this approach, the system must be distributed as many copies of the original also known as instances. These instances make the system. The control of each instance is the responsibility of a different entity. Some users could be running blockchain nodes while others could run an analytics node. Therefore, decentralization is more organizational than technical. However, there needs to be a consensus mechanism to avoid arbitrary decisions by a single participant. The same consensus mechanism is implemented in each instance. The consensus mechanism is pre decided so that all actions do not get affected by malicious intentions of any participant.

A Distributed Ledger system is inherently distributed. If we add distributed governance, it becomes a decentralized trusted system. In such systems, there is no central authority (owner/founder/operator) we need to trust. The mechanisms make the system trustworthy and can be verified.

These systems can be used for creating Govern Chains. The following section provides some examples.

7.3 Govern Chain Examples

When distributed ledger systems are used as a tool to make citizen-government interactions more efficient, they are called Govern Chains. Governments across the world are recognizing the importance of blockchain. There is a need for effective communication between the citizens and governments. E-governance solutions generate effectiveness for and efficiencies in government. Some examples follow.

7.3.1 Public Procurement

When the government buys goods or services, there are chances of corruption. It is, in fact, the most significant source of corruption the world over. A blockchain based system can provide tamper-proof transactions [7]. This will make the system more transparent and uniform. Automated smart contracts can be used. This system will:

(a) Ensure the delivery of same information to all the participants.
(b) Make the process transparent.
(c) Make the process quick.
(d) Remove the need for a pre-bid conference.
(e) Ensure to remove the chances of alteration in the submitted bids.
(f) Lead to cost saving by automating the process.
(g) Ensure proof of identity of the bidders.

A typical blockchain transaction is shown in Fig. 7.1. After the transaction starts, it goes to a blockchain network consisting of multiple nodes. These nodes compete to validate the transaction using their computing power. The node that wins this competition gets some bitcoins in return. Once the transaction is validated, it is combined with other transactions. This becomes a new block and is added to the chain. This completes the transaction.

The blockchain technology is used to secure an already smooth process. If the underlying process is neither smooth nor efficient, blockchain technology will be of no use. Figure 7.2 shows the public procurement process using blockchain. Every stage is recorded and secured in a blockchain.

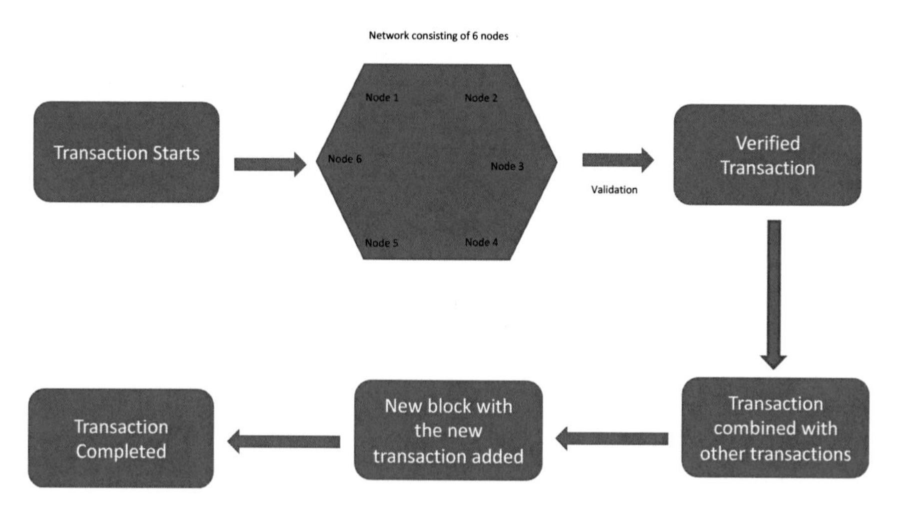

Fig. 7.1 Blockchain transaction

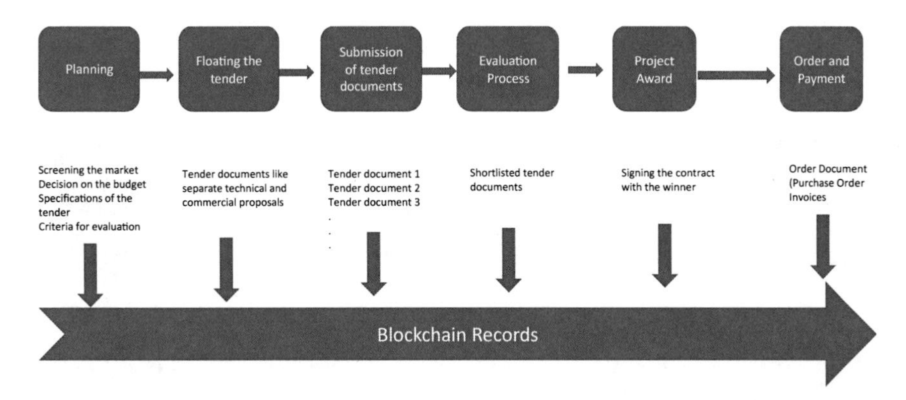

Fig. 7.2 Public procurement process using blockchain

7.3.2 *Land Title Registrations*

Blockchain based land title registrations have started happening worldwide. The first country to register land titles using blockchains in the republic of Georgia. They have developed a system that depends on the cooperation between National Agency of Public Registry (NAPR) and Bitfury, a bitcoin mining company [8–12]. The key to understand this transformation is the fact that the land registration system of Georgia was put through a process of reforms for decades to make it efficient and free of corruption. Blockchain was used after all this work was done. The land register uses a permissioned blockchain that is private. This is administered by NAPR. The role of NAPR here is that of a third-party enforcer. This blockchain includes land titles' sale, new titles' registration, mortgages, rentals, and notary services. Trust and transparency are the key benefits.

In Sweden, the blockchain has a different design. It stores document verification records instead of the actual documents. The actual documents are held by each party that gets into an agreement. The summary of verification records is stored in an external blockchain which is visible to anyone who wishes to verify. Stakeholders such as banks, real estate agents, and the government bodies can access the contracts via another interface. This interface can be integrated with their own systems.

Blockchain can be used if it is adapted to the land registration architecture specific to a country or state. It cannot be applied in its classic form. Legal certainty will be missing in the classic form. For a blockchain supporting land registrations, access to the blockchain system should be limited. The miners should be less in number, and they should be qualified. Liability rules need to be defined and the users should be properly identified [13–16].

Figure 7.3 shows the land title registration process with the help of blockchain technology. The citizens of a state/country can still use the existing interface (web or mobile) for their land registration application. The entire process is secured and validated using blockchain.

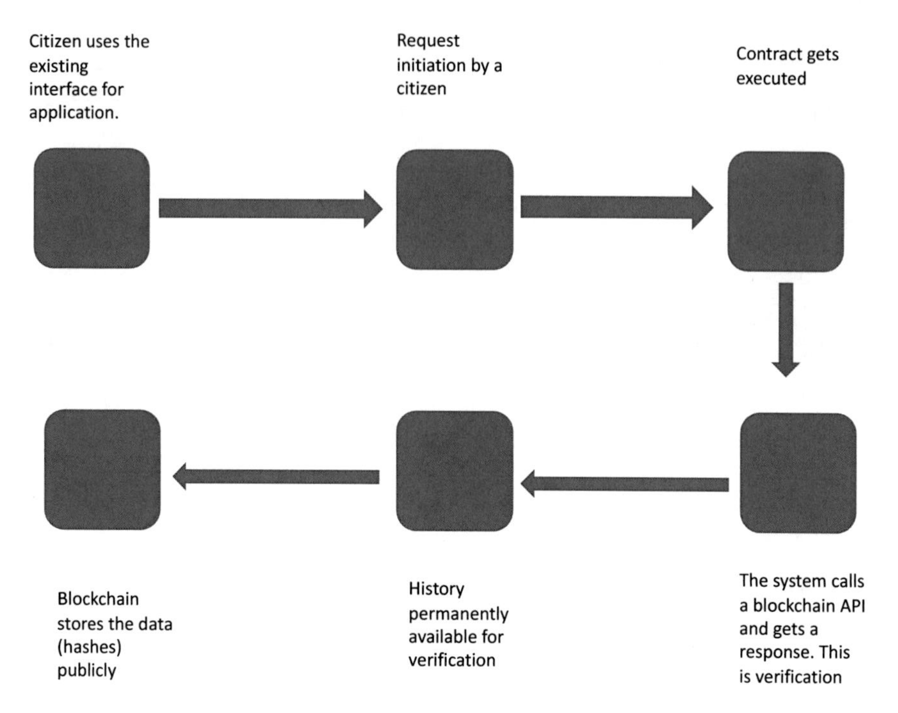

Fig. 7.3 Land title registration using blockchain

7.3.3 Electronic Voting

Electronic voting is a good use case for the blockchain technology as it will provide authentication, anonymity, accuracy, and verifiability. People who are registered to vote will be allowed to vote. It won't allow any link between voters' identities and ballots. Votes will be accurate, and every vote will count. The votes can't be changed, duplicated, or removed.

In this blockchain based voting system, the first transaction will represent the candidate. This transaction will include the candidate's name. Every vote for this candidate will be placed on the top of it. The first transaction will just be the candidate's name. A blank vote would be treated as a protest vote. Each vote is treated as a transaction, and it gets recorded and the blockchain gets updated. The block also carries the previous voter's information. If the blocks are fiddled with, one can find out as all blocks are connected. The actual voting takes place in the blockchain. This blockchain is decentralised. There is no single point of failure. Each vote goes to one of the nodes on the system, and the node automatically adds the vote to the blockchain [17–22].

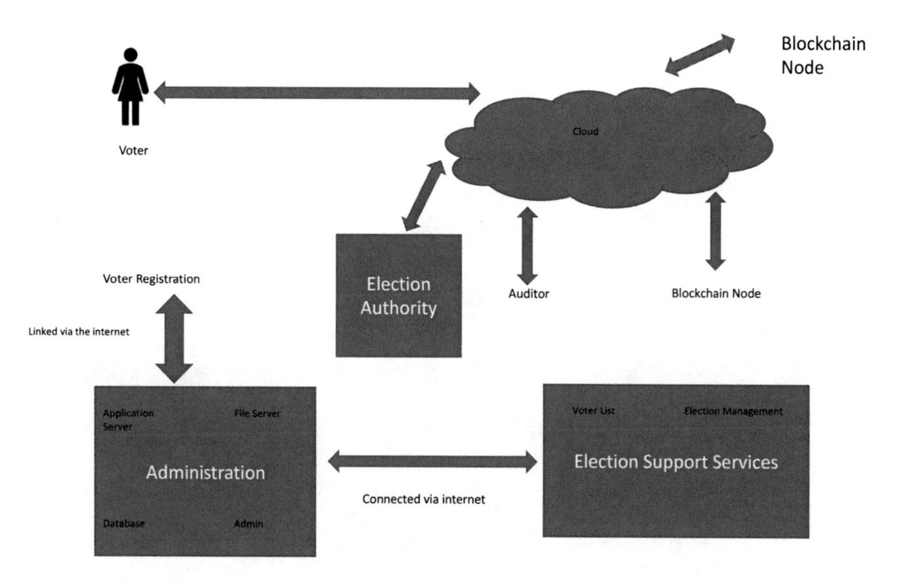

Fig. 7.4 Electronic voting using blockchain

Figure 7.4 shows the electronic voting process using blockchain technology. It starts with the voter registration with the election administration team. The administration relates to the election support services like voter list management and election management through internet.

7.4 Blockchain Applications for Energy Conservation

The requirement for energy conservation in the modern world cannot be overstated. With the global temperatures rising, the weather systems in all parts of the world have gone for a toss. Extreme events like floods, droughts, forest fires, and extreme snow events are commonplace now. There is a big reason to conserve energy. A traditional way to supply electricity is through the electric grid. The electric grid must balance demand and supply of energy. This balance is fueled by data viz. electric consumption data, billing, and trading data. A smart grid is a levelling up of sorts. It is more interoperable and has an open communication framework. The next level is Energy Internet (EI). It is also called Internet of Energy or Smart Energy. The ideal end state for this is sharing energy and data as easily and efficiently as information via the internet. This would require close interactions between participants, autonomy in decision making, fast and efficient access to various distributed and centralized energy resources, energy sharing to maintain the balance, and participants acting as both producers and consumers. The advantages that blockchain technology provides, its usage is ideal for such energy related services. Imagine a scenario where a family generates its own electricity using solar energy and can sell the surplus energy to

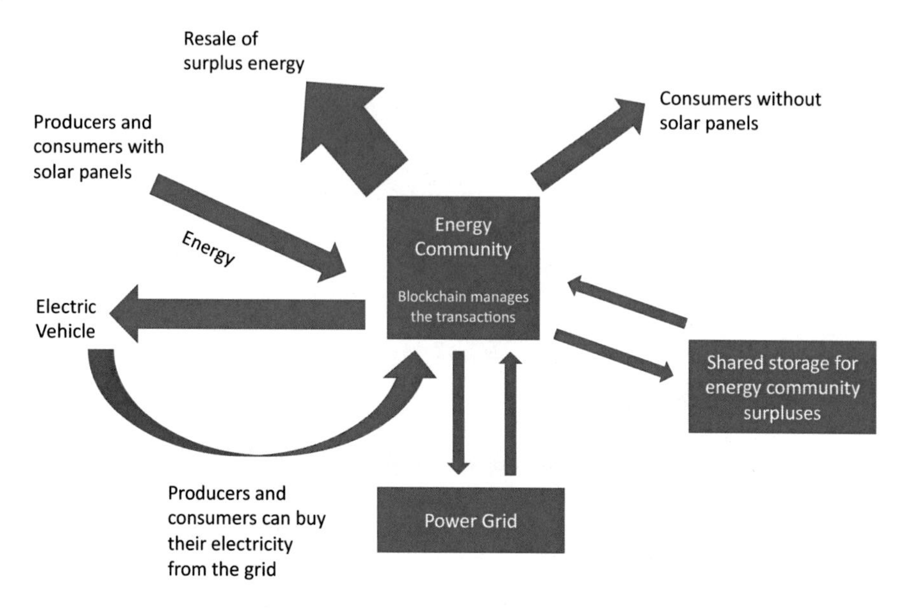

Fig. 7.5 Blockchain for energy conservation

its neighbors. In case of no requirement in the neighborhood, they could even sell it in the energy market. The entire data and all the transactions are recorded in a blockchain. This is possible.

Figure 7.5 on the next page shows the use of blockchain for energy conservation. This has an interesting component. Consumers of solar energy can also become producers. They can trade their excess energy with the community. They can start by supplying to their neighbours.

7.5 Conclusion

Blockchain technology does provide benefits like security, transparency, and immutability. One prerequisite is that the original system that is to be transformed to blockchain must be efficient and effective to start with. Converting an inefficient system to blockchain technology will not solve the systemic problems.

References

1. Yildiz, M.: E-government research: reviewing the literature, limitations, and ways forward. Gov. Inf. Q. **24**(3), 646–665 (2007)
2. May, P., Ehrlich, H.C., Steinke, T.: ZIB structure prediction pipeline: composing a complex biological workflow through web services. In: Nagel, W.E., Walter, W.V., Lehner, W. (eds.) Euro-Par 2006. LNCS, vol. 4128, pp. 1148–1158. Springer, Heidelberg (2006)
3. Foster, I., Kesselman, C.: The Grid: Blueprint for a New Computing Infrastructure. Morgan Kaufmann, San Francisco (1999)
4. Czajkowski, K., Fitzgerald, S., Foster, I., Kesselman, C.: Grid information services for distributed resource sharing. In: 10th IEEE International Symposium on High Performance Distributed Computing, pp. 181–184. IEEE Press, New York (2001)
5. Foster, I., Kesselman, C., Nick, J., Tuecke, S.: The physiology of the grid: an open grid services architecture for distributed systems integration. Technical Report, Global Grid Forum (2002)
6. National Center for Biotechnology Information, http://www.ncbi.nlm.nih.gov
7. Singh, S., Pal, O.: Blockchain technology and its applications in E-governance services. Int. J. Recent. Technol. Eng. **8**(4) (2019)
8. Kaczorowska, M.: Blockchain-based land registration: possibilities and challenges. Masaryk Univ. J. Law Technol. **13**(2), 339–360 (2019)
9. Panda, S.K., Mohammad, G.B., Nandan Mohanty, S., Sahoo, S.: Smart contract-based land registry system to reduce frauds and time delay. Secur. Priv. **2021**, e172 (2021). https://doi.org/10.1002/spy2.172
10. Panda, S.K., Satapathy, S.C.: Drug traceability and transparency in medical supply chain using blockchain for easing the process and creating trust between stakeholders and consumers. Pers. Ubiquit. Comput. (2021). https://doi.org/10.1007/s00779-021-01588-3
11. Niveditha, V.R., Sekaran, K., Amandeep Singh, K., Panda, S.K.: Effective prediction of bitcoin price using wolf search algorithm and bidirectional LSTM on internet of things data. Int. J. Syst. Syst. Eng. **11**(3–4), 224–236
12. Sathya, A.R., Panda, S.K., Hanumanthakari, S.: Enabling smart education system using blockchain technology. In: Panda, S.K., Jena, A.K., Swain, S.K., Satapathy, S.C. (eds.) Blockchain Technology: Applications and Challenges. Intelligent Systems Reference Library, vol. 203. Springer, Cham (2021). https://doi.org/10.1007/978-3-030-69395-4_10
13. Lokre, S.S., Naman, V., Priya, S., Panda, S.K.: Gun tracking system using blockchain technology. In: Panda, S.K., Jena, A.K., Swain, S.K., Satapathy, S.C. (eds.) Blockchain Technology: Applications and Challenges. Intelligent Systems Reference Library, vol. 203. Springer, Cham (2021). https://doi.org/10.1007/978-3-030-69395-4_16
14. Panda, S.K., Daliyet, S.P., Lokre, S.S., Naman, V.: Distributed Ledger Technology in the Construction Industry Using Corda. The New Advanced Society: Artificial Intelligence and Industrial Internet of Things Paradigm. https://doi.org/10.1002/9781119884392.ch2
15. Panda, S.K., Satapathy, S.C.: An investigation into smart contract deployment on ethereum platform using Web3.js and solidity using blockchain. In: Bhateja, V., Satapathy, S.C., Travieso-González, C.M., Aradhya, V.N.M. (eds.) Data Engineering and Intelligent Computing. Advances in Intelligent Systems and Computing, vol. 1. Springer, Singapore (2021). https://doi.org/10.1007/978-981-16-0171-2_52
16. Panda, S.K., Rao, D.C., Satapathy, S.C.: An investigation into the usability of blockchain technology in internet of things. In: Bhateja, V., Satapathy, S.C., Travieso-González, C.M., Aradhya, V.N.M. (eds.) Data Engineering and Intelligent Computing. Advances in Intelligent Systems and Computing, vol. 1. Springer, Singapore (2021). https://doi.org/10.1007/978-981-16-0171-2_53
17. Panda, S.K., Dash, S.P., Jena, A.K.: Optimization of block query response using evolutionary algorithm. In: Bhateja, V., Satapathy, S.C., Travieso-González, C.M., Aradhya, V.N.M. (eds.) Data Engineering and Intelligent Computing. Advances in Intelligent Systems and Computing, vol. 1. Springer, Singapore (2021). https://doi.org/10.1007/978-981-16-0171-2_54

18. Nanda, S.K., Panda, S.K., Das, M., Satapathy, S.C.: Automating vehicle insurance process using smart contract and ethereum. In: Chakravarthy, V.V.S.S.S., Flores-Fuentes, W., Bhateja, V., Biswal, B. (eds.) Advances in Micro-Electronics, Embedded Systems and IoT. Lecture Notes in Electrical Engineering, vol. 838. Springer, Singapore (2022). https://doi.org/10.1007/978-981-16-8550-7_23
19. Panda, S.K., Elngar, A.A., Balas, V.E., Kayed, M. (Eds.): Bitcoin and Blockchain: History and Current Applications, 1st edn. CRC Press (2020). https://doi.org/10.1201/9781003032588
20. Panda, S.K., Jena, A.K., Swain, S.K., Satapathy, S.C. (eds.): Blockchain Technology: Applications and Challenges. Intelligent Systems Reference Library. Springer. https://doi.org/10.1007/978-3-030-69395-4
21. Varaprasada Rao, K., Panda, S.K.: A design model of copyright protection system based on distributed ledger technology. In: Satapathy, S.C., Lin, J.C.W., Wee, L.K., Bhateja, V., Rajesh, T.M. (eds.) Computer Communication, Networking and IoT. Lecture Notes in Networks and Systems, vol. 459. Springer, Singapore (2023). https://doi.org/10.1007/978-981-19-1976-3_17
22. Varaprasada Rao, K., Panda, S.K.: Secure Electronic voting (E-voting) system based on blockchain on various platforms. In: Satapathy, S.C., Lin, J.C.W., Wee, L.K., Bhateja, V., Rajesh, T.M. (eds.) Computer Communication, Networking and IoT. Lecture Notes in Networks and Systems, vol. 459. Springer, Singapore (2023). https://doi.org/10.1007/978-981-19-1976-3_18

Chapter 8
Blockchain in Healthcare: A Review

Rohit Saxena, Deepak Arora, Vishal Nagar, and Satyasundara Mahapatra

Abstract Since Bitcoin introduced the blockchain, research has been conducted to expand its use cases beyond finance. One sector where blockchain is anticipated to have a big influence is healthcare. Researchers and practitioners in health informatics constantly struggle to keep up with the advancement of this field's young but quickly expanding body of research. This chapter provides a thorough review of studies carried out to demonstrate the benefits of blockchain technology that have been utilized in the domain of healthcare, in addition to the pandemic, COVID-19, which led to a massive and pervasive repercussion on healthcare and has significantly accelerated the implementation of digital technology. This chapter also depicts how researchers have presented the use cases for adopting Blockchain technology in the healthcare sector. The state-of-the-art blockchain application development for healthcare has also been described in this chapter, along with any inadequacies and potential future study topics.

Keywords Blockchain · Healthcare · Covid-19 · Ethereum · Hyperledger

8.1 Introduction

With the release of the white paper on Bitcoin in October 2008, blockchain soon became well-known as a distributed ledger technology [1]. Blockchain technology enables the users in a distributed network to exchange digital currency with no necessity for a centralised, trustworthy third party. In the past, the trading of digital currencies between people or corporations required the services of a bank or any other centralized body as a trusted third party (TTP). Depending upon a TTP can be problematic due to several reasons. A TTP can weaken a system by acting as a potential source of collapse because of the risk that it could be intentionally infiltrated

R. Saxena (✉) · V. Nagar · S. Mahapatra
Pranveer Singh Institute of Technology, Kanpur, India
e-mail: rohit.saxenacse@gmail.com

D. Arora
Amity University Uttar Pradesh, Lucknow Campus, Noida, India

S. K. Panda et al. (eds.), *Recent Advances in Blockchain Technology*,
Intelligent Systems Reference Library 237, https://doi.org/10.1007/978-3-031-22835-3_8

to leave the system unworkable. Other issues such as transaction fees and transaction delays could be other disadvantages of TTP. The infamous Bitcoin white paper was released after a year Bitcoin cryptocurrency was introduced. Since it was made accessible as open-source, anyone may modify and enhance it to develop new iterations of Blockchain-based technologies. The initial blockchain technology generation, also known as blockchain 1.0 [2], is made up of early versions of Blockchain-based digital currencies like Bitcoin. The technologies in Blockchain 1.0 comprises Litecoin [3], Monero [4], and Dash [5]. The two aspects of Blockchain 2.0 are smart contracts and smart properties [6]. Smart contracts are computer programs that contain the laws and management guidelines for smart properties. NEO [7], Ethereum [8], Ethereum Classic [7], and QTUM [9] are examples of cryptocurrencies that come under Blockchain 2.0. Building on the aforementioned, Blockchain 3.0, the third iteration of Blockchain technology, is now concentrated on Blockchain's non-financial applications [2]. To this purpose, attempts have been undertaken to expand the technology's potential applications outside of banking, allowing other sectors of the economy and use cases to gain from Blockchain's intriguing qualities. Because of this, Blockchain is regarded as a multipurpose technology [10, 11] with its applications in numerous domains, including insurance, contract administration, dispute resolution, insurance, supply chain management, identity management, and healthcare [10, 12]. A variety of use cases for applying blockchain in the healthcare sector have been established as a result of the increasing interest in blockchain technology and its adoption by numerous businesses and industries.

Organizing the massive transmission of confidential information, which includes contact tracing, vaccination tracking and management, and the issue of COVID-19 health certificates, requires safe, decentralised multipurpose platforms. This is especially true in light of the COVID-19 pandemic [13]. These conditions provide a strong justification for concentrating efforts to boost acceptance and eliminate some of these barriers to the general use of blockchain technology. However, it is unclear whether external one-off factors, like COVID-19 alone, that can promote the use of these technologies would be sufficient to produce long-lasting and sustained outcomes. The entire healthcare sector would need to perform specialized studies in this area and better understand blockchain technology as well as its medicinal applications [14–19].

Nonetheless, there is indeed a number of misleading bits of knowledge, speculative thoughts, and a lack of certainty about the prospective takeaways of Blockchain in the domain of healthcare because it is an innovation but there has been a huge buzz about it in the media and other articles (opinion articles, editorials, blog posts, interviews, etc.). This chapter predominantly concentrates on the use cases for Blockchain in healthcare, the Blockchain-based applications developed for this sector, the obstacles and constraints of such applications, and the approaches taken to address these obstacles and constraints.

8.2 Blockchain Overview

This chapter does not focus on a deep technical study of blockchain technology. However, it is essential to clarify a few blockchain principles, features, and terminologies to further our discussion and help others understand how blockchain technology might be used to address healthcare issues. The most notable advantage of blockchain technology may be its ability to eliminate the requirement for a centralised TTP in distributed systems. The potential cause of the collapse is eliminated by the Blockchain thus enabling the parties to execute transactions in a distributed setting. Figure 8.1 illustrates the distinctions between centralised and decentralised systems [15]. Figure 8.1a depicts a number of ledgers, but the Regional Health Information Organization (RHIO), in this instance, is a centralised repository for all the data. The RHIO essentially keeps the ledger up to date. The RHIO is contacted as the last arbiter to settle a dispute between two nodes. On the other hand, Fig. 8.1b demonstrates that although there is just one ledger, every node has a replica of it and varying degrees of reference to its contents.

While records in Fig. 8.1a are kept in RHIO, there are many ledgers in Fig. 8.1b where all the nodes have some access to the single ledger. The decentralised architecture speeds up transactions, does away with the requirement for a TTP, and does away with the transaction fees which the TTP levies (RHIO). The bulk of the blockchain's functionality is drawn from the cryptographic fundamentals that form its basis. Utilizing public key infrastructure (PKI), each node within the blockchain network offers and creates transactions on behalf of the other participants [20]. The private and the public keys are available to all the participants in the Blockchain network. The address of the user is its public key. For authentication purposes, the user's private key is utilized. The public key of the sender is incorporated inside the transaction, along with the transaction message and public key of the receiver.

The grouping of the recently received legitimate transactions is said to be a block [1]. A transaction proposal that complies with the validation standards is considered valid. This is accomplished, for instance, by confirming that the proposed transaction is legitimate and comes from authenticated users (node). The validation of the

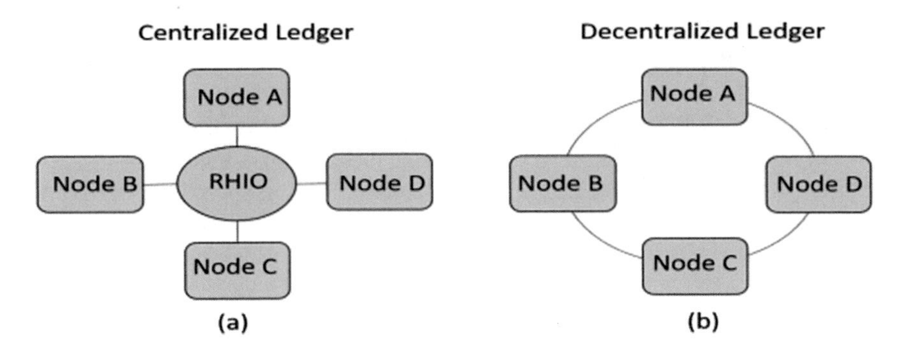

Fig. 8.1 Comparison of centralized system with decentralized system [15]

blocks and their publication to the ledger are decided by the consensus algorithm. The consensus methods in a blockchain network are maintained by specific nodes. The process that involves approving transactions and publishing them to the blockchain is called mining. The term "miners" refers to these specialised nodes. A miner first evaluates if a transaction is valid after receiving a transaction proposal. A block only contains transactions that have been verified. A chained sequence of blocks is formed, after a predetermined amount of time, when a fresh block containing validated transactions is linked to the preceding blocks (or block). Every node in the network possesses a duplicate database or ledger of every network transaction. If a node is free to enter a Blockchain network without the requirement for authorization or access permissions and participate in the process of mining, then such a blockchain implementation is said to be permissionless or public. In a permissioned blockchain, authorization and the proper access permissions are required before a node may join and take part in mining operations. Due to their qualities, permissioned blockchain networks seem to be more compact, quick, and secure than public/permissionless blockchain networks. A blockchain with permissions can also be categorized as private or consortium. The number of nodes allowed to act as miners is what separates private from consortium Blockchains. It is more suitably said to be a private blockchain if just one node is allowed to be a miner. In a consortium blockchain, only the registered users are allowed to take part in mining, Decentralization has its benefits, and the consortium blockchain also benefits from the private blockchain's enhanced security and privacy.

A hash function is another cryptographic essential that chains block together that converts an arbitrary-length message into a fixed-length output called a message digest or digital fingerprint. It is an intriguing feature that the hash function is collision-resistant, indicating it won't produce the same hash value for two different messages. For linking a newer block to the Blockchain network, the header of the new block must include the hash of the previous block header (Fig. 8.2).

Figure 8.2 shows a streamlined illustration depicting the linking of blocks to create a Blockchain. The output of a fixed-length hash is produced by hashing the given transactions within a block, and it is then incorporated into the block header. Every valid block is connected to the end of the blocks that arrived before it through a chain made up of the hash of the preceding block header, which is found in all the blocks. Thus, a "blockchain" is created by connecting each block to its predecessors.

As demonstrated in Fig. 8.2, if a small change is made to any of the transactions in a block, the associated hash output will vary significantly, thereby breaking the chain of the linked blocks in the blockchain. As a result, the network can quickly identify any change to a block's contents in the blockchain. A transaction inside the blocks, as a result, is unchangeable and irreversible. This feature is therefore referred to as immutability. Immutability, a core feature of Blockchain, make sure that the records, once generated, cannot be accessed or modified. Blockchain is also known as an append-only ledger because editing a record over the blockchain necessitates creating a new record. The transactions are time-stamped, leaving a record of what was done and when.

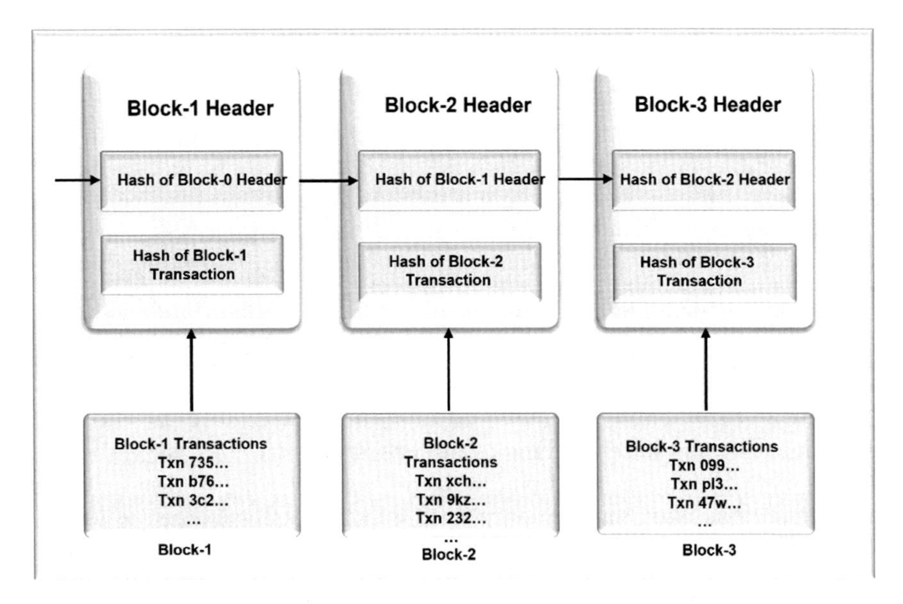

Fig. 8.2 A streamlined view of blockchain

The methods for mining and achieving consensus vary among the various kinds of blockchain. The miners may receive the proposals for a transaction in diverse orders. On a public blockchain, such as Bitcoin Blockchain, the miners may yield various valid blocks with varying orderings of transactions. To make sure that every node has an exact copy of the blockchain, for appending the validated nodes to the blockchain, the miners must concur on the order and only one valid block is appended in every cycle. Depending on the consensus protocol being used, there are various ways to accomplish this. Bitcoin's Proof of Work (PoW) scheme is a well-known example of a consensus protocol [1].

In order to determine which miner's block is allowed for incorporation in the blockchain, miners must complete a computationally difficult challenge and must adhere to the PoW protocol, which is also dependent on cryptographic hash functions. A predetermined pattern functioning as an electronic fingerprint is provided as part of the PoW protocol, and the miners' job is to produce a random number that may be incorporated into transactional messages and then hashed with one another to form the same pattern. The capability to incorporate a block into the blockchain is granted to the miner to complete a cycle's worth of calculations. There are more real-world instances of consensus protocols, but they all aim to preserve the "actual state" of distributed ledgers beyond the need for a centralised, dependable third party. Blockchain is a term used to describe an immutable ledger or repository in which the peers in a network share and maintains chronological records of events or transactions. It is obvious that blockchain offers certain interesting features that are beneficial for applications in the healthcare sector.

Decentralization, which enables the implementation of distributed healthcare apps without depending on a centralized authority, is a key aspect of blockchain making it unquestionably advantageous in the healthcare sector. Furthermore, the fact that the blockchain's data is mirrored all across network nodes promotes openness and transparency, allowing healthcare stakeholders, notably patients, to know how, by whom, when, and also how their data is used. More crucially, as the ledger's data is duplicated across a number of network nodes, the compromised condition of any particular node inside the blockchain network has no impact on the ledger's present state. Because of this, blockchain can protect against potential data losses, damage, and security breaches such as ransomware attacks because of the way it was designed.

Additionally, it is impossible to alter or make amendments to the records that have been added to the Blockchain which perfectly matches the necessity for keeping healthcare records. This is because of the immutability property of the Blockchain. It is crucial to guarantee the validity and integrity of patients' medical records. Furthermore, by using cryptographic techniques for encrypting the data kept on the blockchain, it is made sure that only those users who are authorised to view the data can decode it, optimising both the privacy and security of the stored records. Moreover, because patients' identities in a blockchain are made pseudonymous by the usage of cryptographic keys, healthcare stakeholders can share patient health data without knowing the patients' real names. Smart contracts [2] are another feature of the blockchain that can be utilized to create the rules that give patients choice over how their medical records can be shared and utilised. As a result, blockchain can speed up the creation of an EHR management system that complies with GPDR by encoding in sensitive patient data that should never be exchanged or used without the necessary authorizations, thanks to a set of laws known as smart contracts. Table 8.1 summarises the possible advantages of blockchain with respect to its application in the healthcare sector.

8.3 Review of Blockchain Use Cases

A thorough analysis of the use cases, blockchain-based applications, and blockchain platforms on which all applications have been implemented is provided in this section. Table 8.2 depicts the use cases along with the application areas.

8.3.1 Electronic Health Record (EHR)

The administration of EHR is among the most prominent utilization of blockchain. EHRs are used to create, store, and manage patient personal, medical, or health-related data electronically. Electronic health records (EHR) for patients can be stored and managed using blockchain technology because of their decentralized nature, reliability, immutability, data provenance, resilience, smart contracts, security, and

Table 8.1 Advantages of blockchain in the healthcare sector [19]

Decentralization	A decentralised management structure is necessary for healthcare because of the rapid spread of stakeholders. Blockchain has the capability to become the decentralised health data management foundation that would allow all stakeholders to gain restricted entry to almost all health records without anybody acting as the global health data governing body
Improved privacy and information security	The privacy and security of the patient's health records kept on the Blockchain are considerably enhanced by its immutability attribute, which prevents data from being corrupted, changed, or retrieved after it has been saved. On the blockchain, every piece of health information is time-stamped, encrypted, and added in chronological sequence
Ownership of health records	The ownership of and control regarding the use of a patient's data must be theirs. The patients must be capable to identify instances of data misuse and need to be given the reassurance that their health information is not being used improperly by third parties. Blockchain's robust cryptographic protocols and specified smart contracts help it to meet these needs
Availability of data	The availability of the healthcare data recorded on the blockchain is ensured since records are duplicated across numerous nodes, making the system resistant to data loss, data corruption, as well as some security breaches on data availability
Trust and openness	Blockchain fosters trust in distributed healthcare apps because of its openness and transparency. This enables healthcare stakeholders to accept such applications
Verifiability of data	The legitimacy and integrity of the records kept on the blockchain can be confirmed even without access to their plaintext. This capability is especially helpful in healthcare settings where record verification is necessary, such as in the processing of insurance claims and pharmaceutical supply chains

privacy [41–46]. Blockchain is frequently suggested as an effective technique to strengthen the framework of healthcare to encourage patients in sharing their data, processed, or utilised [18, 19, 47–52].

Guardtime utilises Blockchain-based infrastructure to protect more than a million records of the patient in Estonia, which is noted as a well-known illustration of the use-case application of the blockchain for the governance of EHR [18, 53]. MedRec [21] is another illustration of the use case of EHR that aims to give patients

Table 8.2 Blockchain use case and application using blockchain

Use cases	Blockchain-based applications
Electronic health records	MedRec [21], HealthChain [22], Ancile [23], DPS [24], MedShare [25], BlockHIE [26], FHIRChain [27], MedBlock [28],
Supply chain for drugs and pharmaceuticals	Medium.io AG [29]
Education and research in biomedicine	[30–32]
Remote patient observation	[33–37]
Insurance claims for health	MIStore [38]
Data analytics for health	[39, 40]

control of the data. Through various granular access rights built on blockchain, this application controls who has access to them. Another application use case is the Gem Health Network (GHN), which was developed using the Ethereum blockchain, permits several healthcare professionals to share the same data [21]. Healthbank, a startup in Swiss digital health, allows patients to fully own their data [21] and is employing a blockchain platform. Celesti et al. [19] talks about the Healthcoin project, which wants to create a global EHR system. The blockchain-based platform of the Medicalchain project can make it simpler to share patient health records with global healthcare institutions.

Interoperability among the different Blockchain-based EHR solutions due to a shortage of benchmarks, scalability due to the massive amount of medical data, patient involvement since all patients are not able and willing to control their personal records, privacy and security, and a lack of incentives are some of the challenges faced by the Blockchain-based patient-centric electronic health records [19, 30, 47, 49, 54–56]. HealthChain [22] is an EHR application. HealthChain may provide confidentiality, scalability, and security to health records thanks to the modular Hyperledger Fabric design [12].

Ancile [23] is the Blockchain-based framework that also uses smart contracts to accomplish access control, data protection, privacy, and compatibility of digital health records. Ancile has been developed on the Ethereum Blockchain platform. Two other examples of EHR blockchain implementations using the Ethereum blockchain-based platform include MedRec [21] and the health data preservation system (DPS) created by Li et al. [24]. MeDShare [25], BlockHIE [26], FHIRChain [27], and MedBlock [28] are further blockchain-based EHR applications.

There are certain cryptographic methods that have been suggested for improving the authenticity and security of the EHRs recorded on the Blockchain-based EMRs. To improve security and maximize system performance, Hussein et al. [57] suggested a blockchain-based access management technique for EHRs. There has also been a proposal for an attribute-based signature scheme that involves different authorities, in which a patient can confirm the incorporation of a message to the Blockchain-based application solely on the message's attributes and without giving any private data. It is proven that this protocol is computationally safe and that it can withstand collusion attacks [58]. An identity-based signature (IBS), attribute-based encryption (ABE), and identity-based encryption (IBE) are also suggested in [59] for usage with blockchain. In [60, 61], Zhao et al. suggested the Key Management Scheme to address the security issues in blockchain-based EHR. Later, Zhang and Poslad [62] proposed Granular Access Authorization architecture to support Flexible Queries or GAA-FQ. It enables secure authorization at various granularity levels without the necessity for PKI.

For protecting privacy, a secure EHR system was proposed by Dubovitskaya et al. [52] that stores the authentic EHR across private and consortium blockchains and references to the EHR, respectively. In order to guarantee the availability of the system, this technique similarly uses asymmetric encryption, but it also incorporates conformance testing mechanisms. The authors of [63] suggest MediBchain, a platform that protects the privacy and uses cryptographic operations to identify patient records in blockchain-based EHR applications. Yeu et al. also suggested in [64] an architecture for blockchain-based EHR applications dubbed the Healthcare Data Gateway (HDG) that gives patients privacy-preserving ownership, control, and choice over how to share their data. Regarding the administration and exchange of medical data for diabetic patients, an analogous architecture is put out in [65] that employs multi-signature smart contracts for ensuring privacy and access control to data.

8.3.2 Supply Chain for Drugs and Pharmaceuticals

The administration of the health supply chain, notably in the medical and pharmaceutical business, is another example of a use case for blockchain. The supply of fake or subpar pharmaceuticals can have serious negative effects on patients, but the pharmaceutical business frequently deals with this issue. It has been determined that blockchain technology can solve this issue [18, 45, 53, 55, 66]. In his poll, Engelhardt highlights a few businesses that are investigating how blockchain may be used to spot fraudulent prescription drug sales. Nuco, HealthChainRx, and Scalamed are among the businesses mentioned [46]. The fundamental idea is to log every prescription-related transaction on the blockchain network, which connects all parties involved (manufacturers, distributors, physicians, patients, and pharmacists). A startup named Modum.io AG employs blockchain technology to attain immutability of data and open access to the temperature measurements of pharmaceuticals during transportation, allowing for the verification of their adherence to quality assurance temperature criteria [29]. However, Mackey et al. [67] state that they discovered countless instances of prototypes and studies in relation to leveraging Blockchain for pharmaceutical supply chain management in the literature they consulted. This suggests that despite the fact that few research publications on the topic, commercial blockchain-based products may have been released by industrial players to combat the trade in bogus medicines.

8.3.3 Education and Research in Biomedicine

Blockchain offers an intriguing application in biological research and education. Blockchain can assist in clinical trials to help prevent data fabrication and the under-reporting or exclusion of unwanted clinical research results [45, 68–70]. Due to the inherent anonymization, blockchain makes it simpler for patients to permit the utilisation of their data for clinical trials [55]. The immutability feature of Blockchain also validates the accuracy of data gathered using blockchain for medical research. It is also simpler to duplicate research using blockchain-based data because of the openness and transparency of the technology. These are just a few of the ways blockchain is anticipated to transform biomedical research [41, 46]. Funk et al. [71] provided a rationale for leveraging blockchain to develop a value-based, competency-based, and self-certifying system for the education of health professions.

Clinical trial approval traceability implementation using blockchain technology was demonstrated by [72]. Similar research is presented by Nugent et al. in [31], where they demonstrated the enhancement of transparency of clinical trials through the utilization of smart contracts on the Ethereum Blockchain. Another blockchain-based approach that is suggested to notarize papers obtained from biomedical databases is likewise implemented using the Ethereum platform [32].

8.4 Remote Patient Observation (RPO)

To remotely observe a patient's condition outside of typical healthcare venues like the hospital, biological data must be collected using Body Area Sensors, Internet of Things devices, and Mobile Devices. To store, share, and retrieve remotely obtained biological data, blockchain has been suggested [73–75]. Ethereum Blockchain-based smart contracts offer real-time patient surveillance with the capability to deliver electronic medications in a secure environment, as demonstrated in [33]. In [34], an example of a Hyperledger-based implementation of Blockchain-enabled data sharing and gathering between healthcare stakeholders has been illustrated. In a similar vein, blockchain is utilised to create SMEAD, a mobile-enabled aid for monitoring diabetic patients [35]. The use of mobile devices (smartphones) to communicate information to a blockchain-based application that runs on Hyperledger Fabric is demonstrated in another example application [36]. In order to achieve end-to-end data confidentiality and privacy in an uninterrupted remote patient monitoring application, Uddin et al. [37] developed a blockchain-based patient-centric agent (PCA). Using practical swarm optimization (PSO) for feature optimization and root exploits detection in blockchain-based smart phone healthcare data management was suggested by the authors in [37].

8.5 Insurance Claims for Health

The transparency, decentralisation, immutability, and auditability of the records maintained over the blockchain can be advantageous for processing insurance claims in the healthcare sector [55]. Insurance claim processing is mentioned in several articles as a very potential field for blockchain applications in the healthcare sector [41, 49, 55, 76, 77]. The MIStore, a blockchain-based medical insurance storage solution that has been implemented on the Ethereum blockchain platform, serves as an excellent illustration [38]. Additionally, [46] discusses a project by a business called Pokitdok that seeks to collaborate with Intel for the development of a blockchain-based system to simplify the settlement of medical insurance claims.

8.6 Data Analytics for Health

Blockchain offers a chance to take advantage of the capabilities of numerous cutting-edge technologies to perform data analytics on health records and further the study of precision medicine, including deep learning and transfer learning techniques [78]. This use case is covered in [51] and [41] as well as [39], which provides a full roadmap on how this may be implemented. In an experimental study, Juneja and Marefat [40] used Blockchain in a deep-learning architecture to categorise arrhythmias.

The blockchain-based healthcare applications and the blockchain platforms on which they have been developed are listed in Table 8.3.

8.7 Review of COVID-19 Blockchain-Based Applications

The COVID-19 pandemic has made it necessary to coordinate and manage a lot of data. First off, even though these data are frequently sensitive, they must be controlled and simple to verify [13, 79]. If such data management platforms gave central authorities free access to the data, serious privacy concerns may surface, which could be counterproductive [80, 81]. To overcome these issues, Garg et al. [82] developed a token-based movement pass that runs on a blockchain and uses

Table 8.3 Blockchain-based healthcare applications and the platforms

Blockchain platforms	Healthcare applications
Ethereum	MedRec [21], Ancile [23], DPS [24], GHN [53], FHIRChain [27], MedShare [25], SMEAD [35], Medium.io AG [29], MIStore[38]
Hyperledger fabric	HealthChain [22], MedicalChain [34, 36, 40, 49]
Bitcoin	[72]
Proprietary	Guardtime [76], MedBlock [28], BlockHIE [26, 37]

smart contracts. This pass does away with the requirement for personal information for verification reasons. A functional blockchain platform incorporating IoT data collection was proposed by Xu et al. [83] that can expose a user's identity and geographic location through its hash function, safeguarding the confidentiality of COVID-19 patients as well as the privacy of the general population in a decentralised environment. These illustrations show how blockchain systems can deal with the difficulty of compiling verifiable yet anonymous tracking data. Second, in order to conduct swift, extensive screening and vaccination procedures, it might be required to establish health and immunity certifications. Blockchain technology can provide a secure and decentralised environment for cross-border COVID-negative or immunity state verification [84–86].

Eisenstadt et al. [84] used a consortium, Ethereum-based blockchain architecture along with a mobile application to accomplish rapid verification of tamper-proof test data. To prevent restricting possession of sensitive keys or data, this was accomplished by employing widely dispersed public or private key pairs. The COVID-19 immunisation records of each recipient were stored on-chain on a publicly readable platform in a different, original survey by Chaudhari et al. [87]. They then used an iris extraction technique to authenticate users and locate vaccination records anonymously, enabling them to conceal the input and prevent the leakage of any personally identifiable information. A further study that used the Hyperledger Fabric architecture as well as a distributed system for file storing and accessing Interplanetary File Storage [88] in a simulated setting of a high-travel volume European Member State addressed concerns about scalability, latency, and storage as well [89]. Third, the administration of supply chains for important items like PPE, necessary drugs, and COVID-19 vaccinations is being scrutinized more closely [90–94]. Utilizing IoT, oracles, and programming interfaces, detailed monitoring can be carried out at a granular level for each unique vial or package. In a study by Antal et al. [95], end-to-end surveillance and visibility were achieved by combining IoT sensor devices and self-enforcing smart contracts over an Ethereum blockchain platform, thereby verifying the accuracy of COVID-19 vaccination distribution data. By providing on-chain side effect recording, the proposed approach also addressed another significant worry with such vaccinations. Instead, Ahmad et al. [96] showed the usage of blockchain to control the whole upstream supply chain and dispose of medical goods and equipment relevant to COVID-19. Similar to this, their blockchain was created using Ethereum architecture in conjunction with smart contracts, emphasizing Ethereum's adaptability. Fourth, a peer-to-peer, secure network could be utilized for telemedicine projects like testing kit management [19] and trustworthy stakeholders sharing medical data [97–101]. A consortium network blockchain was developed for the dissemination of COVID-19-related reports, according to a study by Kumar and Tripathi [97] (e.g., chest CT scans). The blockchain was built to recognize and verify such reports before they could be stored on-chain. To do this, it compares each report's perceptual hash to current on-chain perceptual hashes, weeding out reports that are unrelated to COVID-19. Using Proof of Authority consensus, Lee et al. [102] established a global International Patient Summary electronic health record system. Using an accessible Application Programming Interface and Fast Healthcare Interoperability Resources, the system facilitated real-time uploading from clinic electronic medical records systems. They might then support pandemic research, help authorities adopt health-care regulations on time, and provide rapid public updates on the epidemic.

8.8 Limitations and Issues with Blockchain-Based Applications

Patient engagement, scalability, speed, privacy and security, and interoperability are a few of the issues that have been recognized as barriers to the implementation and deployment of blockchain-based systems [55, 103–105].

The lack of an enduring benchmark for developing Blockchain-based healthcare solutions presents an interoperability difficulty because it may be impossible for applications created by various suppliers or on various platforms to work together. Consider the two applications for remote health monitoring that were created, one using the Hyperledger Fabric platform [34] and the other using the Ethereum platform [33]. It would be challenging to transfer data between the two platforms.

Concerns have been raised about the privacy and security of applications in healthcare based on Blockchain. Despite encryption methods used, it may still be feasible to determine patients' identities recorded by connecting the recorded data that is associated with that patient on a public blockchain [45]. Additionally, there is a chance that malicious attacks can be directed targeting the healthcare blockchain by organised criminals or even governments could result in security lapses that endanger patient privacy. The blockchain networks that support various cryptocurrencies have been the target of numerous reported attacks [106–109]. The blockchain's private keys, which are used for cryptographic protocols, are also vulnerable to compromise, which might allow third parties access to the recorded health information.

The "right to be forgotten" provision of the European Union GDPR states that users have the right to request the complete erasure of their personal data [110]. This causes another worry that blockchain's immutability property raises. This provision is in conflict with blockchain's immutability property. When it is desired to erase a patient's medical history, it could be counterproductive due to the immutability of the blockchain, which guarantees that information once stored in the blockchain cannot be removed or altered. Healthcare blockchain-based solutions face a significant scaling difficulty, especially given the amount of data involved. The enormous volume of biomedical data should not be stored on blockchain since this would result in severe performance deterioration, and in some circumstances, it is not even practical. Speed is another problem, as blockchain-based processing might cause noticeable delays. For instance, the Ethereum blockchain platform's current validation method requires that every node in a network take part in the validation process [106, 111–118].

How to involve the patients in the data management process of Blockchain is another difficulty. The administration of patient health data may not be of interest to or feasible for some patients, particularly the young and the elderly [45].

8.9 Strategies Used to Overcome Obstacles and Limitations

For getting over some of the difficulties and restrictions brought on by the employment of blockchain in the healthcare sector, numerous workarounds have been proposed. For instance, it is advised to keep the secure health information "off-chain." as a solution to the scalability issue, leaving the blockchain with only a limited amount of compressed information about the stored data as well as how to access it [49, 50, 116]. Since the real healthcare information recorded off-chain can be completely obliterated, even though the reference to the information on the blockchain can never be deleted, this also solves the GDPR's "right to be forgotten" concern. This preventative measure, nonetheless, has significant drawbacks, such as the partial loss of the blockchain's built-in redundancies, which improves availability of data. Alternative to using the permissionless and public blockchain for healthcare applications, permissioned blockchains like the private or consortium blockchain is employed to further safeguard the data and protect patients' privacy [22].

Additionally, many security vulnerabilities may be avoided by adhering to a strict process for software development and using all known security protections throughout code development. Controls are also implemented in permissioned healthcare blockchains for enabling the ability to undo fraudulent or incorrect transactions [47]. Several rules can be established and programmed to regulate the operating procedures for the healthcare application and management of the patients' data using blockchain-based smart contracts [2].

Moreover, only chosen nodes are allowed to be involved throughout the consensus as well as validation procedures [22] in order to boost system performance and processing speed. This differs from the protocols used by public blockchains like Bitcoin, where any node could participate throughout the consensus or validation procedure.

8.10 Areas for Future Research and Open Research Questions

There is a demand for researchers to create more proofs-of-concept and prototypes because the use of blockchain technology in healthcare is currently in its infancy. This will help researchers better understand the technology and its maturity regarding this application. To determine their advantages and disadvantages, several of the suggested frameworks, ideas, models, and architectures, like [39], need to be put into practice and tested. Open standards are required to provide compatibility between various blockchain products. At the moment, the emphasis is on proving the viability of blockchain prototypes. However, open standards enabling interoperability must be established before blockchain can be completely adopted and implemented in practical healthcare systems. Researchers must begin investigating the problems with interoperability and standardization procedures. Researchers can already submit their contributions to a standards body (ISO/TC 307) [46]. To increase stakeholders' trust in the technology's use and promote its adoption in healthcare, it is necessary to conduct concerted additional research engagements to delve into the unresolved research questions surrounding interoperability, speed, scalability, and data privacy and security that are inherent to blockchain-based healthcare applications.

8.11 Conclusion

Blockchain has evolved since it was first used in Bitcoin to become a multipurpose technology with uses in a range of sectors, including healthcare. This chapter reviews the scope and future of Blockchain-based applications and use-cases in the domain of healthcare. The study's specific goals were to recognise blockchain technology's applications in healthcare, the sample solutions that were created for such use cases, the difficulties and restrictions of these applications, the methods currently used to create these applications, and opportunities for further research. This study demonstrates that blockchain has numerous applications in the healthcare industry, such as the planning and management of electronic health records, managing the supply chains for drugs and pharmaceuticals, conducting biomedical research and education, monitoring patients remotely, and performing health data analytics. A number of blockchain applications for the healthcare industry have been developed as prototypes using cutting-edge blockchain concepts like permissioned blockchain, off-chain storage, smart contracts, etc. To properly comprehend, characterize, and assess the usefulness of blockchain technology in healthcare, more studies must yet be done. Additional study is required in addition to current initiatives to solve the issues of scalability, interoperability, latency, security, and privacy regarding the application of Blockchain technology in healthcare.

References

1. Nakamoto, S., Bitcoin, A.: A peer-to-peer electronic cash system. Bitcoin **4**, 2 (2008). https://bitcoin.org/bitcoin.pdf. Last accessed 15 Apr 2022
2. Swan, M.: Blockchain: blueprint for a new economy. O'Reilly Media, Inc. (2015)
3. Litecoin—Open source P2P digital currency. https://litecoin.org/. Accessed 15 Apr 2022
4. The Monero Project. https://getmonero.org/the-monero-project/. Accessed 12 Apr 2022
5. Dash Official Website|Dash Crypto Currency—Dash. https://www.dash.org/. Accessed 11 Apr 2022

6. NEO Smart Economy (2018). https://neo.org/. Accessed 11 Apr 2022
7. Ethereum classic—a smarter blockchain that takes digital assets further 2018. https://ethereumclassic.org/. Accessed 12 Apr 2022
8. Ethereum Project. https://www.ethereum.org/. Accessed 11 Apr 2022
9. Qtum (2018). https://qtum.org/en. Accessed 15 Apr 2022
10. Burniske, C., et al.: How blockchain technology can enhance electronic health record operability. In: Ark Invest. New York, NY, USA (2016)
11. Jovanovic, B., Rousseau, P.L.: General purpose technologies. In: Handbook of Economic Growth, vol. 1, pp. 1181–1224. Elsevier (2005)
12. Androulaki, E., et al.: Hyperledger fabric: a distributed operating system for permissioned blockchains. In: Proceedings of the Thirteenth EuroSys Conference (2018)
13. Ting, D.S.W., et al.: Digital technology and COVID-19. Nat. Med. **26**(4), 459–461 (2020)
14. Kuo, T.-T., Kim, H.-E., Ohno-Machado, L.: Blockchain distributed ledger technologies for biomedical and health care applications. J. Am. Med. Inform. Assoc. **24**(6), 1211–1220 (2017)
15. Agbo, C.C., Mahmoud, Q.H., Eklund, J.M.: Blockchain technology in healthcare: a systematic review. Healthcare **7**(2) (2019). MDPI
16. Klaine, P.V., et al.: Privacy-preserving contact tracing and public risk assessment using blockchain for COVID-19 pandemic. IEEE Internet of Things Mag. **3**(3), 58–63 (2020)
17. Shamsi, K., Khorasani, K.E., Shayegan, M.J.: A secure and efficient approach for issuing KYC token as COVID-19 health certificate based on stellar blockchain network (2020). arXiv:2010.02169
18. Alsamhi, S.H., et al.: Blockchain for decentralized multi-drone to combat COVID-19 and future pandemics: framework and proposed solutions. Trans. Emerg. Telecommun. Technol. **32**(9), e4255 (2021)
19. Celesti, A., et al.: Blockchain-based healthcare workflow for tele-medical laboratory in federated hospital IoT clouds. Sensors **20**(9), 2590 (2020)
20. Housley, R.: Public key infrastructure (PKI). The internet encyclopedia (2004)
21. Azaria, A., et al.: Medrec: using blockchain for medical data access and permission management. In: 2016 2nd International Conference on Open and Big Data (OBD). IEEE (2016)
22. Ahram, T., et al.: Blockchain technology innovations. In: 2017 IEEE Technology & Engineering Management Conference (TEMSCON). IEEE (2017)
23. Dagher, G.G., et al.: Ancile: privacy-preserving framework for access control and interoperability of electronic health records using blockchain technology. Sustain. Cities Soc. **39**, 283–297 (2018)
24. Li, H., et al.: Blockchain-based data preservation system for medical data. J. Med. Syst. **42**(8), 1–13 (2018)
25. Xia, Q.I., et al.: MeDShare: trust-less medical data sharing among cloud service providers via blockchain. IEEE Access **5**, 14757–14767 (2017)
26. Jiang, S., et al.: Blochie: a blockchain-based platform for healthcare information exchange. In: 2018 IEEE International Conference on Smart Computing (smartcomp). IEEE (2018)
27. Zhang, P., et al.: FHIRChain: applying blockchain to securely and scalably share clinical data. Comput. Struct. Biotechnol. J. **16**, 267–278 (2018)
28. Fan, K., et al.: Medblock: efficient and secure medical data sharing via blockchain. J. Med. Syst. **42**(8), 1–11 (2018)
29. Bocek, T., et al.: Blockchains everywhere-a use-case of blockchains in the pharma supply-chain. In: 2017 IFIP/IEEE Symposium on Integrated Network and Service Management (IM). IEEE (2017)
30. Liu, P.T.S.: Medical record system using blockchain, big data and tokenization. In: International Conference on Information and Communications Security. Springer, Cham (2016)
31. Nugent, T., Upton, D., Cimpoesu, M.: Improving data transparency in clinical trials using blockchain smart contracts. F1000Research 5 (2016)
32. Mytis-Gkometh, P., et al.: Notarization of knowledge retrieval from biomedical repositories using blockchain technology. In: International Conference on Biomedical and Health Informatics. Springer, Singapore (2017)

33. Griggs, K.N., et al.: Healthcare blockchain system using smart contracts for secure automated remote patient monitoring. J. Med. Syst. **42**(7), 1–7 (2018)
34. Liang, X., et al.: Integrating blockchain for data sharing and collaboration in mobile healthcare applications. In: 2017 IEEE 28th Annual International Symposium on Personal, Indoor, and Mobile Radio Communications (PIMRC). IEEE (2017)
35. Saravanan, M., et al.: SMEAD: A secured mobile enabled assisting device for diabetics monitoring. In: 2017 IEEE International Conference on Advanced Networks and Telecommunications Systems (ANTS). IEEE (2017)
36. Ichikawa, D., Kashiyama, M., Ueno, T.: Tamper-resistant mobile health using blockchain technology. JMIR mHealth uHealth **5**(7), e7938 (2017)
37. Uddin, M.A., et al.: Continuous patient monitoring with a patient centric agent: a block architecture. IEEE Access **6**, 32700–32726 (2018)
38. Thomas, C., et al.: Blockchain-based medical insurance storage systems. In: Recent Trends in Blockchain for Information Systems Security and Privacy, pp 219–235. CRC Press (2021)
39. Mamoshina, P., et al.: Converging blockchain and next-generation artificial intelligence technologies to decentralize and accelerate biomedical research and healthcare. Oncotarget **9**(5), 5665 (2018)
40. Juneja, A., Marefat, M.: Leveraging blockchain for retraining deep learning architecture in patient-specific arrhythmia classification. In: 2018 IEEE EMBS International Conference on Biomedical & Health Informatics (BHI). IEEE (2018)
41. Roman-Belmonte, J.M., De la Corte-Rodriguez, H., Rodriguez-Merchan, E.C.: How blockchain technology can change medicine. Postgrad. Med. **130**(4), 420–427 (2018)
42. Liu, W., et al.: Advanced block-chain architecture for e-health systems. In: 2017 IEEE 19th International Conference on e-Health Networking, Applications and Services (Healthcom). IEEE (2017)
43. Magyar, G.: Blockchain: Solving the privacy and research availability tradeoff for EHR data: a new disruptive technology in health data management. In: 2017 IEEE 30th Neumann Colloquium (NC). IEEE (2017)
44. Cunningham, J., Ainsworth, J.: Enabling patient control of personal electronic health records through distributed ledger technology. Stud. Health Technol. Inform. **245**, 45–48 (2018)
45. Radanović, I., Likić, R.: Opportunities for use of blockchain technology in medicine. Appl. Health Econ. Health Policy **16**(5), 583–590 (2018)
46. Engelhardt, M.A.: Hitching healthcare to the chain: an introduction to blockchain technology in the healthcare sector. Technol. Innov. Manage. Rev. **7**(10) (2017)
47. Alhadhrami, Z., et al.: Introducing blockchains for healthcare. In: 2017 International Conference on Electrical and Computing Technologies and Applications (ICECTA). IEEE (2017)
48. Patel, V.: A framework for secure and decentralized sharing of medical imaging data via blockchain consensus. Health Inform. J. **25**(4), 1398–1411 (2019)
49. Gordon, W.J., Catalini, C.: Blockchain technology for healthcare: facilitating the transition to patient-driven interoperability. Comput. Struct. Biotechnol. J. **16**, 224–230 (2018)
50. Esposito, C., et al.: Blockchain: a panacea for healthcare cloud-based data security and privacy? IEEE Cloud Comput. **5**(1), 31–37
51. Roehrs, A., Da Costa, C.A., da Rosa Righi, R.: OmniPHR: a distributed architecture model to integrate personal health records. J. Biomed. Inf. **71**, 70–81 (2017)
52. Dubovitskaya, A., et al.: Secure and trustable electronic medical records sharing using blockchain. In: AMIA Annual Symposium Proceedings, vol. 2017. American Medical Informatics Association (2017)
53. Mettler, M.: Blockchain technology in healthcare: the revolution starts here. In: 2016 IEEE 18th International Conference on E-health Networking, Applications and Services (Healthcom). IEEE (2016)
54. Kamau, G., et al.: Blockchain technology: is this the solution to emr interoperability and security issues in developing countries? In: 2018 IST-Africa Week Conference (IST-Africa). IEEE (2018)

55. Kamel Boulos, M.N., Wilson, J.T., Clauson, K.A.: Geospatial blockchain: promises, challenges, and scenarios in health and healthcare. Int. J. Health Geograph. **17**(1), 1–10 (2018)
56. Rifi, N., et al.: Towards using blockchain technology for eHealth data access management. In: 2017 Fourth International Conference on Advances in Biomedical Engineering (ICABME). IEEE (2017)
57. Hussein, A.F., et al.: A medical records managing and securing blockchain based system supported by a genetic algorithm and discrete wavelet transform. Cogn. Syst. Res. **52**, 1–11 (2018)
58. Guo, R., et al.: Secure attribute-based signature scheme with multiple authorities for blockchain in electronic health records systems. IEEE Access **6**, 11676–11686 (2018)
59. Wang, H., Song, Y.: Secure cloud-based EHR system using attribute-based cryptosystem and blockchain. J. Med. Syst. **42**(8), 1–9 (2018)
60. Zhao, H., et al.: Efficient key management scheme for health blockchain. CAAI Trans. Intell. Technol. **3**(2), 114–118 (2018)
61. Zhao, H., et al.: Lightweight backup and efficient recovery scheme for health blockchain keys. In: 2017 IEEE 13th International Symposium on Autonomous Decentralized System (ISADS). IEEE (2017)
62. Zhang, X., Poslad, S.: Blockchain support for flexible queries with granular access control to electronic medical records (EMR). In: 2018 IEEE International Conference on Communications (ICC). IEEE (2018)
63. Al Omar, A., et al.: Medibchain: a blockchain based privacy preserving platform for healthcare data. In: International Conference on Security, Privacy and Anonymity in Computation, Communication and Storage. Springer, Cham (2017)
64. Yue, X., et al.: Healthcare data gateways: found healthcare intelligence on blockchain with novel privacy risk control. J. Med. Syst. **40**(10), 1–8 (2016)
65. Cichosz, S.L., et al.: How to use blockchain for diabetes health care data and access management: an operational concept. J. Diab. Sci. Technol. **13**(2), 248–253 (2019)
66. Tseng, J.-H., et al.: Governance on the drug supply chain via gcoin blockchain. Int. J. Environ. Res. Public Health **15**(6), 1055 (2018)
67. Mackey, T.K., Nayyar, G.: A review of existing and emerging digital technologies to combat the global trade in fake medicines. Expert Opin. Drug Saf. **16**(5), 587–602 (2017)
68. Benchoufi, M., Ravaud, P.: Blockchain technology for improving clinical research quality. Trials **18**(1), 1–5 (2017)
69. Shae, Z., Tsai, J.J.P.: On the design of a blockchain platform for clinical trial and precision medicine. In: 2017 IEEE 37th International Conference on Distributed Computing Systems (ICDCS). IEEE (2017)
70. Angeletti, F., Chatzigiannakis, I., Vitaletti, A.: The role of blockchain and IoT in recruiting participants for digital clinical trials. In: 2017 25th International Conference on Software, Telecommunications and Computer Networks (SoftCOM). IEEE (2017)
71. Funk, E., et al.: Blockchain technology: a data framework to improve validity, trust, and accountability of information exchange in health professions education. Acad. Med. **93**(12), 1791–1794 (2018)
72. Benchoufi, M., Porcher, R., Ravaud, P.: Blockchain protocols in clinical trials: transparency and traceability of consent. F1000Research **6** (2017)
73. Dey, T., et al.: HealthSense: a medical use case of Internet of Things and blockchain. In: 2017 International Conference on Intelligent Sustainable Systems (ICISS). IEEE (2017)
74. Zhang, J., Xue, N., Huang, X.: A secure system for pervasive social network-based healthcare. IEEE Access **4**, 9239–9250 (2016)
75. Weiss, M., et al.: Blockchain as an enabler for public mHealth solutions in South Africa. In: 2017 IST-Africa week conference (IST-Africa). IEEE (2017)
76. Angraal, S., Krumholz, H.M., Schulz, W.L.: Blockchain technology: applications in health care. Circ. Cardiovasc. Qual. Outcomes **10**(9), e003800 (2017)

77. Gatteschi, V., et al.: Blockchain and smart contracts for insurance: is the technology mature enough? Future Internet **10**(2), 20 (2018)
78. Shae, Z., Tsai, J.: Transform blockchain into distributed parallel computing architecture for precision medicine. In: 2018 IEEE 38th International Conference on Distributed Computing Systems (ICDCS). IEEE (2018)
79. Mahmood, S., et al.: Global preparedness against COVID-19: we must leverage the power of digital health. JMIR Public Health Surveill. **6**(2), e18980 (2020)
80. Bay, J., et al.: BlueTrace: a privacy-preserving protocol for community-driven contact tracing across borders. Government Technology Agency-Singapore, Tech. Rep 18 (2020)
81. Idrees, S.M., Nowostawski, M., Jameel, R.: Blockchain-based digital contact tracing apps for COVID-19 pandemic management: Issues, challenges, solutions, and future directions. JMIR Med. Inf. **9**(2), e25245 (2021)
82. Garg, C., Bansal, A., Padappayil, R.P.: COVID-19: prolonged social distancing implementation strategy using blockchain-based movement passes. J. Med. Syst. **44**(9), 1–3 (2020)
83. Xu, H., et al.: BeepTrace: blockchain-enabled privacy-preserving contact tracing for COVID-19 pandemic and beyond. IEEE Internet Things J. **8**(5), 3915–3929 (2020)
84. Eisenstadt, M., et al.: COVID-19 antibody test/vaccination certification: there's an app for that. IEEE Open J. Eng. Med. Biol. **1**, 148–155 (2020)
85. Hasan, H.R., et al.: Blockchain-based solution for COVID-19 digital medical passports and immunity certificates. IEEE Access **8**, 222093–222108 (2020)
86. Bansal, A., Garg, C., Padappayil, R.P.: Optimizing the implementation of COVID-19 "immunity certificates" using blockchain. J. Med. Syst. **44**(9), 1–2 (2020)
87. Chaudhari, S., et al.: Framework for a DLT based COVID-19 passport. In: Intelligent Computing, pp. 108–123. Springer, Cham (2021)
88. Bieri, C.: An overview into the InterPlanetary File System (IPFS): use cases, advantages, and drawbacks. Communication Systems XIV; University of Zurich: Zurich, Switzerland 78 (2021)
89. Hernández-Ramos, J.L., et al.: Sharing pandemic vaccination certificates through blockchain: case study and performance evaluation. In: Wireless Communications and Mobile Computing 2021 (2021)
90. Alexander, G.C., Qato, D.M.: Ensuring access to medications in the US during the COVID-19 pandemic. JAMA **324**(1), 31–32 (2020)
91. Gordon, D.E., et al.: A SARS-CoV-2 protein interaction map reveals targets for drug repurposing. Nature **583**(7816), 459–468 (2020)
92. Ho, D.: Addressing COVID-19 drug development with artificial intelligence. Adv. Intell. Syst. **2**(5), 2000070 (2020)
93. Khurshid, A.: Applying blockchain technology to address the crisis of trust during the COVID-19 pandemic. JMIR Med. Inform. **8**(9), e20477 (2020)
94. Kovács, G., Falagara Sigala, I.: Lessons learned from humanitarian logistics to manage supply chain disruptions. J. Supply Chain Manage. **57**(1), 41–49 (2021)
95. Antal, C., et al.: Blockchain platform for COVID-19 vaccine supply management. IEEE Open J. Comput. Soc. **2**, 164–178 (2021)
96. Ahmad, R.W., et al.: Blockchain-based forward supply chain and waste management for COVID-19 medical equipment and supplies. IEEE Access **9**, 44905–44927 (2021)
97. Kumar, R., Tripathi, R.: A secure and distributed framework for sharing COVID-19 patient reports using consortium blockchain and IPFS. In: 2020 Sixth International Conference on Parallel, Distributed and Grid Computing (PDGC). IEEE (2020)
98. Christodoulou, K., et al.: Health information exchange with blockchain amid COVID-19-like pandemics. In: 2020 16th International Conference on Distributed Computing in Sensor Systems (DCOSS). IEEE (2020)
99. Panda, S.K., Mohammad, G.B., Nandan Mohanty, S., Sahoo, S: Smart contract-based land registry system to reduce frauds and time delay. Secur. Priv., e172 (2021). https://doi.org/10.1002/spy2.172

100. Panda, S.K., Satapathy, S.C.: Drug traceability and transparency in medical supply chain using blockchain for easing the process and creating trust between stakeholders and consumers. Pers. Ubiquit. Comput. (2021). https://doi.org/10.1007/s00779-021-01588-3
101. Niveditha, V.R., Sekaran, K., Singh, K.A., Panda, S.K.: Effective prediction of bitcoin price using wolf search algorithm and bidirectional LSTM on internet of things data. Int. J. Syst. Syst. Eng. **11**(3–4), 224–236
102. Lee, H.-A., et al.: Global infectious disease surveillance and case tracking system for COVID-19: development study. JMIR Med. Inf. **8**(12), e20567 (2020)
103. Sathya, A.R., Panda, S.K., Hanumanthakari, S.: Enabling smart education system using blockchain technology. In: Panda, S.K., Jena, A.K., Swain, S.K., Satapathy, S.C. (eds.) Blockchain Technology: Applications and Challenges. Intelligent Systems Reference Library, vol. 203. Springer, Cham (2021). https://doi.org/10.1007/978-3-030-69395-4_10
104. Lokre, S.S., Naman, V., Priya, S., Panda, S.K.: Gun tracking system using blockchain technology. In: Panda S.K., Jena A.K., Swain S.K., Satapathy S.C. (eds.) Blockchain Technology: Applications and Challenges. Intelligent Systems Reference Library, vol. 203. Springer, Cham (2021). https://doi.org/10.1007/978-3-030-69395-4_16
105. Panda, S.K., Daliyet, S.P., Lokre, S.S., Naman, V.: Distributed ledger technology in the construction industry using Corda. In: The New Advanced Society: Artificial Intelligence and Industrial Internet of Things Paradigm. https://doi.org/10.1002/9781119884392.ch2
106. Yli-Huumo, J., et al.: Where is current research on blockchain technology?—a systematic review. PloS One **11**(10), e0163477 (2016)
107. Panda, S.K., Satapathy, S.C.: An investigation into smart contract deployment on ethereum platform using Web3.js and solidity using blockchain. In: Bhateja, V., Satapathy, S.C., Travieso-González, C.M., Aradhya, V.N.M. (eds.) Data Engineering and Intelligent Computing. Advances in Intelligent Systems and Computing, vol. 1. Springer, Singapore (2021). https://doi.org/10.1007/978-981-16-0171-2_52
108. Panda, S.K., Rao, D.C., Satapathy, S.C.: An investigation into the usability of blockchain technology in internet of things. In: Bhateja, V., Satapathy, S.C., Travieso-González, C.M., Aradhya, V.N.M. (eds.) Data Engineering and Intelligent Computing. Advances in Intelligent Systems and Computing, vol. 1. Springer, Singapore (2021). https://doi.org/10.1007/978-981-16-0171-2_53
109. Panda, S.K., Dash, S.P., Jena, A.K.: Optimization of block query response using evolutionary algorithm. In: Bhateja, V., Satapathy, S.C., Travieso-González, C.M., Aradhya, V.N.M. (eds.) Data Engineering and Intelligent Computing. Advances in Intelligent Systems and Computing, vol. 1. Springer, Singapore (2021). https://doi.org/10.1007/978-981-16-0171-2_54
110. Foglia, M.: Patients and privacy: GDPR compliance for healthcare organizations. Eur. J. Priv. L. Tech. 43 (2020)
111. Nanda, S.K., Panda, S.K., Das, M., Satapathy, S.C.: Automating vehicle insurance process using smart contract and ethereum. In: Chakravarthy, V.V.S.S.S., Flores-Fuentes, W., Bhateja, V., Biswal, B. (eds.) Advances in Micro-Electronics, Embedded Systems and IoT. Lecture Notes in Electrical Engineering, vol. 838. Springer, Singapore (2022). https://doi.org/10.1007/978-981-16-8550-7_23
112. Panda, S.K., Elngar, A.A., Balas, V.E., Kayed, M. (eds.): Bitcoin and Blockchain: History and Current Applications, 1st edn. CRC Press (2020). https://doi.org/10.1201/9781003032588
113. Blockchain technology: applications and challenges. In: Panda, S.K., Jena, A.K., Swain, S.K., Satapathy, S.C. (eds.) Springer, Intelligent Systems Reference Library. https://doi.org/10.1007/978-3-030-69395-4
114. Varaprasada Rao, K., Panda, S.K.: A design model of copyright protection system based on distributed ledger technology. In: Satapathy, S.C., Lin, J.C.W., Wee, L.K., Bhateja, V., Rajesh, T.M. (eds.) Computer Communication, Networking and IoT. Lecture Notes in Networks and Systems, vol. 459. Springer, Singapore (2023). https://doi.org/10.1007/978-981-19-1976-3_17
115. Varaprasada Rao, K., Panda, S.K.: Secure electronic voting (e-voting) system based on blockchain on various platforms. In: Satapathy, S.C., Lin, J.C.W., Wee, L.K., Bhateja,

V., Rajesh, T.M. (eds.) Computer Communication, Networking and IoT. Lecture Notes in Networks and Systems, vol. 459. Springer, Singapore (2023). https://doi.org/10.1007/978-981-19-1976-3_18

116. Xia, Q., et al.: BBDS: Blockchain-based data sharing for electronic medical records in cloud environments. Information **8**(2), 44 (2017)
117. Zhang, A., Lin, X.: Towards secure and privacy-preserving data sharing in e-health systems via consortium blockchain. J. Med. Syst. **42**(8), 1–18 (2018)
118. Firdaus, A., et al.: Root exploit detection and features optimization: mobile device and blockchain based medical data management. J. Med. Syst. **42**(6), 1–23 (2018)

Chapter 9
Blockchain: A Study of New Business Model

K. Varaprasada Rao, Dileep Kumar Murala, and Sandeep Kumar Panda

Abstract One of the most amazing technological advancements of the twenty-first century is blockchain. Blockchain technology is most famously used in the creation and management of cryptocurrencies like bitcoin and Ethereum. Blockchain is being studied outside of the financial services sector in areas like taxation, supply chain management, company operations, and governance. However, not all pertinent fields of research have investigated blockchain in depth. Current research in this area mostly focuses on designing business processes for the blockchain or on execution engines that can implement processes using smart contracts. On the other hand, research into whether and how blockchains can help with business process monitoring has received less attention. With the use of blockchain technology, collaborative business operations involving untrustworthy parties can be carried out in a decentralized setting. Business Process Management (BPM) provides a good context in which these technologies can be used since blockchain and distributed ledger technologies are particularly suitable to build trusted settings when players do not trust one another. Several research proposals have shown that it is possible to create collaborative business processes for blockchains using high-level notations like the Business Process Model and Notation (BPMN) and then have the platform automatically generate the code artifacts needed to run those processes. The study shows how businesses and regulators may use Blockchain to uplevel corporate operations, improve efficiency, and cut costs. The main blockchain disadvantages that users should be aware of before implementing the technology are also emphasized. The chapter also considers how businesses may use blockchain to fully capitalize on the fourth industrial revolution. This work attempts to close this gap by outlining a collection of pertinent research

K. V. Rao (✉) · D. K. Murala
Department of Computer Science and Engineering, Faculty of Science and Technology (IcfaiTech), ICFAI Foundation for Higher Education, Hyderabad, Telangana, India
e-mail: varaprasad.fst@ifheindia.org

S. K. Panda
Department of Data Science and Artificial Intelligence, Faculty of Science and Technology (IcfaiTech), ICFAI Foundation for Higher Education, Hyderabad, Telangana, India
e-mail: sandeeppanda@ifheindia.org

S. K. Panda et al. (eds.), *Recent Advances in Blockchain Technology*,
Intelligent Systems Reference Library 237, https://doi.org/10.1007/978-3-031-22835-3_9

difficulties that result from the implementation of blockchain technologies in business process monitoring solutions, describing the current methodologies to address these challenges, and presenting a reference architecture for doing so.

Keywords Blockchain · BPM · Trust · Business model · Blockchain technology · Innovation strategy · Smart contract · Consensus · Business processes

9.1 Introduction

Over the past few decades, consensus methods in a conventional distributed system have been the subject of extensive research. Following the popularity of Bitcoin, the first cryptocurrency that launched in early 2009 [1], blockchain technology has drawn interest from the academic community and business sectors [2]. Blockchain applications are currently starting to appear. Includes a wide range of subjects outside of cryptocurrencies, like insurance [3], medicine [4], and economics [5]. Software engineering (SE) [6], the Internet of Things (IoT), the supply chain, etc. The basis for any Blockchain application uses the consensus mechanism for information replication and exchange. Exchanges between participants, as well as broadcasting participant states. As a result, support for consensus processes has increased. There has been a resurgence of attention recently [7].

In decentralized contexts, there are numerous ways to implement consensus, and some of them are still under consideration. Blockchain is a compelling alternative for businesses looking to modernize their business processes because of its core qualities of integrity, resilience, and transparency [8]. To foster cooperation, information sharing, and group decision-making, it is essential to adopt flexible business processes in open environments [9]. Technology development opens up a wide range of possibilities for automating, exchanging information, and altering enterprises [10], particularly as IoT devices proliferate [11]. From this vantage point, it is possible to include recently developed Blockchain technology as a crucial component for business processes to address the problems with security, distribution, openness, cost-effectiveness, and most importantly trust. The cutting-edge business process known as "business process 4.0" uses contemporary technologies to pursue the ultimate objectives of interoperability, automation, trust, and transparency.

Collaboration between different organizations is essential to achieve greater, common goals. Think about a supply chain, for instance, where the cooperation of various businesses results in a product through the various production and distribution phases [12]. As of right now, considerable information sharing is needed for the integration of each party's associated procedures [13]. This makes it challenging to design and manage these inter-organizational business processes. Additionally, the huge volume of message exchanges results in redundant data and a lack of complete awareness of the how, when, and where of operations. These factors contribute to the continued reliance of businesses on approved third parties to mediate and oversee the execution of inter-organizational business operations.

9.2 Blockchain and Functionalities

9.2.1 Blockchain and Its Usage

A blockchain is a shared distributed database or ledger between computer network nodes. A blockchain serves as an electronic database for storing data in digital form. The most well-known use of blockchain technology is for preserving a secure and decentralized record of transactions in cryptocurrency systems like Bitcoin. The innovation of a blockchain is that it fosters confidence without the necessity for a reliable third party by ensuring the fidelity and security of a record of data. The way the data is organized in a blockchain differs significantly from how it is typically organized. In a blockchain, data is gathered in groups called blocks that each includes sets of data. Blocks have specific storage capabilities, and when filled, they are sealed and connected to the block that came before them to create the data chain known as the blockchain. Every additional piece of information that comes after that newly added block is combined into a brand-new block, which is then added to the chain once it is full. A blockchain, as its name suggests, arranges its data into pieces (blocks) that are strung together, whereas a database typically organizes its data into tables. When used in a decentralized way, this data structure creates an irreversible chronology of data by design. When a block is completed, it is irrevocably sealed and added to the timeline. When a block is added to the chain, it receives a precise timestamp. When a block is filled, it is sealed and linked to the block before it to form the data chain known as the blockchain. Blocks have predefined storage capacities. When the chain is complete, a new block is created from each piece of information that follows that just-added block and added to the chain. A database normally arranges its data into tables, whereas a blockchain, as its name suggests, divides its data into parts (blocks) that are tied together. This data structure intentionally produces an irreversible chronology of data when employed decentralized. A completed block is permanently sealed and put to the timeline (Fig. 9.1).

This straightforward concept introduces innovation. Immutability, transparency, and security are some of its primary characteristics. Blockchain is built on a decentralized approach that has undergone a lot of changes in recent years and has applications in a variety of fields, including finance, the internet of things, reputation management, and BPM [14]. Figure 9.2 depicts several applications of these domains. BPM (BPM) is a new field that uses blockchain technology. To support industry 4.0, this article is introducing the features of blockchain technology that model, monitor, and effectively carry out business activities.

Each activity in a business process has a clear input and output. These business procedures are necessary to carry out daily tasks. Any business process has three essential components: modeling, monitoring, and execution [15]. Different business process modeling languages, such as EPC (Event-driven Process Chain) and BPMN (Business Process Modelling Notation), make business process modeling easier [8]. Monitoring the company process is another crucial step in preventing conflicts and determining the current situation. Execution of the business process must take into

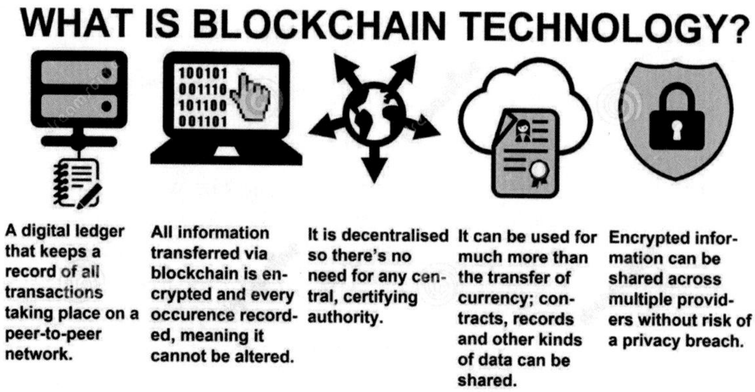

WHAT IS BLOCKCHAIN TECHNOLOGY?
100101
001110
101100
001101
A digital ledger that keeps a record of all transactions taking place on a peer-to-peer network.
All information transferred via blockchain is encrypted and every occurence recorded, meaning it cannot be altered.
It is decentralised so there's no need for any central, certifying authority.
It can be used for much more than the transfer of currency; contracts, records and other kinds of data can be shared.
Encrypted information can be shared across multiple providers without risk of a privacy breach.

Fig. 9.1 Blockchain technology

Use cases of blockchain
Finance
Online payment
Remittance
Digital assets
Reputation systems
Academics
Judicial systems
Internet of Things
E-business
Privacy
Business Processes management
Process monitoring
Process execution

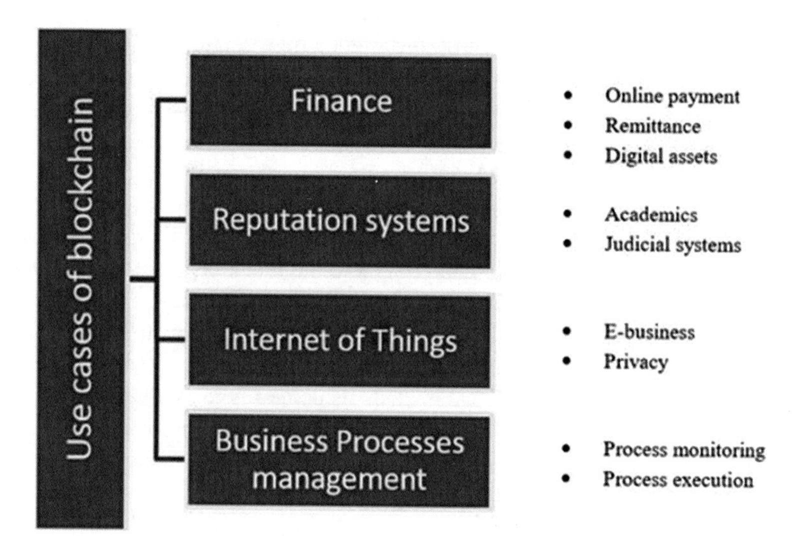

Fig. 9.2 Blockchain use-cases

account a variety of criteria, including those for trust, scalability, service quality, and cost-effectiveness [11]. Blockchain has the potential to alter both intra- and inter-organizational BPM. Blockchain technology's reputation for reliability makes it an excellent network contender for the BPM solution, which uses it to address peer trust issues. The blockchain network's consensus algorithm is used to verify every transaction that the BPM solution executes. Consensus works well because it also ensures that the nodes contribute to making the network function. A feature like this enables the development of more adaptable and dynamic inter-relationships amongst the participating entities. A blockchain is a promising tool for business process monitoring in particular because it allows participants to share and trust information about the state of an ongoing operation [16].

9.2.2 Features of Blockchain Technology

Blockchain systems are characterized by several elements. In the sections that follow, we highlight crucial traits related to deployments, implementation, and attributes.

(1) Blockchain with Private, Public, and Permissions: The method of ledger sharing and who is permitted to join in a system distinguish these various Blockchains from one another [17]. Ledgers are shared and verified by a predetermined group of nodes in a private blockchain. Nodes that want to join the system must initiate or be validated by the system [18]. Consensus maintenance is the responsibility of authorized nodes. Private Blockchain is appropriate for closed networks with complete node trust. The ultimate power to restrict access to approved nodes is the owner. On the other hand, public blockchains like Bitcoin, Ethereum, and others let anybody access and maintain the distributed ledger with permissions to verify the ledger's accuracy through the use of a consensus mechanism. Anybody can join, participate in, and exit a public Blockchain network freely. It is entirely open and distributed. As a result, this system runs on unidentified and untrusted nodes [19].

The major nodes are initially and rigorously chosen in permission blockchains, which are a combination of private and public blockchains. Permissioned blockchain is appropriate for semi-closed systems made up of a few businesses that are frequently arranged into consortiums. The degree of data openness varies; often, access controls are used to limit access to participants' information and data inside Blockchains, as established by the consortium. Despite the system's partial closure, decentralization's advantages can still be realized. In the case that certain nodes act maliciously, the system, for instance, has some false tolerance. The permission Blockchain implementations Hyperledger Fabric [20], Ripple [16], and Stellar [21] are a few examples.

The following characteristics are shared by all blockchain variations notwithstanding their diverse settings when it comes to the benefits that blockchain technology provides. They have the following three features: (1) They operate on a peer-to-peer (P2P) network, which provides some degree of decentralization; (2) Consensus mechanisms are used by multiple nodes to maintain the integrity of the ledger; and (3) Data is stored in Blockchain, which offers immutability, even if some nodes are broken or malicious [22].

(2) Centralization and Decentralization: In conventional centralized database systems, transactions are validated by centrally located, reputable intermediates who are trusted to be reliable. When employing central servers, this results in increased expenses and performance problems. The distributed transaction management issues [23] that arise when transactions are carried out amongst peers in a P2P network are being addressed by blockchain technology. Since public blockchains operate in a completely decentralized environment, it is possible to create trust in transactions between previously unknown or untrusted nodes. Private Blockchains operate in a secure, controlled environment and make use of access control methods to maintain the same position. The selection of nodes under an owner's control indicates the degree of decentralization [24]. The environment here is comparable to a conventional distributed database system. Permissioned Blockchains operate in a trusted environment like private Blockchains do, but they are more decentralized. The membership status is given to nodes by consortium policies. Blockchain data is maintained by all nodes, and no single party has complete control over the system. Different degrees of decentralization across all types of Blockchains have the same positive effects on single points of failure and data integrity [21].

(3) Persistency: A Blockchain ledger's transactions are deemed persistent because they are dispersed over the network, where each node retains and manages its records. Persistency is constantly maintained as long as the vast majority of nodes are benign. This property gives rise to several qualities, such as transparency and immutability (temper resistance). Blockchains are auditable due to their transparency and immutability [25].

(4) Validity: Blockchains do not require executions from each node, in contrast to other distributed systems. Other nodes in a Blockchain system would verify the transactions or blocks that were broadcast. Therefore, any falsification might be immediately found. The three main roles in this system are proposers, who put forth a value, acceptors, who evaluate and select a value, and learners, who agree with the selected value [7].

(5) Anonymity and Identity: Public Blockchains' primary distinguishing feature is anonymity. This system's identity can be separated from a user's real-world identity. To prevent identity disclosure, one person can acquire many identities [4]. Private information does not require to be maintained by a central organization. As a result, based on the transaction information, it is impossible to determine the real-world identity, while maintaining some kind of privacy. Contrarily, systems that are run and managed by well-known entities, such as private and permission Blockchains, typically require identity.

(6) Auditability: One may easily check and trace earlier records through nodes in a Blockchain network thanks to record timestamps and durable information. The sorts of Blockchain technologies and how they are implemented affect how auditable a system is. Public blockchains have the best auditability. Nodes are decentralized, followed by permission blockchains, which have the lowest because nodes are managed by a single entity. Private Blockchain has the lowest auditability because nodes are managed by a single entity [10].

(7) Closedness and Openness: Opened Blockchains depend on public nodes to keep track of transaction logs. As a result, by adhering to a set of guidelines, anyone can publish a transaction and join the system, and the data contained in this Blockchain is open to the general public. Because nodes must first be pre-specified or verified before joining, permission blockchains are regarded as semi-opened [16]. Between public and private Blockchains, they exist. The policies of the consortium, which can regulate whether the information is fully opened, partially opened, or closed, control the information inside this Blockchain. Similar to permission

Blockchains, private Blockchains use regulations to regulate the selection of nodes and the level of data transparency. But they are dependent on a single organization or owner [16].

9.3 Business Process Management

9.3.1 BPM Is What, Exactly?

BPM is a structured method for enhancing the procedures that firms employ to complete tasks, provide for their customers, and produce a profit. A business process is an action or series of actions that aid in achieving the objectives of an organization, such as boosting revenue or fostering diversity in the workforce. BPM employs a variety of techniques to enhance a business process, including analysis, modeling how it functions in various circumstances, adjustments, monitoring the new process, and ongoing improvement of its capacity to produce desired business objectives and results [26]. BPM is adjusting by incorporating ground-breaking technological developments as digital transformation methods gain popularity. There are countless options for BPM to advance, from low code platforms to robotic process automation. BPM is essential to a company's long-term viability and financial success. All business operations carried out with proper BPM experience increased levels of productivity, efficiency, and effectiveness. Businesses can utilize BPM to improve control over their business processes, acquire a competitive advantage, and avoid being overwhelmed by the workforce management process. Businesses may better serve and keep customers by improving control and consistency. Businesses can make more tough decisions and are more likely to come up with novel solutions thanks to the openness and agility that BPM offers [25, 27, 28]. Businesses become leaders in sustainability activities when waste and bottlenecks are reduced, resulting in leaner management capabilities. BPM is a process that is continuously carried out; it is not a one-time process of predetermined operations that will address business process inefficiencies. BPM has a wide range of advantages for organizations, but it may be challenging to set up and keep running. For firms to maintain a competitive advantage, continuous process improvement is necessary and involves several steps. Figure 9.3 presents the life cycle of BPM. The six primary steps of the BPM life cycle are as follows:

1. Analysis: Existing business processes are in-depth studied during the analysis BPM life cycle step. Making sure that current procedures are in line with business goals and strategy is a key goal of the analysis life cycle stage [17, 29]. Qualitative and quantitative analysis are the two basic types of analysis performed. While qualitative analysis is not reliant on numbers, quantitative analysis is.

2. Design: New processes are built or current processes are changed during the design stage. If the analytical life cycle stage deems it important, make sure to use process design and process modeling. Future efforts to create new business processes

Fig. 9.3 BPM life cycle

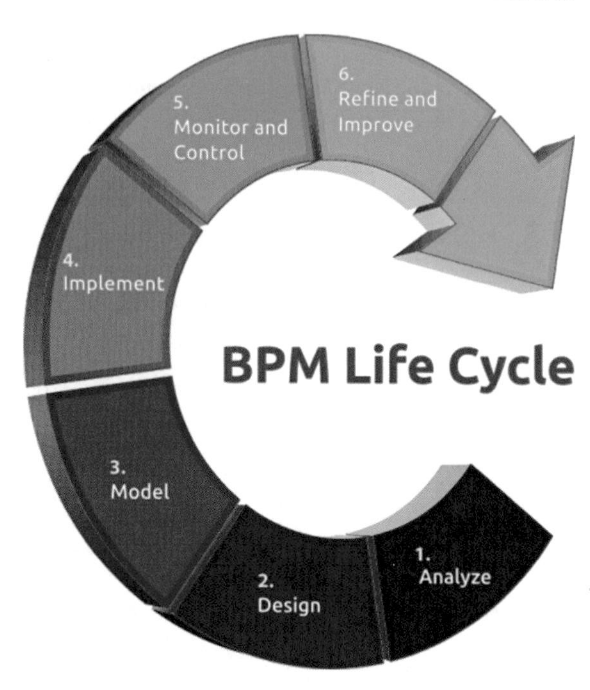

and evaluate the performance of existing ones can both benefit from the business process model developed during the design life cycle stage [20, 30, 31].

3. Model: A thorough understanding of both corporate strategy and objectives is necessary for the process modeling BPM life cycle stage. Business processes can be divided into three categories: management, secondary, and primary. Primary processes are fundamental procedures that directly benefit customers. While supporting primary operations, secondary processes do not directly benefit customers. Monitoring and measuring primary and secondary processes are done through the use of management techniques [32–35].

4. Implementation: Both newly built and previously designed processes are carried out throughout the BPM life cycle's implementation stage. Systemic or non-systemic implementations are also possible. A non-systemic implementation lacks the specialized BPM software and technology that is used in a systemic implementation. Make sure to take into account both the available money and business processes when choosing between systemic and non-systemic implementation [22, 36].

5. Monitoring and controlling: Process performance is assessed throughout this step to ascertain whether any further action is required. Consistent monitoring, measuring, and regulating are essential for process improvement [37–40].

6. Refining and improving: Processes are regularly modified and re-engineered during the refinement stage. Businesses optimize business processes to the greatest extent possible using the data obtained throughout the entire BPM process.

9.3.2 Advantages and Drawbacks (BPM)

Businesses and organizations strive for both ongoing growth and immediate success. This makes sense since when it comes to providing excellent service; success is not the goal. Multitasking, collaborative accounting, targeted campaigns, and other practices or tactics are used to speed up development. BPM, or BPM, is one particular subject that is applied globally [23]. For effective growth and return on investment, BPM emphasizes consistency, scalability, and visibility (ROI). All crucial organizational processes are examined by BPM. To ensure that these processes are implemented effectively and smoothly within the workforce, effective methodologies are devised.

9.3.2.1 Advantages of BPM

Knowing BPM's benefits and drawbacks will help one comprehend it better. The following are some benefits of BPM:

Streamlining Problem Solving: Because work processes are evenly divided among the best personnel for the job, BPM's operational structure is simplified. If there are issues that need to be fixed or improved, BPM kicks in. Employees may address problems in their sectors with more concentration and resources thanks to BPM. It is not the typical executive meeting where all issues are discussed and managers argue over who should handle which issue. Because businesses can quickly identify issues and find solutions, this benefit is commonly referred to as agility [24].

Lower Risks: This one has a strong connection to the BPM attribute of problem-solving. Employees who are strategically placed reduce organizational risks and pitfalls to the extent that focus is well-maintained and resources are efficiently managed. Organizations can make better judgments for foreseen problems as a result of the risks being decreased [1].

Measurability: Visibility is one of the primary goals of BPM, as was previously stated. Early on in the process, the discipline can demonstrate observable and quantifiable results. Organizations can evaluate what is working and what is not working by carefully laying out and interpreting analytics or statistical data. Several tools or applications that are available today assist staff in keeping track of and documenting important numbers [2].

Superiority in Both Customers and Employees: Customer focus and employee growth are two benefits of BPM. Customers benefit from superior IT support, customer service, and preferences for the goods or services they want. Their requirements are considered. Professionals can advance in their careers at a difficult but rewarding stage thanks to the BPM framework. BPM is, in essence, a win–win situation [3].

Increasing the Workforce with Technology: Many workers are concerned about their job security due to the growing danger of automation. The human workforce—true employees—remains in control of BPM. Instead of smart computers, a firm, corporation, or organization that uses BPM needs competitive employees. Both are maximized in BPM to guarantee the caliber of the task or service [4].

9.3.2.2 BPM's Disadvantages

Poor management and money loss: Only when the BPM framework is appropriately implemented will its advantages be fully realized. You do not want to lose significant amounts of money and other resources as a result of failing to follow the approaches. BPM is not inexpensive, despite being reasonably priced. You must still fork over a sizable sum of cash. If done correctly, you will quickly recoup that money or experience an increase in your ROI. If not, that will cause serious issues for your company. Analyses that are poorly distributed can likewise be poorly analyzed [8].

Lack of dialogue: The majority of the time, communication is the key. Some businesses are concerned that BPM's approach to separating workflows and processes could lead to poor communication among staff members working in various departments [11].

Briefly stated: Many firms, organizations, and companies continue to use BPM despite its drawbacks. On the other hand, despite its benefits, some people continue to question its effectiveness. This is because BPM can cause a business to lose money, clients, sales, intellectual property, and other things if it is not used properly. The majority of firms require these approaches to be productive in fields like accounting, ICT, engineering, web design, and more, thus it is clear that BPM manages to accomplish the job well even with its drawbacks [5].

We advise you to contact us at Proceso.pro if you are interested in BPM. Process management services like workflow acceleration, multimedia engagement, analytics reports, and more are our areas of expertise. BPM is always a wise decision. We can assist you in making it beneficial to your business, your personnel, and your clients. Call us right away.

BPM focuses on establishing a standardized, automated process for common transactions and interpersonal interactions. Reducing waste and duplication of effort and raising team productivity, aids in lowering the operational costs of the company. Businesses using BPM can decide to use one of the several BPM approaches, such as Six Sigma and Lean [13].

9.3.3 What Is not BPM

BPM is not a piece of software. It is possible to develop standardized and automated business processes with the aid of BPM tools. For instance, HappyFox Workflows assists companies in automating intricate, multi-step, and monotonous business procedures. However, BPM is not a standalone piece of software [41]. Task management is not BPM. Managing or coordinating a group of activities is what task or project management is all about. To organize tasks and ad hoc projects, use a project management tool like Microsoft Project, Jira, Asana, or Trello. On the other hand, BPM (BPM) is primarily concerned with continuing, repetitive procedures that adhere to a defined pattern [6].

9.3.4 What Kinds of BPM Systems Are There?

BPM systems can be grouped according to the function they perform. The three categories of BPM are as follows:

System-focused BPM (or Integration-centric BPM): With little to no human participation, this kind of BPM system manages operations that mostly rely on already-in place business systems (such as HRMS, CRM, and ERP). To build quick and effective business processes.

System-centric BPM: This software offers wide integrations and API access. Online banking is an illustration of a process that is integration-centric and may involve the integration of multiple software systems [12].

Human-focused BPM: Human-centric BPM puts the needs of the people first, with the help of several automation features. These are tasks that are largely carried out by people, and automation cannot simply take their place. These frequently require numerous permissions and separate activities. Customer support, handling complaints, employee onboarding, running e-commerce, and filing expense reports are a few examples of human-centric procedures [14].

9.3.5 Why Do People Utilize BPM?

BPM sounds like a term that should be easy to define: after all, it refers to controlling business processes. But that's too straightforward. BPM is a layered technique that focuses on codifying, optimizing, and continually improving operations or processes, particularly those that are typically thought of as ad hoc solutions or the kind of "institutional IQ" that leaves the building when employees depart [15]. BPM is the administration of business processes, which are frequently hidden in manuals, regulations, laws, and worksheets as well as in people's thoughts. Any firm would unavoidably produce and maintain [these], usually without adequate governance for long-term upkeep, according to Jim Tyrrell, these procedures are occasionally developed naturally, occasionally by legislation, and occasionally they are simply concocted on the spot. These can be useful when considering the function BPM performs within your company or when describing it to others. After that, we'll examine various BPM instances and go further into the connection between BPM and IT [18].

9.3.6 What Are the Advantages of Putting BPM into Practice?

Organizations can complete their digital transformation and achieve their larger organizational goals with the aid of BPM. The following are some major advantages of implementing BPM in your company:

1. Increased Business Agility: To adapt to changing market conditions, an organization's business processes must be changed and improved. Businesses can pause business processes, make adjustments, and then restart them thanks to BPM. Workflow alterations, reuse, and customization make business processes more responsive and provide the company with a greater understanding of the implications of process changes [19].

2. Lower Expenses and Higher Revenues: By eliminating bottlenecks, BPM technology gradually lowers costs. Reduced lead times for product sales, which give clients immediate access to services and goods, can have the effect of increasing sales and improving revenue. BPM systems can also be used to allocate and monitor resources to decrease waste, which can help lower costs and increase profitability [7].

3. Greater Efficiency: The integration of corporate processes offers the possibility of process efficiency enhancement from beginning to end. Process owners can carefully monitor delays and, if necessary, assign more resources with the proper information. More efficiency in the business process is gained through automation and the elimination of repeated processes [42].

4. Greater Visibility: Automation is made possible by BPM software, which also ensures real-time monitoring of crucial performance indicators. Better management and the ability to efficiently change structures and procedures while monitoring results are the results of this increased openness [21].

5. Compliance, Safety, and Security: A thorough BPM ensures that businesses abide by regulations and keep up with legal developments. By correctly documenting procedures and simplifying compliance, BPM may also support safety and security measures. As a result, businesses can urge their employees to protect the company's assets, such as confidential data and tangible assets, from misuse, theft, or loss [43].

9.3.7 When Ought Businesses Start Using BPM?

Here are a few business process illustrations where deploying BPM will yield a significant return on investment.

1. Dynamic processes that call for modifications to regulatory compliance, such as alterations to consumer information management in response to modifications to financial or privacy legislation.
2. Complex business processes that call for coordination and orchestration among numerous functional departments, business divisions, or workgroups.

3. Mission-critical processes that can be measured and can notably raise key performance indicators [9].
4. Business procedures that must be completed using one or more legacy apps.
5. Manually operated business operations that occasionally call for speedy responses.

9.3.8 BPM (BPM) Versus Business Process Automation (BPA)

BPM and BPA are similar and even complementary in some aspects, but they are not the same. Process automation is the focus of BPA, whereas process management, which may or may not include automation, is the focus of BPM. Simply said, not every BPM may involve BPA, but all BPA can be thought of as a sort of BPM.

9.3.9 Business Process Characteristics

The following list provides a summary of business process features.

(1) Transient and Persistent: Workflow in business is referred to as transient or ad hoc if it occurs temporarily. In a typical use scenario, the workflow instance ends when its objectives are met. When services are available in plenty and can be obtained when needed, this trait is evident. It routinely engages in brief interactions with previously undiscovered and typically distributed, autonomous, and diverse services. On the other hand, persistent workflow repeatedly performs the same tasks to achieve the same goals. This trait can be traced back to classic ERP systems used in industries like payroll and the automotive industry. Modern ERP systems, which have few modifications to business workflows but allow services to be dynamically chosen to execute workflow tasks, also exhibit persistent characteristics.

(2) Static and Dynamic Workflows: The frequency of changes determines whether a workflow is static or dynamic. A static workflow has minimal or no changes throughout workflow implementation, whereas dynamic workflows experience ongoing modifications [29]. Changes in this context could be made to the branching and task sequences inside a workflow structure or to the services that provide features for particular tasks. The latter presents a significant challenge for service selection, composition, and replacement in open environments because Services may offer equivalent functionalities with varied QoS qualities; (1) they commonly reside in diverse dispersed domains where modifications to the services are unclear; (2) they may cause performance and security instabilities; and (3) they may provide comparable functionalities.

(3) Workflow Creation and Implementation: Two phases can be distinguished in a workflow process. The first step is workflow formation, which establishes organization across phrase planning and design. The enactment phase occurs as workflows are carried out. The enactment tasks typically involve process scheduling, service contact, monitoring, and real-time service assessment and replacement. Dynamism and uncertainty frequently arise during the execution of workflows.

9.3.10 Blockchain and Business Process Characteristics

By relating the qualities of Blockchain and its features to those of business processes as they are outlined in the preceding part, this section explains how they can enhance business processes. Benefits from blockchain technology can be seen in many different ways in business operations.

(1) Transient and Persistent: Business process workflows need services in both cases to provide features for their internal duties. Blockchain technology can be used in two ways to enhance company procedures that frequently rely on distributed and autonomous services.

- Using blockchain to construct services can increase trust because persistency, validity, and auditability can be acquired natively. To choose a service, identity is required.
- To retain the QoS qualities of services and offer access to QoS information for service selection purposes, specialized Blockchain systems can be employed. Where several organizations are in charge of maintaining QoS ledgers, permission blockchain is an appropriate option. Depending on the policies of the consortium, the QoS information may be opened or closed [32].
- Private blockchains have trust issues, whereas public blockchains have a problem with inconsistent confirmation settlement times (also known as transaction finality).

(2) Dynamic and Static: In open environments, firms frequently make modifications to their operations to improve their efficacy and efficiency. When several firms are engaged, modern business processes become more dynamic and changes are highly challenging to solve. The trust and QoS that blockchain technology offers have been used to solve the dynamic nature of services. However, when adjustments made by one organization have an impact on others, it can create a new problem with trust. It is necessary to open cross-organizational business processes by making their business logic visible to all parties involved. Smart contracts can be used, where a business process can be encoded into smart contract transactions, to achieve Blockchain benefits for process executions [17]. The business logic is only accessible to those concerned. In this situation, where the amount of decentralization fluctuates based on the number of participants involved, permission blockchain makes sense. But it is possible to obtain a certain level of validity, persistency, and auditability.

(3) Workflow Formation and Enactment: The step of business workflow construction using process definitions is called workflow formation. Execution of processes during the implementation stage. Blockchain technology has two uses when multiple entities are involved.

(1) The repository for business process descriptions ensures that the workflow structure and flow controls are valid. This process is straightforward because all that is required to record static descriptions is a specific kind of transaction. Additionally, process executions will be generated and encoded into smart contracts [17, 20] when a business process workflow has been instantiated as an instance, which ensures operation validity according to the workflow descriptions.

(2) Since smart contract code is installed and honestly executed by outside parties to modify the state of a Blockchain ledger, abstraction of business process descriptions and instances, which are kept as transactions, would directly benefit from blockchain advantages. Specifically, smart contracts running on a Blockchain infrastructure can coordinate a process workflow made up of actions carried out by many services. Additionally, it allows users to keep tabs on the overall status of process instances.

(4) Decentralized Versus Centralized Management: Centralized workflow management is having a difficult time balancing the competing demands of scalability, security, openness, and cost as corporate processes grow more decentralized. Cross-organizational business processes that operate together and are integrated rely on distributed, autonomous, and heterogeneous services to complete tasks. The trustworthiness of services is a security concern of decentralized administration, nevertheless, and as already indicated, Blockchain technology plays a crucial role in establishing trust. In either case, centralized management may be able to benefit from blockchain technology by configuring a private blockchain, but decentralized management's design is more amenable to permission or public blockchain.

In conclusion, cross-organizational business procedures usually involve a coalition of businesses. For this situation, a permission blockchain is a good option. Since it is Blockchain-implemented, companies (nodes) participating in a Blockchain network may be able to achieve some degree of the advantages of persistency, validity, and auditability. Different levels of closeness and transparency are required by consortium policies as well as other elements like local legislation and regulations. For authentication and authorization between the relevant organizations, identity must be explicit.

9.4 Using Blockchain to Implement BPM

Every corporation that wishes to progress its operations must make the switch to new technical developments and revolutionary discoveries as the business environment changes quickly. Businesses already utilize BPM (BPM) platforms and their powerful features. However, combining several new, developing technologies could have a substantial impact on how well businesses function.

9.4.1 Blockchain Networks Power Business Process Automation

There are a huge number of transactions between business entities while the companies are operating in the rapidly shifting economic environment. A blockchain network that streamlines value exchanges effectively and affordably can facilitate these transactions. Blockchain network integration can fuel BPM (BPM) platforms, ensuring the transparency of transactions that take place outside of a company's internal systems. A blockchain network, for instance, ensures the veracity of information flowing from one organization to another. In other words, a new peer-to-peer BPM system is introduced by blockchain technology. The shareholders may be able to collaborate in these processes more effectively thanks to this new peer-to-peer BPM system, which may also result in a value-added chain that is more efficient.

9.4.2 How Smart Contracts Promote Collaboration in the Enterprise

By executing smart contracts in a decentralized network to which corporate counterparts are connected, several enterprise blockchain solutions can be created. With the execution of smart contracts on the network, blockchain solutions are implemented in businesses. According to Wikipedia, smart contracts help businesses exchange assets when specific requirements are met. Companies can specify their own business rules and penalties for breaching agreements using this option. The system's requirement to operate by predetermined guidelines and carry out specific steps to complete a process validates that all participants in the network execute accurately, preserving the consistency of workflows. However, blockchain technology offers various levels of security, trust, and transparency amongst various players in a business context.

Smart contracts also enable real-time auditing of workflows and processes. The contract's developer can code some rules and assess a process's quality. Organizations can obtain automated functions using Low-Code BPM platforms, which are created within a blockchain network, without having any prior expertise in blockchain or smart contract creation. Only the BPM suppliers develop and integrate these contracts based on the requirements of each company. For businesses looking to increase their network efficiency and collaboration with partners, Comidor offers a place to start. Overall, the market's existing BPM systems are given essential functionality by blockchain and smart contracts. They enable businesses and dispersed entities to work together transparently and effectively inside a certain ecosystem. By using the strength of blockchain networks and cutting-edge technical advancements, the digital business processes and workflows that are automated by Low-Code BPM systems can be brought to the next level.

9.4.3 Blockchain Implementation Difficulties

The blockchain technology sector has its difficulties and limitations, much like any other sector. The next section will go over the typical issues that users of blockchain technology encounter when implementing it:

A Scarcity of Knowledgeable Blockchain Developers: This is a significant issue with the use of blockchain technology. Smaller businesses are compelled to provide competitive incentives to attract and retain the industry's restricted pool of highly trained blockchain professionals because the market is still seeing a slow inflow of expertise.

Diverse and Inconsistent Blockchain Regulation: Regulators that will direct the application of blockchain technologies are yet unable to be agreed upon by various blockchain industry participants. The difficulty is not in adopting rules; rather, it is in determining where and to what extent the legislation should apply to the operations of using blockchain technology. This is why everyone involved in the blockchain industry needs to urge the government and other regulatory bodies to adopt a rule that will work for everyone.

Insufficiency in Scalability: This is another significant obstacle to the adoption of blockchain technology. Many blockchain developers are currently confronted with finding a solution to the conflict between preserving the standard advantages of DLT and yet making sure that its scope is enormous at high speeds. The blockchain technology sector's full potential is still mostly unscalable.

Security Concerns: This could be seen as the most alarming problem with blockchain technology. It should be emphasized that security formed the cornerstone of blockchain technology. However, the blockchain business is exposed to serious dangers from hackers as a result of the growing practical implications of the 51 percent assault theory.

Consensus Protocols That Consume Energy: In actuality, this is one of the biggest difficulties with implementing blockchain technology. When it was discovered that in 2017, the energy used for bitcoin mining was equivalent to the amount used by Denmark's whole population, this challenge was given further consideration. You must find this surprising, don't you? The development of novel and creative modalities of consensus for forward-thinking decision-making actions by stakeholders is the answer to energy-intensive consensus protocols. It has been hypothesized that a solution to this problem will trigger an evolutionary process that could advance DLT.

9.5 Business Process Monitoring

Business process monitoring seeks to ascertain how well-performing processes are about performance metrics and goals. Depending on the instruments and information at hand, a business process platform can monitor the instances of the running processes, and find abnormalities during their execution, such as behavior that differs from what is anticipated of it. This section briefly discusses the key traits of business process monitoring platforms in terms of potential monitoring objectives (i.e., the why), available monitoring approaches (i.e., the how), and the monitoring topic (i.e., the what).

9.5.1 Why to Monitor

Various factors warrant a monitoring platform be made known. The process owner first wants to know If the procedure to recognize critical situations is functioning as intended such as bottlenecks, stalemates, and departures from the intended execution [18]. Even if design-time verification methods It is possible to use simulation and validation to develop a sound procedure because only during runtime is a true check achievable. when the procedure is carried out in a genuine environment where Unexpected occurrences could affect a successful execution. Finally, when multiple players are involved in the same business process, business process monitoring is important [15]. In this instance, the 'choreographed process' refers to the protocol of communication between the various participants. After that, business process monitoring techniques are in charge of ensuring that this protocol is implemented correctly. The features of a business process monitoring platform are concerned with (i) runtime monitoring, (ii) execution data postprocessing, and (iii) runtime control to accomplish these goals. Runtime monitoring seeks to continuously monitor both the choreography and the process execution. To provide the process owner or the choreography participants with insights, the execution data post-processing runs analysis on top of the data gathered at run time. Such insights rely on a few indicators to explain the behavior of the process instances that have already been completed, such as Key Performance Indicators (KPIs) or Process Performance Indicators (PPIs) [14]. The runtime control then employs an active strategy to avert potential critical circumstances. It accomplishes this by utilizing techniques for process intelligence, which, among other things, could predict how the process will go. To be more clear, this study largely focuses on the passive method, or how the blockchain might be used to increase the trust in monitoring data, regardless of the indicators to be computed or the potential actions to be performed to resolve critical scenarios [19].

9.5.2 How to Monitor

The creation of events relating to the execution of a particular process instance is known as event data logging. A blockchain would be useful for event data logging since it would make it easy to trace the history of events. Events would also be permanently and irrevocably stored. The distributed nature of a blockchain can also be used as a communication method, as stated in [22], making it possible to readily transmit events relating to inter-organizational business operations. Each organization would make sure that events were recorded on the blockchain and spread to the other parties with access to it. The events generated by the other participants would also be automatically received similarly. Runtime performance evaluation (RPE) and business activity monitoring (BAM) BAM [9] analyses real-time information on the actions being carried out and is also known as "monitoring" [7]. (e.g., response time and failure rate). The effectiveness of activities can be assessed by measuring KPIs pertinent to the process using this technique. The adoption of a blockchain would be advantageous for BAM and runtime performance analysis. In fact, by recording the parameters used by the algorithms that analyze event logs on the blockchain, the participants might agree on how the analysis should be carried out. Furthermore, a sophisticated programmable blockchain would automatically ensure that the analysis is carried out as agreed by the participants. These blockchains would indeed enable the use of analysis algorithms in smart contracts [24].

9.5.3 What to Monitor

Depending on the monitoring technique and the underlying representation of the activity to monitor, different sorts of events must be logged for the monitoring to be reliable. The starting or termination of activities, as well as the sending and receiving of messages among participants, are frequently signaled by events in conformity checking techniques [23]. More sophisticated events that also identify when or by whom an artifact was edited are often required by BAM, runtime performance analysis, and compliance testing approaches (e.g., starting an activity or modifying an artifact). Different outcomes can result from combining the three main corners those are: (i) monitoring an organization's actions, (ii) all of the activities conducted by all of the organizations involved in the multi-party business process, or (iii) participant involvement.

9.6 Existing Approaches

9.6.1 Overview and Design Principles of the Lorikeet System

A Model-Driven Engineering (MDE) tool called Lorikeet is used to create blockchain applications for asset management and business processes. The initial "killer application" of the blockchain is thought to be the administration of assets, starting with fungible assets like cryptocurrency and tokens. An interesting area of use for blockchains is business procedures that handle non-fungible assets (such as by transferring titles to vehicles or properties). Non-fungible assets, in contrast to fungible ones, have the potential to be extensively customized, which might result in counterparty risks by introducing inefficiencies and uncertainty. The centralized trustworthy authority has usually been used to administer non-fungible assets, which again raises trust concerns. Therefore, a system that allows business activities to be automated on the blockchain shouldn't ignore features of asset management. Lorikeet can automatically generate blockchain smart contracts using asset data schemata and business process models. The BPMN translation techniques from both [18, 19] are implemented in Lorikeet, along with the registry editor Regenerator [15]. The process model is the artifact or blueprint that is implemented in a conventional BPMS. In contrast, Lorikeet develops the implementation-related code, which is then put into action. Before execution, the generated code can be examined, modified, or improved. This feature aids in the understanding of the code by technical specialists and supports the future necessity for establishing trust in the created smart contracts.

In both business and academia, generating blockchain smart contracts is done with the help of the highly regarded program Lorikeet. The architecture of Lorikeet is presented in Fig. 9.4, which comprises the BPMN translator, Registry generator, and Blockchain trigger on the back end, as well as a modeler user interface [19]. To enable users to create business process and registry models, the modeler UI component is delivered as a web application. BPMN 2.0 is used to model business processes [13]. To facilitate representation of and interaction with registries in the BPMN process paradigm, Lorikeet expands that standard. The update includes two new features in the form of new graphical notations and new XML attributes, namely Registry Reference and Action Invocation. An Action Invocation displays the asset registry action to be invoked, whereas a Registry Reference represents asset data stored on a blockchain [13].

The back-end parts, which adhere to a microservice-based architecture, are individually deployed as Docker containers and include the BPMN translator, Registry generator, and Blockchain trigger. From the aforementioned BPMN models, the

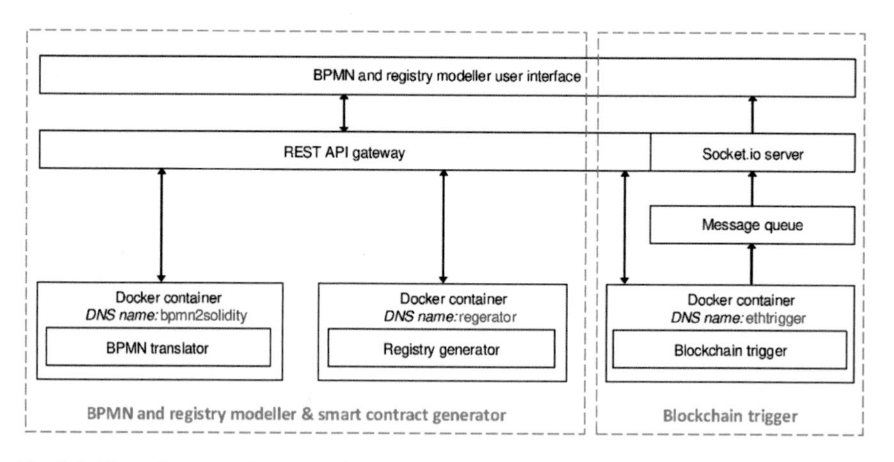

Fig. 9.4 The software architecture of Lorikeet [19]

BPMN translator creates Solidity smart contracts automatically. The information needed to invoke registry operations, as well as to create and run the process model, is contained in the smart contracts. Based on registry models that detail the data structure, registry types, as well as fundamental and sophisticated functions, the Registry generator also generates Solidity smart contracts. The smart contracts can then be implemented by users on the blockchain. The Blockchain trigger manages compilation, deployment, and interaction with Solidity smart contracts, as well as communication with an Ethereum Blockchain node.

Through an API gateway, the BPMN and registry modeler UI communicate with the back-end microservices. The API gateway routes API requests from the modeler UI to the appropriate microservice, such as translating a BPMN model. The Lorikeet business process modeler user interface is shown in Fig. 9.4. It is divided into two panels, one on the left for BPMN process modeling and one on the right for source code display. The Solidity smart contract code corresponding to those changes is modified at design time as the user makes changes to the BPMN model. The registry modeler UI resembles the business process modeler UI in terms of design and concept. Lorikeet has been utilized in cross-border partnerships with academic and business partners.

9.6.2 *The System of Caterpillar*

The goal of Caterpillar is to make it possible for its users to create native blockchain applications that will ensure that collaborative business processes are executed correctly, starting with a BPMN process model. The term "native" in this context means that all of the execution logic recorded in the process model is encoded in code artifacts distributed on the blockchain. Caterpillar specifically attempts to uphold the three design tenets listed below [13].

Figure 9.5 displays the Caterpillar architecture [25]. The core of Caterpillar is comprised of the modules inside the dashed rectangle, which also includes the Execution Engine, Work Item Manager, Event Monitor, and repositories for process models, metadata, service interfaces, and runtime data (specifically, process instance data and work item data). The entire mapping from BPMN to Solidity is implemented in the Compilation Tools module. This module creates a smart contract in Solidity from a BPMN model that is in standard XML format and contains the workflow routing logic of the process model. The smart contract specifically includes scripts to update the state of a process instance once a job is finished or an event takes place, as well as variables to encode the state of a process instance. Caterpillar features complex BPMN control flow elements like subprocesses, multi-instances, and event handling in addition to supporting fundamental ones like tasks and gateways. The Modelling Tool, built on top of Camunda's BPMN modeler, is connected to the Compilation Tools module. The Caterpillar Execution Engine offers operations for deploying a process model (or deploying the smart contracts produced by the Compilation Tools), creating instances of a deployed process model, choosing which tasks are enabled,

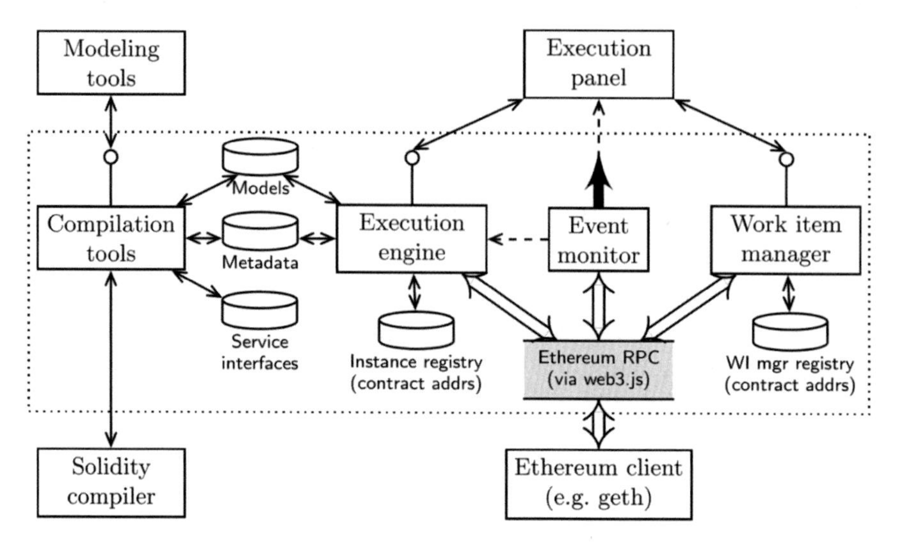

Fig. 9.5 Caterpillar architecture [25]

and controlling and recording their execution. When a service task is enabled, the execution engine is also in charge of carrying out automatic script tasks and initiating external service requests [15].

The Event Monitor keeps the other Caterpillar components informed by listening to events produced by the blockchain and producing notifications whenever a transaction about an instance of a process is validated on the blockchain. Additionally, Caterpillar offers a default Work Item Manager and an Execution Panel (to create and track process instances) (to handle user tasks). The Caterpillar engine's functionality is made available via a REST API, enabling programmers to create their execution panel or job list handler in place of the pre-installed ones.

9.7 Certain General Advice and Learned Lessons

You can start your blockchain adventure by taking into account several factors and using several resources. Identifying a section of your business process and the number of organizations or entities involved should come first. There is a greater possibility of improvement the more parties involved because there is a stronger tendency for bureaucracy. Within the confines of one organization, implementing blockchain technology has limited utility. Second, you consider whether identity, privacy, and trust are essential elements for your business network. Is it crucial to hunt down and verify a work's or asset's origins, or is it not? Is consensus-based record-keeping that requires participants to agree on the record's contents and current status beneficial? Is the provenance, or the origin of a work or object, its tracing and is verification important? Is typical retaining records that call for participant agreement on the contents of the record and its current state helpful to you? The solutions to You can make a decision using these queries. whether you should select a public using a private, permission-based blockchain Firstly, blockchain is not the solution to every issue, some businesses Problems don't lend themselves to blockchain technology. Starting by utilizing your business procedures, a design or a method in the engineering process will guarantee that you'll be on the proper path to accomplish the objective of completing the project. Otherwise, you risk drowning in the sea of technical and semantic words associated with the blockchain, such as bitcoin, ethereum, hyperledger, and ripple [44].

Blockchain is useful for tracking the ownership and status of valuable assets. These assets can range widely, for instance, from a work of art to an industrial component or an engineering service contract. An intangible object, such as a piece of intellectual property or a digital right, can also be considered an asset. It makes it possible for parties that don't know or trust each other to conduct business effectively and faithfully. By using Smart Contracts, it automates the contractual agreements between parties in a network. Whether you should construct your blockchain application on a public network or a private network, there are numerous options available. Bitcoin is an example of an unpermitted type of a public blockchain network, in contrast to the permission and private nature of blockchain for business in the context of company-to-business transactions. Identity is valued above anonymity. Unlike bitcoin, which is used for cryptocurrencies, it is used to track assets. What degree of privacy and permission is necessary is one of the main factors to be taken into account. In this example, the boundary for the blockchain technology applications was a section of a value chain for a corporation and a specific procedure [45].

There are various ways to develop a consensus mechanism. Proof of Work, Proof of Stake, Byzantine Agreement, Tendermint, and Federated Byzantine Agreement are a few examples of the current consensus processes. While the Hyperledger Fabric blockchain is built on the Public Byzantine Fault Tolerant algorithm, the Bitcoin and Ethereum blockchains use Proof of Work. A technology manager may or may not need to be familiar with the consensus method, but they may need to have some understanding of the blockchain system.

There are a lot of educational materials accessible for learning about blockchain technology. Blockchain training and certification programs are available from MIT. Blockchain education is available through the online education provider edX. For IEEE members looking for information on blockchain technology, application examples, business references, and opportunities to participate in working groups for Blockchain conferences, Blockchain IoT, standards and certifications, training, and community engagement, IEEE Blockchain is the go-to resource.

If your business has the funds and blockchain is a top priority, you might even think about hiring industry advisors. For instance, in San Francisco, New York, Toronto, London, Munich, Dubai, Tokyo, and Singapore, IBM provides quick start "Blockchain Garage" services. Identifying the areas of your organization where blockchain offers the most value is one of the problems. There is a need for coverage of issues like who should foot the bill for the creation and maintenance of the blockchain network, where the network peer will be located, when and how new participants can join, the network confidentiality rules, who should be in charge of the Smart Contract processes, and the trusted forms of identity. This is only a preliminary list of challenges that businesses looking for a more extensive blockchain implementation would encounter.

9.8 Conclusion

There is no question that Blockchain technology has the potential to fundamentally alter how business is conducted on a global scale. However, before we can witness the legal, economic, and technological viability of this technology in the operation of various business applications, there are several adoption and viability difficulties with Blockchain technology that need to be addressed to fully realize its enormous potential. The tremendous potential of Blockchain technology to transform company operations has emerged. Modern corporate processes can be significantly enhanced to achieve digitization, automation, and transparency thanks to the features of persistency, validity, auditability, and disintermediation that Blockchain offers. The work done to integrate Blockchain into business operations is still in its infancy, though. This article defines Blockchain in the first place and looks into the properties of Blockchain in business processes. This paper provides a qualitative assessment of the research on the use of Blockchain technology in commerce and related fields. Although the study is qualitative and is based on a review of the literature, publication and sample selection bias still exist. Additionally, the Blockchain architecture for the complete business process is not integrated or presented. The researchers could do additional research on the financial measurement of cost, revenue, profit, and investment relevant to various business and Blockchain-related fields. This will make it easier for them to comprehend how Blockchain technology fits into business administration.

9.9 Future Directions for Research

Blockchain technology will radically alter how we handle transactions in general and, as a result, how businesses run their internal business operations. Our examination of the difficulties relating to the BPM lifecycle and beyond identifies seven key areas for further research. For some of these, we anticipate that practical insights will surface sooner than for others. The arrangement just suggests how quickly these revelations might occur.

(1) Creating a variety of blockchain execution and monitoring solutions. Study on this we will need to show that employing blockchains in process-aware information systems is feasible. Here, design science and algorithm engineering

will be necessary among other things. Information from distributed systems and software engineering will be useful.

(2) Developing novel analytical techniques and creating blockchain-based business processes. This area of study will need to look into the most effective ways to specify and use blockchain-based procedures. To examine this subject, formal research techniques and design science will be necessary. Here, lessons learned from database and software engineering studies will be useful.

(3) Redesigning procedures to make use of the blockchain's opportunities. Research in this area will need to look into how blockchain technology can make it possible to collaborate with outside parties and reimagine particular processes. This technology has the potential to revitalize the entire field of choreographies. Design science will be necessary here, among other things. Organizational science and operations management insights will be instructive.

(4) Establishing suitable processes for evolution and adaptability. The potential assurances that may be possible for specific forms of evolution and adaptation must be examined in this field of study. Formal research techniques will be necessary in this case, among others. It will be instructive to conclude theoretical computer science and verification.

(5) Creating methods for locating, discovering, and examining pertinent procedures for implementing blockchain technology. The properties of blockchain technology that best fulfill the needs of particular processes must be the subject of further study. Empirical research techniques and design science will be necessary, among other things. Here, lessons learned from management science and innovation research will be helpful.

(6) Knowing how blockchains affect strategy and governance, especially with new business and governance models made possible by revolutionary innovation based on blockchain. Which procedures in a company will need to be studied for this topic's research? Using blockchain, a setting could be organized differently, with these implications brings. Empirical research techniques will be needed, among others, to study this subject. It will be useful to conclude organizational science and business studies.

(7) Examining the cultural trend toward transparency in corporate management and operations. Processes, as well as hiring and upskilling personnel as necessary. Studies on this subject will need to look into how the adoption of Blockchain affects business culture. And how this is different from adopting other technology. Empirical techniques will be necessary, among others, for this kind of research. It will be useful to conclude organizational science and business studies.

The information systems and BPM communities have a special chance to influence this. An essential change to support inter-organizational communication toward a dispersed, reliable infrastructure process. With this study, we hope to make the research challenges more distinct, focused, and energetic that have arrived.

References

1. Abadi, J., Brunnermeier, M.: Blockchain economics. https://doi.org/10.3386/w25407. Working Paper, National Bureau of Economic Research, No. 25407 (2018)
2. Pal, A., Tiwari, C.K., Haldar, N.: Blockchain for business management: applications, challenges and potentials. J. High Technol. Manag. Res. https://doi.org/10.1016/j.hitech.2021.100414
3. Agrawal, H.: What is the bitcoin Mempool & why does it matters? https://coinsutra.com/bitcoin-mempool/ (2019). Accessed 10 May 2020
4. Aguilera, R.V., Grøgaard, B.: The dubious role of institutions in international business: a road forward. J. Int. Bus. Stud. **50**(1), 20–35 (2019)
5. Ahmed, S.: Cryptocurrency & robots: how to tax and pay tax on them. South Carol. Law Rev. **69**(3), 697–740 (2018)
6. Ainsworth, R.T., Alwohaibi, M., Cheetham, M.: UK & KSA VATs: a cutting-edge proposal—mini-blockchain and VATCoin, from SSRN: https://doi.org/10.2139/ssrn.3574381; https://ssrn.com/abstract=3574381 (2020a). Accessed 3 May 2020
7. Murala, D.K., Panda, S.K., Swain, S.K.: Secure dynamic groups data sharing with modified revocable attribute-based encryption in cloud. Int. J. Recent Technol. Eng. (IJRTE) **8**(Issue4) (2019). ISSN: 2277-3878
8. Ahluwalia, S., Mahto, R.V., Guerrero, M.: Blockchain technology and startup financing: a transaction cost economics perspective. Technol. Forecast. Soc. Change **151**(February), 119854 (2020). https://doi.org/10.1016/j.techfore.2019.119854
9. Rahimi, F., Moller, C., Hvam, L.: BPM, and IT management: the missing integration. Int. J. Inf. Manag. **36**(1), 142–154 (2016)
10. Ariouat, H., Hanachi, C., Andonoff, E., Benaben, F.: A conceptual framework for social BPM. Proc. Comput. Sci. **112**, 703–712 (2017)
11. Ahmed, K.: YouTube podcast on blockchain and the banking industry. https://youtu.be/rq3SW-KUKhg (2020). Accessed 7 May 2020
12. Ainsworth, R.T., Alwohaibi, M., Cheetham, M., Tirand, C.: A VATCoin proposal following on the 2017 EU VAT proposals—MTIC, VATCoin, and BLOCKCHAIN, Boston University, School of Law. Law and Economics Research Paper No. 18-09. Available at SSRN: https://ssrn.com/abstract=3151465 (2017). Accessed 4 May 2020
13. Ainsworth, R.T., Alwohaibi, M.: Blockchain, bitcoin, and VAT in the GCC: the missing trader example. Boston University School of Law, Law and Economics Research Paper No. 17-05, from SSRN: https://doi.org/10.2139/ssrn.2919056; https://ssrn.com/abstract=2919056 (2017a). Accessed 1 May 2020
14. Ainsworth, R.T., Alwohaibi, M., Cheetham, M., Tirand, C.: A VATCoin proposal following on the 2017 EU VAT proposals—MTIC, VATCoin, and BLOCKCHAIN, Boston University School of Law. Law and Economics Research Paper No. 18-09, from SSRN: https://ssrn.com/abstract=3151465. Accessed 6 May 2020
15. Devine, A., Jabbar, A., Kimmitt, J., Apostolidis, C.: Conceptualising a social business blockchain: the coexistence of social and economic logics. Technol. Forecast. Soc. Chang. (2021). https://doi.org/10.1016/j.techfore.2021.120997
16. García-Bañuelos, L., Ponomarev, A., Dumas, M., Weber, I.: Optimized execution of business processes on blockchain. In: Carmona, J., Engels, G., Kumar, A. (eds.) BPM. BPM 2017. Lecture Notes in Computer Science, vol. 10445. Springer, Cham (2017)
17. Haarmann, S.: Estimating the duration of blockchain-based business processes using simulation. In: 11th ZEUS Workshop, ZEUS 2019, Bayreuth, Germany, 14–15, February (2019). http://ceur-ws.org/Vol-2339
18. Di Ciccio, C., Cecconi, A., Dumas, M., García-Bañuelos, L., López-Pintado, O., Lu, Q., Mendling, J., Ponomarev, A., Binh Tran, A., Weber, I.: Blockchain support for collaborative business processes (2019). https://doi.org/10.1007/s00287-019-01178-x
19. Kimani, D., Adams, K., Attah-Boakye, R., Ullah, S.: Blockchain, business and the fourth industrial revolution: whence, whither, wherefore and how? Technol. Forecast. Soc. Chang. (2020). https://doi.org/10.1016/j.techfore.2020.120254

20. Morkunas, V.J., Paschen, J., Boon, E.: How blockchain technologies impact your business model. (2019). https://doi.org/10.1016/j.bushor.2019.01.009
21. Murala, D.K., Panda, S.K., Swain, S.K.: A novel hybrid approach for providing data security and privacy from malicious attacks in the cloud environment. J. Adv. Res. Dyn. Control Syst. **11**, 1291–1300 (2019)
22. Viriyasitavat, W., Da Xu, L., Bi, Z., Sapsomboon, A.: Blockchain-based BPM (BPM) framework for service composition in industry 4.0. J. Intell. Manuf. 1–12 (2018)
23. Viriyasitavat, W., Hoonsopon, D.: Blockchain characteristics and consensus in modern business processes. J. Ind. Inf. Integr. (2018). https://doi.org/10.1016/j.jii.2018.07.004
24. Li, Y., Luo, Z., Yin, J., Xu, L.D., Yin, Y., Wu, Z.: Enterprise pattern: integrating the business process into a unified enterprise model of a modern service company. **11**(1) (2015). https://doi.org/10.1080/17517575.2015.1053415
25. López-Pintado1, O., García-Bãnuelos1, L., Dumas, M., Weber, I.: Caterpillar: A Blockchain-Based BPM System
26. Meidan, J., Garcia-Garcia, A., Escalona, M.J., Ramos, I.: A survey on business processes management suites. Comput. Stand. Interfaces **51**, 71–86 (2017)
27. Panda, S.K., Mohammad, G.B., Nandan Mohanty, S., Sahoo, S.: Smart contract-based land registry system to reduce frauds and time delay. Secur. Privacy e172 (2021). https://doi.org/10.1002/spy2.172
28. Panda, S.K., Satapathy, S.C.: Drug traceability and transparency in medical supply chain using blockchain for easing the process and creating trust between stakeholders and consumers. Pers. Ubiquit. Comput. (2021). https://doi.org/10.1007/s00779-021-01588-3
29. Niveditha, V.R., Sekaran, K., Singh, K.A., Panda, S.K.: Effective prediction of bitcoin price using wolf search algorithm and bidirectional LSTM on internet of things data. Int. J. Syst. Syst. Eng. **11**(3–4), 224–236
30. Sathya, A.R., Panda, S.K., Hanumanthakari, S.: Enabling smart education system using blockchain technology. In: Panda, S.K., Jena, A.K., Swain, S.K., Satapathy, S.C. (eds.) Blockchain Technology: Applications and Challenges. Intelligent Systems Reference Library, vol. 203. Springer, Cham (2021). https://doi.org/10.1007/978-3-030-69395-4_10
31. Lokre, S.S., Naman, V., Priya, S., Panda, S.K.: Gun tracking system using blockchain technology. In: Panda, S.K., Jena, A.K., Swain, S.K., Satapathy, S.C. (eds.) Blockchain Technology: Applications and Challenges. Intelligent Systems Reference Library, vol. 203. Springer, Cham (2021). https://doi.org/10.1007/978-3-030-69395-4_16
32. Viriyasitavat, W., Hoonsopon, D.: Blockchain characteristics and consensus in modern business processes. J. Ind. Inf. Integr. **13**, 32–39 (2019)
33. Panda, S.K., Daliyet, S.P., Lokre, S.S., Naman, V.: Distributed Ledger technology in the construction industry using Corda. In: The New Advanced Society: Artificial Intelligence and Industrial Internet of Things Paradigm. https://doi.org/10.1002/9781119884392.ch2
34. Panda, S.K., Satapathy, S.C.: An Investigation into Smart Contract Deployment on Ethereum Platform Using Web3.js and Solidity Using Blockchain. In: Bhateja, V., Satapathy, S.C., Travieso-González, C.M., Aradhya, V.N.M. (eds.) Data Engineering and Intelligent Computing. Advances in Intelligent Systems and Computing, vol. 1. Springer, Singapore (2021). https://doi.org/10.1007/978-981-16-0171-2_52
35. Panda, S.K., Rao, D.C., Satapathy, S.C.: An investigation into the usability of blockchain technology in internet of things. In: Bhateja, V., Satapathy, S.C., Travieso-González, C.M., Aradhya, V.N.M. (eds.) Data Engineering and Intelligent Computing. Advances in Intelligent Systems and Computing, vol. 1. Springer, Singapore (2021). https://doi.org/10.1007/978-981-16-0171-2_53
36. Panda, S.K., Dash, S.P., Jena, A.K.: Optimization of block query response using evolutionary algorithm. In: Bhateja, V., Satapathy, S.C., Travieso-González, C.M., Aradhya, V.N.M. (eds.) Data Engineering and Intelligent Computing. Advances in Intelligent Systems and Computing, vol. 1. Springer, Singapore (2021). https://doi.org/10.1007/978-981-16-0171-2_54
37. Viriyasitavat, W., Xu, L., Bi, Z.M., Sapsomboon, A.: Blockchain-based BPM (BPM) framework for service composition in industry 4.0. J. Intell. Manuf. online published (2018). https://doi.org/10.1007/s10845-018-1422-y

38. Nanda, S.K., Panda, S.K., Das, M., Satapathy, S.C.: Automating vehicle insurance process using smart contract and Ethereum. In: Chakravarthy, V.V.S.S.S., Flores-Fuentes, W., Bhateja, V., Biswal, B. (eds.) Advances in Micro-Electronics, Embedded Systems and IoT. Lecture Notes in Electrical Engineering, vol. 838. Springer, Singapore (2022). https://doi.org/10.1007/978-981-16-8550-7_23
39. Panda, S.K., Elngar, A.A., Balas, V.E., Kayed, M. (eds.): Bitcoin and Blockchain: History and Current Applications, 1st edn. CRC Press (2020). https://doi.org/10.1201/9781003032588
40. Blockchain technology: applications and challenges. In: Panda, S.K., Jena, A.K., Swain, S.K., Satapathy, S.C. (eds.) Intelligent Systems Reference Library. Springer. https://doi.org/10.1007/978-3-030-69395-4
41. Ainsworth, R.T., Alwohaibi, M.: The first real-time blockchain VAT—GCC Solves MTIC Fraud (July 24, 2017). Boston University School of Law. Law and Economics Research Paper No. 17-23, from SSRN: https://doi.org/10.2139/ssrn.3007753; https://ssrn.com/abstract=3007753 (2017b). Accessed 3 May 2020
42. Murala, D.K., Panda, S.K., Swain, S.K.: A survey on cloud computing security and privacy issues and challenges. J. Adv. Res. Dyn. Control Syst. **11**, 1276–1290 (2019)
43. Silva, D.: Decentralized Enforcement of Business Process Control Using Blockchain
44. Varaprasada Rao, K., Panda, S.K.: A design model of copyright protection system based on distributed ledger technology. In: Satapathy, S.C., Lin, J.C.W., Wee, L.K., Bhateja, V., Rajesh, T.M. (eds.) Computer Communication, Networking and IoT. Lecture Notes in Networks and Systems, vol. 459. Springer, Singapore (2023). https://doi.org/10.1007/978-981-19-1976-3_17
45. Varaprasada Rao, K., Panda, S.K.: Secure electronic voting (E-voting) system based on blockchain on various platforms. In: Satapathy, S.C., Lin, J.C.W., Wee, L.K., Bhateja, V., Rajesh, T.M. (eds.) Computer Communication, Networking and IoT. Lecture Notes in Networks and Systems, vol. 459. Springer, Singapore (2023). https://doi.org/10.1007/978-981-19-1976-3_18

Chapter 10
Understanding the Blockchain Technology Adoption in Transportation Management: Application in Trucking Industry

Krishna Kumar Dadsena

Abstract Over the past year the transportation industry realizes the changes, expectation and challenges with the growing digitally-driven buy-and-sell ecosystem. Therefore, the transporter plan for the best possible way to provide service to their clients by not only delivering their products/goods safely but also ensure the security of their data and information involved in the transaction. Hence, in order to scale the modern transport business, managers need to ensure a secure platform for recording and maintaining the confidential data in their business operation. Blockchain technology (BT) can help in redesigning of transportation operations and supports in transforming towards more efficient business environment. The BT enables the effective information management and data security of business operations, which can also help in building the transparent, efficient, reducing physical paperwork, and trustworthy business environment. Therefore, this chapter presented recent development in transportation industry considering the blockchain technology. We have further discussed the scope of blockchain technology adoption to enhance the security, trust and operational performance of the transportation business. In addition to this, potential benefits of blockchain technologies in payment, immutability, flexibility and security has been highlighted. We have presented the role of BT considering the Indian trucking industry. This chapter helps in strategic planning and decision making to the transport managers in effective implementation of the BT. At the end of this chapter the future scope of BT adoption in trucking industry have been discussed.

Keywords Blockchain technology (BT) · Transportation management · Trucking industry · Security · Payment · Immutability · Flexibility

K. K. Dadsena (✉)
Indian Institute of Management, Ranchi, India
e-mail: Krishnakumar.dadsena@iimranchi.ac.in

S. K. Panda et al. (eds.), *Recent Advances in Blockchain Technology*,
Intelligent Systems Reference Library 237, https://doi.org/10.1007/978-3-031-22835-3_10

10.1 Introduction

In today's fast- paced global market, transportation plays a major role in logistics management, which has become backbone of supply chain management. An efficient and reliable transportation system is the demand of the time to sustain in the global dynamic market. Transportation system contributes not only in economic growth but it also supports the social growth of the country. Recent days, there have been several trends like online shopping, e-marketplace which force transport managers to rethink and strategize their future planning considering the ongoing digital transformation [1]. Beside these trends, now a days transport managers became more attentive to diverse challenges such as costs cutting, theft, delay, and trust issues which needs to be handle efficiently [2]. Therefore, in view of current trend of digitization, the role of data and information security has emerged as an important characteristics in the business operation to ensure the data privacy, immutability and confidentiality of the all transactions among different stakeholders [3]. Thus, transportation industry must focus on more transparent, efficient and reliable data management system [4]. Considering the distributed freight logistics markets which demands the technologies to support the managers in handling challenges related to data privacy and security. In this regards blockchain characteristics can support a transportation business by providing a reliable data recording and management system to the transport manager. Considering the expanding global trading of domestic and international commodity, the transport industry needs to deal with sensitive information's, for which the blockchain can be very vital.

Despite of realization of important benefits and support in handling existing challenges like security, logistic inefficiency, uncertainty, the pace of implementation BT in the transport industry is still very slow. In the transport business the adoption of BT not only help in tracking and tracing of their consignment but also helps in enhancing the operational performance their business [5]. A blockchain technology helps in standardize auditing, data integrity in information management to the organization [6]. The blockchain can be defined as a safe and secure record of transactions, collected into different blocks chained in specific sequence, and distribute across a different server to create reliable provenance [7].

BT allows execution of smart contracts, which helps in controlling the digital assets, formulate the collaboratives commitment to reduce the cost and risk [6]. This also, ensure secure flow of information and cash on the agreement of stakeholders without a third party involvement [8]. In addition to this, blockchain technology serve as enabling forces for economic, social and business transformation by predicting the challenges and opportunities for value addition in the transportation business [9]. Furthermore, the encrypted and distributed data structure of blockchain and absence of a central server reduce the risk of cyberattack [10]. Considering the several advantages in operational, economic benefits the integration of blockchain in transportation management is expected to become popular in near future [11]. Based on the above discussion this study tries to cover the recent development, challenges and opportunities of BT adoption in transportation management. In this chapter, specifically we will discuss the challenges and opportunity in Indian trucking industry. The remaining of the chapter is organized as follows, Sect. 10.2, presented the recent development of BT in transportation industry. Challenges in BT adoption in trucking industry has been discussed in Sect. 10.3. Section 10.4 discussion of the study has been highlighted and future opportunity has been discussed in Sect. 10.5. Section 10.6 presented the key takeaways of the study.

10.2 Application of Blockchain Technology in Transportation Management

The transport sector has become integral part of a highly divers and complex supply chain management. As the competition in this globalized market, where avoidance of adverse events must consider in order to sustain in current business scenario. [1] discussed the role of digital technologies in transportation management and its impact on overall SCM, in which authors indicated that the technological development has become the demand of an hour in modern transportation system. The role BT in this growing competitive environment to improve the overall performance of the transportation industry cannot be ignored. Recent days, integration of BT in transportation industry is well appreciated from the practitioners and researchers' perspective to sustain in the current growing market. BT implementation overcome challenges related to recording and managing the sensitive information in the operational process of the transportation industry [12]. Which makes the BT more popular among the transport managers, practitioners and scholars. The BT implementation motivates the transport managers to rethink and modify their business activities to enhance transparency and trust in their business. In Table 10.1 the relevant existing research in BT implementation in transportation industry has been presented.

Blockchain is a widely acceptable for inherently fragmented industries, like transportation, where coordination with multiple stakeholders is essential.[1] This technology helps in solving the hurdles and disputes among stakeholders through digitizing peer-to-peer collaboration and payments [26], and it also expand the trade opportunities by providing the reliable services and infrastructure for transportation businesses.

Based on the aforementioned discussion the role of BT in different modes of freight transportation, namely, air, water, and road has been presented in Table 10.1. Further, the application of BT can be categories based on the potential scope in transportation industry namely as, immutability, security, payment and flexibility as shown in Fig. 10.1. The detailed discussion is presented in subsequent section.

[1] Blockchain drives transformation in transportation. https://www.ibm.com/thought-leadership/institute-business-value/report/blockchain-expedited-delivery, accessed date 04 May 2022.

Table 10.1 Relevant application of Blockchain technology in transportation management

Author	Mode of transportation	Application
[13]	Road	Road transport companies adopt blockchain in storing transactions for smooth function of their business
[14]	Cargo transportation	The transactions among all the stakeholders were carried out by smart contracts using blockchain application
[15]	Road	Tracking and tracing of operational activity through blockchain to enhance the readiness level (RL)
[16]	Air	In order to manage the flow of information and financial transaction air cargo company used the blockchain technology
[17]	Road	To manage, access and trace the information in blood transportation management a blockchain technology has been implemented
[18]	Water	Maritime industry integrated a blockchain technology to attain lean process and to ensure a high level of information security
[3]	Road	To improve the scalability and privacy related issues the trucking industry integrated the blockchain technology
[19]	Road freight	Transport managers managed their information related to charging points for electric vehicle through blockchain technology implementation
[20]	Road freight	The ride-sharing business, the car owners used the blockchain technology to collect and store the data in a crypto ledger after verification
[21]	Railway	Railway freight supply chain used the blockchain to improve their information management system and financial activity
[22]	Road freight	To enhance trust, transparency in transactions a car riding company integrated the blockchain in their business process
[23]	Railway	To collect the historical driving data and to stablish the smart contract heavy haul railway industry used the blockchain technology
[24]	Water	The blockchain technology has been adopted by maritime industry to enhance the operational efficiency, visibility, and transparency
[25]	Container transportation	In order to improve the safety and traceability in the container transportation, managers implemented the blockchain technology

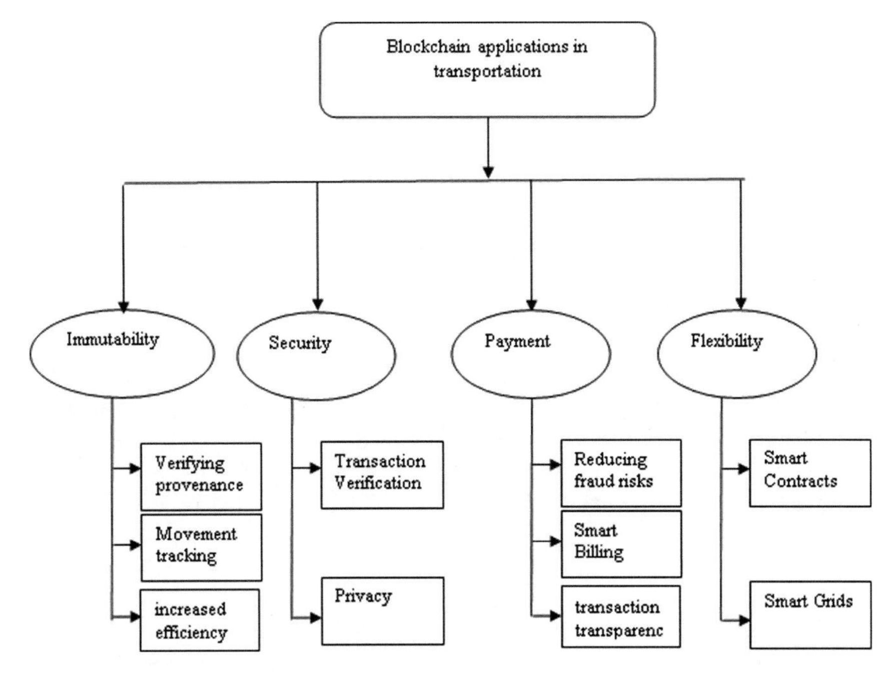

Fig. 10.1 Blockchain applications in different domains of transportation management

10.3 Potential Challenges and Scope of BT Adoption in Indian Trucking Industry

The road transport provides service nearly every industry, retailer, direct customers and other resource base sector of the economy. The demand for freight transportation service via trucks has been increasing over the years and seeks considerable attention from the practitioners as well as academia. It is interesting that nearly 70% of the US community serves via trucks,[2] 65% of Indian community served by trucks [27] which is mostly unorganized, and "freight trucking industry in China has been increasing at 13.8% per year" (www.joc.com/trucking-logistics). According to a report published ATA 2016 the US trucking industry is the lifeblood of the economy as $726.4 billion in gross freight revenue only from the primary shipment from the trucking industry. Similarly for the Indian trucking industry valued at $130 billion, and for China trucking industry generates revenue $94 billion as per 2016 reports.[3]

[2] Federal Highway Administration, https://www.fhwa.dot.gov/policy/2015cpr/es.cfm, accessed date 14 June 2022.

[3] https://www.ibisworld.com, accessed date 13 June 2022.

Table 10.2 A comparative statistics of freight movement via road (Report published by Novonous, 2015)

Country	Total freight movement by road (%)	Total freight movement by rail (%)
US	60	10
EU-28	74.9	18.2
Africa	76	10
India	63	27

10.3.1 Background of Indian Trucking Industry

As the current growing market and use of road freight transportation become an indispensable part of logistics supply chain. The road transportation has significant contribution in freight movement and economic development of any country. In the recent years, roads as compared to rail, transport lions-share of the freights as they can reach each and every corner of the country. Table 10.2, indicates the role of road freight transportation industry over the rail freight transport. Freight transport by truck is more preferable over railways due to its ability to reach beyond geographical barriers, quantity barriers, and ease of availability of trucks.

The Indian transportation (Air, Sea, Rail & Road Freight) market is approximately worth US$300 billion, which is contributing around 13–14% of the country's GDP.[4] However, the road-freight transport (by means of trucks) is valued at US$150 billion per annum and is growing at a CAGR of around 12% [28]. As, the Indian trucking industry is highly fragmented, because it largely owned by private sector and it is dominating by the small truck owners. Around 90% of which belongs to micro truck owners, having less than twenty trucks, making it totally fragmented and unorganized, this might be one of the main reasons in its inefficient which functioning [29, 30]. Considering all these facts the trucking industry, like many other industries, is also encountering many uncertain undesirable events such as low concentration, low technology innovation, low efficiency etc., which have adverse impact on its overall performance of the trucking business.

Despite of importance of trucking industry, the current performance level is very low due to inadequacies of the current infrastructure and operations to meet the growing demand for freight [31]. Paniati [32] investigated the role of data, measure, information and decision in development of performance measurement index for the transport systems. The author suggested that setting measures may help in achieving mission and vision of the organization and to manage what already exists by effective use of information transparency and data security. Li et al. [33] emphasize that BT is being applied as a key tool for providing better decision making through efficient data management system.

Modern, trucking industry experiencing the benefit of digitization, by integrating global positioning system (GPS), mobile and internet, Internet of thing (IoT) etc.,

[4] https://www.businessworld.in/article/India-S-Road-Freight-Market-Pegged-At-150-Billion-Truxapp-/03-10-2017-127398/, accessed date 24 June, 2022.

however there is still need to encourage adoption of BT to ensure the trust and information transparency among the stakeholder [13, 57, 58]. Considering these facts, practitioners/transporters must understand the potential effects of digital transformation to attain the efficient operations of the trucking business. Therefore, we will briefly discuss on scope of BT adoption in Indian trucking industry in next section.

10.3.2 Challenges and Scope of BT Implementation in Indian Trucking Industry

The rapid technological progressions and its impact on transportation business creating more opportunities and business environment. In the trucking industry where the aim is to provide the efficient service to their customers by providing the right time delivery, transparent and accurate information with minimum cost. The Indian trucking industry has its unique and important role in economic and social development of the country. As this business comprises different stakeholder, hence the flow and accuracy of information matters a lot to perform efficiently and effectively.

The Indian trucking industry, experiencing vast scope to adopt BT, as it assists them in several ways such as helps in minimizing the fraud, enhancing the transparency in sharing the important information's, and providing protected record keeping.

Blockchain technology emerged as an important technology in tucking business, as it helps in providing better and trustworthy culture in the business environment [3]. The blockchain is basically helps the transporters in maintaining the distributed database of records of all transactions and communication among the stakeholders in the network.

Despite of a vital role of the blockchain technology in trucking industry, this business facing several challenges which needs to be consider. The exiting academic literature, trade journal, magazine and reports show that present situation of our Indian trucking industry is wretched condition, and discussed about the technological challenges which effect the India trucking industry. Because of the complexity in the process of trucking industry, where each stage involves the crucial information. Hence, this section, we present an overview of the challenges in Indian trucking industry and scope to tackle those challenges using BT implementation. The scalability is actually needed for blockchain to obtain widespread approval across industry sectors and users [3, 59, 60]. As BT can be used to identify trends, which may help the managers to prepare to response the future demand. BT supports the trucking business form the scalability perspective, in which managers delas with the complex and large set of data [13].

Although the gaining importance of digitization in transportation industry, there are administration and implementation challenges in BT adoption. In this chapter we have tried to identify the challenges in Indian trucking industry and map them with potential scope (security, transparency, flexibility, immutability) of BT adoption. The summary of the important challenges has been shown in Table 10.3.

In adoption of BT information and data management lack of trust are some important challenges. Rodrigues et al. [48] emphasizes on a proper communication and data management system in trucking business may help in overall performance of the industry. Similarly, the fragmented nature of transportation business led to the challenges related to trust, transparency and transaction [49, 50].

10.4 Scope of BT Implementation in Trucking Industry

Despite of its economic and social contributions in all around the world, the trucking industry still suffers from many challenges, which directly affects the trust issue in collaboration among the stakeholders [13, 61, 62]. Hence this chapters tries to highlight the important role of blockchain technology to enhance the transparency and trust in in trucking industry. In this section we have discussed the potential application of BT considering the broad scope for immutability, security, payment and flexibility in the trucking industry.

10.4.1 Immutability

In recent years, the demand of an efficient freight transportation service has been increasing by growth in international trade, growing public accountability and dynamic environment. However, bottlenecks in the system are exposing the inadequacies of current operations management to meet the growing freight demand.

According to, The National Cooperative Freight Research Program (NCFRP) report the decision at strategic, operational and investment on digital transformation will be necessary to maintain performance of transportation sector [51, 63, 64]. Hence, the decision making on adoption of new technologies to develop trustworthy, transparent and systematic data management became an essential part in transportation planning and management. In order to improve the data recording and its management transporters can take the help advanced technique by integrating immutability feature in blockchain.

Immutability of the data ensure that the originality of recorded data, which provides provenance of assets means exact information. Blockchain immutability helps the operations in several ways, such as minimizing the risks, like human error, risk of data leakage etc. The blockchain protocols are design in such a way that transporters can achieve true immutability by cryptography technique. The algorithms developed for this technology ensure that data and information stored on a network cannot manipulated. Which helps in creating a single source of truth for the industry which is very fragmented in nature, specifically trucking industry. In transportation business, the immutable nature of blockchain beneficial for verification and

Table 10.3 Potential challenges and scope for BT adoption

Potential Challenges	Affects	Source	Decision-making level	BT application	
				BT capability	Covers the scope
Capacity utilization[a]	Empty truck/Unused capacity	[34]	Strategic	Smart contract	Flexibility
Operating cost	Cost and profitability	[35–38]	Operational	Removing the middleman and cutting out the costs	Immutability
Shipment delay	Economic loss and Reputation damage	[39, 40]	Operational	Tracking and tracing system	Immutability
Information management	Uncertainty in information about the load	[40, 41]	Tactical	Advance demand information system,	Immutability
In transit theft	Economic loss, reputation loss	[42]	Operational	Tracking	Security
risks of misuse of sensitive information	Trust and collaboration issues	[35]	Strategic	Smart contract	Security
Fragmented nature of business	Lack of integration between carriers, unstructured freight rate	[27, 35])	Strategic	BT helps in minimizing the fraud in transaction	Payment
Improper communication channel	Communication gap/delay	[43]	Operational	Development of a database	Flexibility
Unethical practice (Bribery)	Economic loss/delay in receipt the amount	[44, 45]	Strategic	Transparent transaction	Payment
Rigid infrastructure,	Emission tracking, truck condition	[36, 46, 47]	Strategic	Improvement of road safety, signalling process	Flexibility

[a]Ministry of road transport & highways, government of India (MORTH), (2011). Report of the Sub-Group on Passenger and Freight Traffic Assessment and Adequacy of Fleet and Data Collection and Use of IT in Transport Sector in the Twelfth Five Year Plan (2012–2017).

tracking of goods movement through theft and fraud activities. This technology may also help in improve accountability compliance, which benefits the transporters and minimize the inefficiency of the business.

10.4.2 Security

Although security is an important issue in freight transportation, it is also a vital to develop secure business environment. To reduce the impact of fraud, theft transporters need to think about the reduction of such issues through advanced technologies. Recent days, there are technologies such as IoT, blockchain, demonstrated its impact on minimizing such challenges. Thus, it is very important to use an advanced technology which helps in strategic decision making considering the confidentiality of data storage and security in transportation management [69, 70].

The BT become more important when it comes to the industry which is more volatile and fragmented industry, like trucking industry. In this business the involvement of different stakeholders tries to provide the efficient and trustworthy service to their customers. However, the considering the issues related to improper documentation of record, absence of single source database management system, fake set of documentation theft in goods are still a big challenge for the managers [42]. In addition to this improper communication and lack of proper information about the load for the transporter leads to inefficiency in operation of trucking business [26]. BT ensure the security and safety improvement trough proper verification of all transaction and maintaining the security which lead to operation cost reduction in transportation management [52].

10.4.3 Payment

As trucking industry involves the several stakeholders which take in very frequent regular basis financial transactions in their business operations. Sometimes transporters face challenges such as delay in payment, poor data management and recording of financial transaction, which may lead to dispute among the partners. Considering the size of the trucking industry and the complexity of its operations, it becomes important to have a systematic and secure way of performing the financial transaction. This has motivated to take up this issue and implementing the BT to ensure the reliable and secure system to streamline the financial transactions in transportation business [21, 65, 66].

As the huge amount of monetary transaction in daily business and the fragmentated nature of trucking business there are million dollars are associated in dispute payment [53]. Industry demands a secure and authentic payment method to ensure the trustworthy business environment ecosystem. The digitization of payment reduces the paperwork through electronic transaction which drives the efficient transaction by reducing the execution time for settlement and payment of money. BT helps in minimizing the fraud in transaction, since the entire business structure demand the validation in the operation. The evolving trucking industry need to be prepared and plan an efficient and effective payment method through reliable and informed data management system. In streamlining the payment method, the BT has been found very effective.[5] This can be achieved through blockchain technology adoption as it provides smart billing and effective transaction management system for trucking industry [54].

10.4.4 Flexibility

The road transportation sector needs more flexible, robust, resilient and responsive against the uncertainties in the operations. The trucking industry is exposed to several issues related to trust among the partners and transparency in information management. In order to overcome such issues transports, need to develop a reliable system to maintain all transactions [53]. Hence, now a days transport managers realizes that requirement reliable technology such as BT, which help in developing the smart grid to maintain a secure data base management system through advanced technology [55, 67, 68]. BT enables the smart contracts among the untrusted parties. However, its potential outcome lies in capacity of streamlining the whole process which start from order placing to payment, and also improve the confidence and benefits of the all stakeholders.[6]

In addition to this, transportation industry is facing vital change due to the depletion of fossil fuels, which force the practitioners to think about the non-polluting and efficient energy source.[7] A BT enables the secure energy trading in transportation management considering the important data and information management such as required energy, dynamic pricing, record of connectivity and all the record which is very important for the transporters in operation [55]. This technological advancement enables the transportation operations more flexible and trustworthy.

[5] Blockchain Technology in Trucking https://www.rtsinc.com/articles/blockchain-technology-trucking, accessed date 23 May 2022.

[6] https://www.freightwaves.com/news/is-blockchain-finally-ready-for-prime-time, accessed on 23 April 2022.

[7] Smart Grid's Low-Tech Savings: Fewer Truck Rolls https://www.greentechmedia.com/articles/read/smart-grids-low-tech-savings-fewer-truck-rolls, accessed date 23 April 2022.

10.5 Discussion

The emerging trend such as shopping from e-marketplace, which makes transportation business more challenging in handling and maintaining the data and information efficiently. The modern transportation system demands a trustworthy, secure and transparent business environment to handle such challenges. The evolving digital transformation in the transportation management shows its effectiveness in terms of security, trust, flexibility to attain the goal of an organization. In recent year freight matching through online platforms gained popularity in operation of trucking industry. Which acts as centralized e-marketplaces, where sharing information and data digitally become risky in terms of misuse of confidential information. For such scenario BT has proven its effectiveness in handling such challenges in business operations for transportation industry. Which is also emerged as one of the widely used and most appropriate tool to enhance the security, transparency flexibility and immutability of the transportation system.

This chapter discussed about the recent trend and research in adoption of BT in transportation system. Based on the existing literature it can be observed that BT adoption in transportation industry is getting more attention from the managers perspective to develop a transparent and trustworthy business ecosystem. BT adoption also supports the transportation managers to manage their operations, which helps in overall operational performance improvement through efficient, accurate data recording and management system. The proposed study mainly focuses on mapping the different scopes (immutability, security, payment and flexibility) with transportation management strategies.

10.6 Future Scope

The BT has broad applications across diverse industries such as supply chain management, media and entertainment industry, insurance sector etc. Although the application of BT mainly lies in security domain, the full potential application of this technology is still not explored in the transportation management.

In current scenario the BT is one of the most popular/interesting topics from the transportation managers perspective. As it indicated the tremendous future opportunity of BT implementation in the transportation sector. Specifically considering the fragmented nature of trucking industry this technology needs further exploration for right purpose. Some of the important future scope is listed below:

- Transporters seeks to manage and improve their performances despite of several external forces such as globalization, security issues, transparency and technological influence. Hence, BT can be effectively used to track the performance of individual carrier, and it also allow to monitor the performance of individual trucks regularly [5], which helps in identifying the key areas for improvement.

- In future the application of BT in development of smart contract and collaboration in among the transporters can be further investigated. The information and data involve in the operation from different stakeholder's have become game changer in transportation industry[8]. Hence, BT would help in formation of smart contract by providing an efficient and effective data and information management system to the transporters.
- Considering the involvement of different stakeholders in transportation industry the security of data become very important criteria. In transportation industry there is scope of developing an intelligent transportation in view of different layers such as government, industry [56].
- In the era of growing competitive market, transporters' need to manage their stability under the competitive umbrella. Traditional trucking industry generally concerns the pursuit of their clearly defined vision and mission which is primarily focused on the economic concern. However, with the traditional practice, transporters often struggling to manage transparency and trust in managing their business. In future there is wide scope to study about the integrated effect of BT with other technologies such as big data analytics, cloud computing and Artificial intelligence (AI) tools to attain the sustainable criteria in trucking industry.
- In future management perception on urgency of implementation of BT can be considered as one of the important areas of study for transportation management.

10.7 Key Takeaways

Future scope directed that there is wide scope of exploring the use of BT in transportation industry. Specifically, in the trucking industry where perceived benefits of BT implementation can be further tested through empirical analysis against the risk in BT adoption. It is also imperative that different transporters and stakeholders must think and related their mutual benefits for a particular operational process so that collaborative initiatives can be made in BT implementation. The trucking companies must think and start the adoption of BT to address some important challenges in their operation the transport manager is experiencing. Precisely, it would be interesting to investigate the impact of different variables such as government intervention, competitive pressure in the market, ongoing industry trends, support from top management in BT implementation.

---

[8] The use of Blockchain in the transportation and logistics industry, 2021. https://www.globalialogisticsnetwork.com/blog/2021/07/14/the-use-of-blockchain-in-the-transportation-and-logistics-industry/, accessed date 20 April 2022.

References

1. Dong, C., Akram, A., Andersson, D., Arnäs, P.-O., Stefansson, G.: The impact of emerging and disruptive technologies on freight transportation in the digital era: current state and future trends. Int. J. Logist. Manag. (2021)
2. Dadsena, K.K., Sarmah, S., Naikan, V., Jena, S.K.: Optimal budget allocation for risk mitigation strategy in trucking industry: an integrated approach. Transp. Res. Part A: Policy Pract. **121**, 37–55 (2019)
3. Alacam, S., Sencer, A.: Using blockchain technology to foster collaboration among shippers and carriers in the trucking industry: a design science research approach. Logistics **5**, 37 (2021)
4. Aslam, J., Saleem, A., Khan, N.T., Kim, Y.B.: Factors influencing blockchain adoption in supply chain management practices: a study based on the oil industry. J. Innov. Knowl. **6**, 124–134 (2021)
5. Kuhi, K., Kaare, K., Koppel, O.: Ensuring performance measurement integrity in logistics using blockchain. In: 2018 IEEE International Conference on Service Operations and Logistics, and Informatics (SOLI), pp. 256–261. IEEE (Year)
6. Orji, I.J., Kusi-Sarpong, S., Huang, S., Vazquez-Brust, D.: Evaluating the factors that influence blockchain adoption in the freight logistics industry. Transp. Res. Part E: Logist. Transp. Rev. **141**, 102025 (2020)
7. Angelis, J., Da Silva, E.R.: Blockchain adoption: a value driver perspective. Bus. Horiz. **62**, 307–314 (2019)
8. Chang, Y., Iakovou, E., Shi, W.: Blockchain in global supply chains and cross border trade: a critical synthesis of the state-of-the-art, challenges and opportunities. Int. J. Prod. Res. **58**, 2082–2099 (2020)
9. Morkunas, V.J., Paschen, J., Boon, E.: How blockchain technologies impact your business model. Bus. Horiz. **62**, 295–306 (2019)
10. Irannezhad, E.: Is blockchain a solution for logistics and freight transportation problems? Transp. Res. Procedia **48**, 290–306 (2020)
11. Mollah, M.B., Zhao, J., Niyato, D., Guan, Y.L., Yuen, C., Sun, S., Lam, K.-Y., Koh, L.H.: Blockchain for the internet of vehicles towards intelligent transportation systems: a survey. IEEE Internet Things J. **8**, 4157–4185 (2020)
12. Zhou, Z., Wang, M., Huang, J., Lin, S., Lv, Z.: Blockchain in big data security for intelligent transportation with 6G. IEEE Trans. Intell. Transp. Syst. (2021)
13. Haouari, M., Mhiri, M., El-Masri, M., Al-Yafi, K.: A novel proof of useful work for a blockchain storing transportation transactions. Inf. Process. Manage. **59**, 102749 (2022)
14. Baygin, M., Yaman, O., Baygin, N., Karakose, M.: A blockchain-based approach to smart cargo transportation using UHF RFID. Expert Syst. Appl. **188**, 116030 (2022)
15. Callefi, M.H.B.M., Ganga, G.M.D., Godinho Filho, M., Queiroz, M.M., Reis, V., dos Reis, J.G.M.: Technology-enabled capabilities in road freight transportation systems: a multi-method study. Expert. Syst. Appl. **203**, 117497 (2022)
16. Poleshkina, I.: Blockchain in air cargo: challenges of new World. In: MATEC Web of Conferences, pp. 00021. EDP Sciences (Year)
17. Pradhan, N.R., Singh, A.P., Kumar, V.: Blockchain-Enabled Traceable, Transparent Transportation System for Blood Bank. Advances in VLSI, Communication, and Signal Processing, pp. 313–324. Springer (2021)
18. Pu, S., Lam, J.S.L.: Blockchain adoptions in the maritime industry: a conceptual framework. Marit. Policy Manag. **48**, 777–794 (2021)
19. Fu, Z., Dong, P., Ju, Y.: An intelligent electric vehicle charging system for new energy companies based on consortium blockchain. J. Clean. Prod. **261**, 121219 (2020)
20. Humayun, M., Jhanjhi, N., Hamid, B., Ahmed, G.: Emerging smart logistics and transportation using IoT and blockchain. IEEE Internet Things Mag. **3**, 58–62 (2020)
21. Xie, P., Chen, Q., Qu, P., Fan, J., Tang, Z.: Research on financial platform of railway freight supply chain based on blockchain. Smart Resilient Transp. (2020)

22. Syed, T.A., Siddique, M.S., Nadeem, A., Alzahrani, A., Jan, S., Khattak, M.A.K.: A novel blockchain-based framework for vehicle life cycle tracking: an end-to-end solution. IEEE Access **8**, 111042–111063 (2020)
23. Hua, G., Zhu, L., Wu, J., Shen, C., Zhou, L., Lin, Q.: Blockchain-based federated learning for intelligent control in heavy haul railway. IEEE Access **8**, 176830–176839 (2020)
24. Czachorowski, K., Solesvik, M., Kondratenko, Y.: The Application of Blockchain Technology in the Maritime Industry. Green IT Engineering: Social, Business and Industrial Applications, pp. 561–577. Springer (2019)
25. Wang, X., Shi, H.: Research on container transportation application based on blockchain technology. In: Proceedings of the Asia-Pacific Conference on Intelligent Medical 2018 & International Conference on Transportation and Traffic Engineering 2018, pp. 277–281 (Year)
26. Choi, T.-M.: Blockchain-technology-supported platforms for diamond authentication and certification in luxury supply chains. Transp. Res. Part E: Logist. Transp. Rev. **128**, 17–29 (2019)
27. Raghuram, G.: An overview of the trucking sector in India: significance and structure (2015)
28. Pundhir, A., Shukla, A., Goel, A.D., Pundhir, P., Gupta, M.K., Parashar, P., Varshney, A.M.: Exploring unsafe sexual practices among truck drivers at Meerut District, India: a cross-sectional study. Afr. Health Sci. **21**, 547–556 (2021)
29. Parikh, J., Khedkar, G.: The impacts of diesel price increases on India's trucking industry. Int. Inst. Sustain. Dev. (2013)
30. Dadsena, K.K., Naikan, V.N., Sarmah, S.: A methodology for risk assessment and formulation of mitigation strategies for trucking industry. Int. J. Perform. Eng. **12** (2016)
31. Juntunen, S.-M.: Key performance Indicators of transportation category management: on-time delivery performance (2017)
32. Paniati, J.: 4th International Transportation Systems Performance Measurement Conference. Federal Highway Administration (2011)
33. Li, Y., Ouyang, K., Li, N., Rahmani, R., Yang, H., Pei, Y.: A blockchain-assisted intelligent transportation system promoting data services with privacy protection. Sensors **20**, 2483 (2020)
34. Garza-Reyes, J.A., Villarreal, B., Kumar, V., Molina Ruiz, P.: Lean and green in the transport and logistics sector–a case study of simultaneous deployment. Prod. Plan. Control **27**, 1221–1232 (2016)
35. Ho, W., Zheng, T., Yildiz, H., Talluri, S.: Supply chain risk management: a literature review. Int. J. Prod. Res. **53**, 5031–5069 (2015)
36. Rodrigues, V.S., Piecyk, M., Mason, R., Boenders, T.: The longer and heavier vehicle debate: a review of empirical evidence from Germany. Transp. Res. Part D: Transp. Environ. **40**, 114–131 (2015)
37. Sanchez Rodrigues, V., Stantchev, D., Potter, A., Naim, M., Whiteing, A.: Establishing a transport operation focused uncertainty model for the supply chain. Int. J. Phys. Distrib. Logist. Manag. **38**, 388–411 (2008)
38. Tacken, J., Sanchez Rodrigues, V., Mason, R.: Examining CO2e reduction within the German logistics sector. Int. J. Logist. Manag. **25**, 54–84 (2014)
39. Shams, K., Asgari, H., Jin, X.: Valuation of travel time reliability in freight transportation: a review and meta-analysis of stated preference studies. Transp. Res. Part A: Policy Pract. **102**, 228–243 (2017)
40. McKinnon, A.C.: Benchmarking road freight transport: review of a government-sponsored programme. Benchmarking: Int. J. **16**, 640–656 (2009)
41. Liu, R., Jiang, Z., Fung, R.Y., Chen, F., Liu, X.: Two-phase heuristic algorithms for full truck-loads multi-depot capacitated vehicle routing problem in carrier collaboration. Comput. Oper. Res. **37**, 950–959 (2010)
42. Radivojević, G., Gajović, V.: Supply chain risk modeling by AHP and Fuzzy AHP methods. J. Risk Res. **17**, 337–352 (2014)
43. Sanchez-Rodrigues, V., Potter, A., Naim, M.M.: The impact of logistics uncertainty on sustainable transport operations. Int. J. Phys. Distrib. Logist. Manag. **40**, 61–83 (2010)

44. Giannakis, M., Papadopoulos, T.: Supply chain sustainability: a risk management approach. Int. J. Prod. Econ. **171**, 455–470 (2016)
45. Schneider, K.C., Johnson, J.C.: Professionalism and ethical standards among salespeople in a deregulated environment: a case study of the trucking industry. J. Pers. Sell. Sales Manag. **12**, 33–43 (1992)
46. Arbués, P., Baños, J.F.: A dynamic approach to road freight flows modeling in Spain. Transportation **43**, 549–564 (2016)
47. Demir, E., Huang, Y., Scholts, S., Van Woensel, T.: A selected review on the negative externalities of the freight transportation: modeling and pricing. Transp. Res. Part E: Logist. Transp. Rev. **77**, 95–114 (2015)
48. Rodrigues, V.S., Piecyk, M., Potter, A., McKinnon, A., Naim, M., Edwards, J.: Assessing the application of focus groups as a method for collecting data in logistics. Int J Log Res Appl **13**, 75–94 (2010)
49. Koh, L., Dolgui, A., Sarkis, J.: Blockchain in Transport and Logistics–Paradigms and Transitions, vol. 58, pp. 2054–2062. Taylor & Francis (2020)
50. Galvez, J.F., Mejuto, J.C., Simal-Gandara, J.: Future challenges on the use of blockchain for food traceability analysis. TrAC, Trends Anal. Chem. **107**, 222–232 (2018)
51. Proctor, G.: NCFRP Report 10: Performance Measures for Freight Transportation. Transportation Research Board of the National Academies, Washington, DC (2011)
52. Tob-Ogu, A., Kumar, N., Cullen, J., Ballantyne, E.E.: Sustainability intervention mechanisms for managing road freight transport externalities: a systematic literature review. Sustainability **10**, 1923 (2018)
53. Wang, S., Qu, X.: Blockchain applications in shipping, transportation, logistics, and supply chain. In: Smart Transportation Systems 2019, pp. 225–231. Springer (2019)
54. Cole, R., Stevenson, M., Aitken, J.: Blockchain technology: implications for operations and supply chain management. Supply Chain. Manag.: Int. J. (2019)
55. Chaudhary, R., Jindal, A., Aujla, G.S., Aggarwal, S., Kumar, N., Choo, K.-K.R.: BEST: Blockchain-based secure energy trading in SDN-enabled intelligent transportation system. Comput. Secur. **85**, 288–299 (2019)
56. Tsai, S.-B., Gupta, B., Agrawal, D.P., Wu, W., Liu, A.: Recent advances in intelligent transportation systems for cloud-enabled smart cities. J. Adv. Transp. 2021, (2021)
57. Panda, S.K., Mohammad, G.B., Nandan Mohanty, S., Sahoo, S.: Smart contract-based land registry system to reduce frauds and time delay. Secur. Priv. **2021**, e172 (2021). https://doi.org/10.1002/spy2.172
58. Panda, S.K., Satapathy, S.C.: Drug traceability and transparency in medical supply chain using blockchain for easing the process and creating trust between stakeholders and consumers. Pers. Ubiquit. Comput. (2021). https://doi.org/10.1007/s00779-021-01588-3
59. Niveditha, V.R., Sekaran, K., Amandeep Singh, K., Panda, S.K.: Effective prediction of bitcoin price using wolf search algorithm and bidirectional LSTM on internet of things data. Int. J. Syst. Syst. Eng. **11**(3–4), 224–236
60. Sathya, A.R., Panda, S.K., Hanumanthakari, S.: Enabling smart education system using blockchain technology. In: Panda, S.K., Jena, A.K., Swain, S.K., Satapathy, S.C. (eds.) Blockchain Technology: Applications and Challenges. Intelligent Systems Reference Library, vol. 203. Springer, Cham (2021). https://doi.org/10.1007/978-3-030-69395-4_10
61. Lokre, S.S., Naman, V., Priya, S., Panda, S.K.: Gun tracking system using blockchain technology. In: Panda, S.K., Jena, A.K., Swain, S.K., Satapathy, S.C. (eds.) Blockchain Technology: Applications and Challenges. Intelligent Systems Reference Library, vol. 203. Springer, Cham (2021). https://doi.org/10.1007/978-3-030-69395-4_16
62. Panda, S.K., Daliyet, S.P., Lokre, S.S., Naman, V.: Distributed Ledger Technology in the Construction Industry Using Corda. The New Advanced Society: Artificial Intelligence and Industrial Internet of Things Paradigm. https://doi.org/10.1002/9781119884392.ch2
63. Panda, S.K., Satapathy, S.C.: An investigation into smart contract deployment on ethereum platform using Web3.js and solidity using blockchain. In: Bhateja, V., Satapathy, S.C., Travieso-González, C.M., Aradhya, V.N.M. (eds.) Data Engineering and Intelligent Computing.

Advances in Intelligent Systems and Computing, vol. 1. Springer, Singapore (2021). https://doi.org/10.1007/978-981-16-0171-2_52
64. Panda, S.K., Rao, D.C., Satapathy, S.C.: An investigation into the usability of blockchain technology in internet of things. In: Bhateja, V., Satapathy, S.C., Travieso-González, C.M., Aradhya, V.N.M. (eds.) Data Engineering and Intelligent Computing. Advances in Intelligent Systems and Computing, vol. 1. Springer, Singapore (2021). https://doi.org/10.1007/978-981-16-0171-2_53
65. Panda, S.K., Dash, S.P., Jena, A.K.: Optimization of block query response using evolutionary algorithm. In: Bhateja, V., Satapathy, S.C., Travieso-González, C.M., Aradhya, V.N.M. (eds.) Data Engineering and Intelligent Computing. Advances in Intelligent Systems and Computing, vol. 1. Springer, Singapore (2021). https://doi.org/10.1007/978-981-16-0171-2_54
66. Nanda, S.K., Panda, S.K., Das, M., Satapathy, S.C. (2022). Automating vehicle insurance process using smart contract and ethereum. In: Chakravarthy, V.V.S.S.S., Flores-Fuentes, W., Bhateja, V., Biswal, B. (eds.) Advances in Micro-Electronics, Embedded Systems and IoT. Lecture Notes in Electrical Engineering, vol. 838. Springer, Singapore. https://doi.org/10.1007/978-981-16-8550-7_23
67. Panda, S.K., Elngar, A.A., Balas, V.E., Kayed, M. (Eds.): Bitcoin and Blockchain: History and Current Applications, 1st edn. CRC Press (2020). https://doi.org/10.1201/9781003032588
68. Panda, S.K., Jena, A.K., Swain, S.K., Satapathy, S.C. (Eds.) Blockchain Technology: Applications and Challenges. Intelligent Systems Reference Library. Springer. https://doi.org/10.1007/978-3-030-69395-4
69. Varaprasada Rao, K., Panda, S.K.: A design model of copyright protection system based on distributed ledger technology. In: Satapathy, S.C., Lin, J.CW., Wee, L.K., Bhateja, V., Rajesh, T.M. (eds.) Computer Communication, Networking and IoT. Lecture Notes in Networks and Systems, vol. 459. Springer, Singapore (2023). https://doi.org/10.1007/978-981-19-1976-3_17
70. Varaprasada Rao, K., Panda, S.K.: Secure electronic voting (E-voting) system based on blockchain on various platforms. In: Satapathy, S.C., Lin, J.CW., Wee, L.K., Bhateja, V., Rajesh, T.M. (eds.) Computer Communication, Networking and IoT. Lecture Notes in Networks and Systems, vol. 459. Springer, Singapore (2023). https://doi.org/10.1007/978-981-19-1976-3_18

Chapter 11
MediBlock: A Pervasive Way to Create Healthcare Value in Secured Manner for Personalized Care

Vaidik Bhatt and Samyadip Chakraborty

Abstract These days, thanks to the fast development of new technologies like the internet of things and wearable gadgets, the healthcare industry is expanding at a breakneck pace. These gadgets saw widespread use as a means of facilitating remote patient monitoring. On the other end blockchain is a highly acclaimed technology that offers a wide variety of interesting characteristics, provides more trust and security. Information pertaining to patients' health is subject to stringent regulatory and safety regulations. The implementation of a distributed architecture is required in order to successfully overcome these problems and remain in compliance with applicable security standards. Blockchain has attracted a lot of attention as a potentially useful advanced technology because of its decentralized design and the security benefits it promises to provide for Internet of Things-based applications. These considerations served as the impetus for this research, which presents Medi-Block, a protected healthcare system that combines Blockchain technology with the Internet of Things. It is possible to monitor patients remotely using this method. The study proposes an extra security level in a high value creation process of personalized healthcare through remote patient monitoring. This chapter proposes the framework on how patient layer, physician layer and technological layer interacts to provide better value of healthcare with the maximum privacy and security.

Keywords Healthcare · Internet of Things · Blockchain · MediBlock · Security · Personalized care

V. Bhatt (✉) · S. Chakraborty
Department of Operations & IT, ICFAI Business School (IBS) Hyderabad, IFHE University, Hyderabad, India
e-mail: vaidik.bhatt@ibsindia.org

S. Chakraborty
e-mail: samyadip@ibsindia.org

S. K. Panda et al. (eds.), *Recent Advances in Blockchain Technology*,
Intelligent Systems Reference Library 237, https://doi.org/10.1007/978-3-031-22835-3_11

11.1 Introduction

The provision of medical care is an essential component of human civilization. Healthcare is the process of diagnosing, treating, and preventing injury or illness, whether it be physical or mental to improve or maintain health [1]. This process is carried out by health service providers, such as doctors, support staff, hospital systems, and so on. The development of healthcare sector is crucial for the growth and development of sustainable communities and thereby country. As healthcare sector showed its progressive nature in newer technology adoption, the entire service landscape observed a change. Technology helped healthcare sector to enhance its reach towards last mile [2]. Healthcare observed this paradigm shift due to adoption of various technologies like telehealth, teleconsulting, mHealth and IoMT (Internet of Medical Things). These technologies helped to change the focus from merely a cure of illness towards overall care and wellbeing of patients and community at a large [3], and thereby helping to create sustainable community for a developing nation [4, 5].

Today's healthcare service encounter process is qualitatively different from the it was a decade ago [6]. Adoption of state-of-the-art technologies in healthcare empowered patients to create value for themselves [7], by having access to own healthcare and vitals data [8], now patients are also participating in healthcare value creation for themselves. Technology helps them to co-create the value from the healthcare services obtained for themselves [9], now they are involved in joint decision making with the physician [10] and enhancing their engagement in healthcare service delivery [11]. Now patients show much more compliance towards the treatment and temporal protocols placed by the physicians [12], as it was a joint decision [13]. Now patients can connect to their physicians irrespective of place and time when they require healthcare services [14–16].

On one side, the technology helps to provide a ubiquitous and sustainable care to the society, where every citizen can have an access towards quality and affordable care, on the other side complexity of a healthcare sector makes it difficult to provide healthcare services due to involvement of multiple connections and network actors [1]. Healthcare sector is much more complex and distinguished than it looks. Healthcare sector involves multiple organizations like hospitals, clinics, nursing home, pharmacies, pharmaceutical companies, ambulatory centers, radiology and pathology labs, critical care units etc. These multiple organizations attract multiple parties like doctors, nurses, pharmacist, radiologist, pathologist, nutritionist etc. However, all these attributes do not carry any intrinsic value as they are just value propositions [17, 18]. As per the principles of service dominant logic, the value is co-created between the service provider and receiver. In this case, to obtain the value from healthcare services exercised, patients have to be treated or the patients' health should be maintained in good health [19]. It is important that healthcare sector should work in an orchestrated manner to provide better outcomes and value [20, 21].

Although, the technological innovations can help in creating an orchestrated network, where various network partners acts together at just correct point in time and converts lower degree underachieving process to higher degree over achieving process to create better and socially sustainable healthcare systems [20–22], adoption of such technologies are negatively affected by issues like privacy [23–26], Trust [27–29], data integrity [1, 30] and Technology Anxiety [31]. These issues regarding the trust, data privacy and integrity impacts the healthcare service delivery on a larger extent as users does not want that their personal health information is revealed. However, the decentralized ledger technology known as blockchain is already here and ready to take the internet to a whole new level by removing friction from three crucial areas: control, trust, and value [32].

11.2 MediBlock: A Possible Solution?

The blockchain is a kind of distributed ledger technology that may be used for both the storing and transmission of transactions. The information is kept in a ledger that is constructed up of several blocks. To create a chain of blocks, each block is connected to the one that came before it. A peer-to-peer network is used to assure the delivery of data. As a result, the blockchain is a distributed ledger that is shared in a way that is both safe and decentralized [33]. After the idea given by [33], researchers across the globe are in a heed of exploring the multiple usage of blockchain technologies in various fields.

Reference [34] investigated the potential impact that blockchain technology could have on patient-centered care delivery by analysing the ecosystem of blockchain-based solutions that are currently being developed for key stakeholders. These stakeholders include patients, pharmaceutical companies, hospital systems, and insurance providers. Reference [35] presented a system for exchanging electronic medical records which are seamlessly integrated across the network actors based on the blockchain network. Reference [36] studied how blockchain technology might help to value creation in the healthcare industry by investigating 33 firms selected from a sample of 2404 businesses that were formed through professional incubators and accelerators.

Reference [37] developed an application that is built on blockchain and semantic web technologies that enables users to reach to an agreement with a Health Insurance Organization on their health status and their reimbursement in the event of an accident or illness. Reference [38] provided a paradigm that is powered by blockchain technology that makes it possible for patients and healthcare providers to share their medical information in a way that is advantageous to both parties. Whereas Reference [39] provided a scoping review of blockchain technology in healthcare and concluded that because there is a dearth of research directing healthcare organizations toward the implementation of blockchain design types that are public, private, and hybrid, it will be essential for future work to provide data relating to variations that exist among the various blockchain designs.

A MediBlock is a trustworthy healthcare platform that incorporates Internet of Things and Blockchain technology. Reference [40] focused on the three trust objectives are driven by three important antecedents: the blockchain certificate, structural assurance, and member credibility. The technology was developed to provide remote patient monitoring, which is particularly useful in the case of chronic conditions that call for ongoing observation [30]. Reference [41] discovered that the combination of blockchain technology with big data would result in the provision of real-time data analytics, which would make it possible for providers to settle transactions in a more expedient manner. Reference [42] addresses the concerns with the authentication of healthcare IoT data and the identification of IoT devices. Due to its immense capability of confidentiality, integrity and authenticity it keeps the data safe as well as real time updated. Now each of the service provider have access to the data which are the concern area for them.

The majority of applications for blockchain technology are found in the fields of healthcare and Internet of Things (IoT), specifically for the purpose of enhancing data management operations and, more specifically, data security, which includes data integrity, access control, and the preservation of privacy [43]. A road plan for a blockchain-enabled, decentralized personal health data ecosystem was developed. The goal of the roadmap was to facilitate deep learning for the purposes of medication discovery, biomarker creation, and preventive healthcare [44]. Reference [45] also discussed the implementation of blockchain-based security solutions in addition to addressing the security needs of Internet of Things (IoT)-enabled smart healthcare systems [46, 47].

Owing to the fact that its ledger is decentralized, it has a source provenance, and it is tamper-proof, communities of practices are focusing on Medi-Block to provide quality healthcare services while keeping the data safe and temper proof [48]. In this view, [1] provided the use of Blockchain Technology to have a significant influence on policy-based decision making for the purpose of ameliorating the predicament of the Indian healthcare ecosystem.

Out of many more possible usages of blockchain in healthcare, this chapter concentrates on providing personal care and recommendations to the patients using the state-off-the-art blockchain technology.

11.3 Building MediBlock Network

A secure MediBlock network securely transfer the medical data of the patients to the physicians and other allied network actors and again provide their recommendations to the patients for their well-being. In a way, the secured blockchain network closes a feedback loop in the service encounter process which creates value for services from the lenses of patients [49–51], and healthcare service provider network actors [7, 52–64] by creating the temporal displacement of care [12]. This secure MediBlock can be divided in to three different components of patient layer, healthcare service provider layer and technology layer as shown in Fig. 11.1.

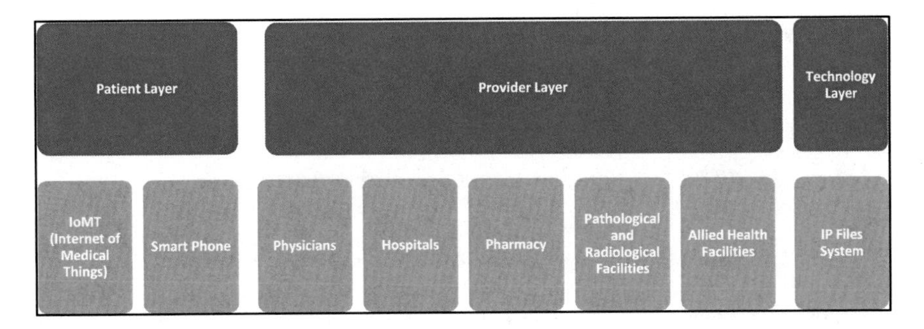

Fig. 11.1 Layers of MediBlock chain

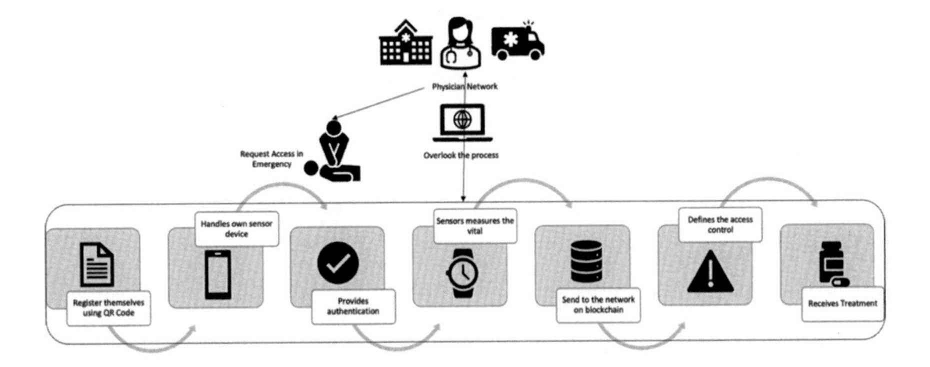

Fig. 11.2 Overview of treatment process

11.3.1 Patient Layer

Every patient has their own individual set of Internet of Things (IoT) medical gadgets as well as electronic wearables that have integrated sensors. These sensors are able to record a variety of readings of vital indicators, as well as keep track of a person's physical activity and sleep habits, among other things [8]. As a result, the function of these devices is to gather patient health information and then transmit that information to the treating medical team through the patient's smartphone. The owner of each of these devices (the patient) registers it on the blockchain, and it is distinguished from other devices in the network by a pair of distinctive values consisting of its MAC address and the identification of its owner.

The smartphone acts as a connection point between Internet of Things (IoT) medical equipment and the medical staff. Therefore, the node is what gives other Internet of Things devices permission to access the blockchain. Because of its restricted capabilities, the smartphone won't be able to store the whole block chain; rather, it will only have access to it through a decentralized application (DAPP). When the patient establishes his profile, the DAPP also generates a Blockchain account for him. Additionally, a pair of keys that will serve as the patient's unique identification are produced at this time.

11.3.2 Healthcare Service Provider/Physician Layer

This component consists of medical professionals, hospitals, pharmaceutical labs, and organizations focused on public health. Through the use of a Blockchain network, these organizations are linked with patients in order to have access to the patients' health data. Through their desktop computers or mobile devices, doctors are able to participate in the Blockchain network as light nodes. The function of full nodes in the Blockchain, which may keep a copy of the Blockchain and participate in the consensus process, is played by hospitals. The other entities are allowed to hold a copy of the Blockchain, but they are not permitted to take part in the process of reaching consensus. The information gathered from the patients will be utilized for purposes like research, remote monitoring, and analysis.

11.3.3 Technology Layer

On the technology side, the encrypted health data is stored in an off-chain database that is built on top of the IP File System (IPFS), which is a peer-to-peer distributed file system. IPFS was chosen as our method of data storage rather than Blockchain for two reasons. On the one hand, this is due to the fact that IPFS is capable of managing vast amounts of data that are produced by a big number of devices. The storage of such a large volume of data in the Blockchain has the potential to alter its growth and will need the use of special full nodes. As a result, the Blockchain is exclusively used for purposes of access control and protecting the integrity of data. On the other hand, even if the data were encrypted, it would not be a good idea to keep sensitive information on the blockchain for practical reasons. Since it is common knowledge that data are saved in the Blockchain indefinitely, it follows that in the event that the encrypted system is compromised at some point in the future, all nodes will be able to decode the information and get access to it.

11.4 Building a Security for a MediBlock Network

In most cases, the safety of a block chain is assured by a combination of various factors, the most important of which are cryptography, immutability, replication, traceability, and consensus method. In addition to these characteristics, the MediBlock system makes use of smart contracts for the purpose of access control and data encryption. Because of the security features and data immutability and state-of-the-art device encryption features, patients' data are secured in close as well as open network of service providers including physicians, nurses, ward boys, pharmacists, lab technicians across the network of clinics, hospitals, radiology and pathological facilities, ambulatory centers and other allied facilities.

11.4.1 Actual Working Mechanism of MediBlock Network

After installing the decentralized application (DAPP), a patient will be unable to build his identification until he has received confirmation from his attending physician. Therefore, the only individual who is authorized to register his patients by producing a QR code for each of them is the attending physician. Within this QR code, the address of the attending physician is included in addition to some other information on the Blockchain network, such as its Medi-Block ID. The patient may next establish his account by creating a unique identity after scanning the QR code provided. When the Blockchain network is finally accessed, the patient will be able to carry out a variety of duties by using a variety of different DAPP interfaces. In addition to this, the smart contract is the engine that runs the back end for DAPP and provides the many features.

Each individual patient is required to first register all of his or her medical gadgets in order to gather various data. A one-of-a-kind combination of variables, including the MAC address of the device and the location of its owner, will be used to identify each individual device on the blockchain. This ensures that the only person who can add a device is the owner, and only data from devices that have been registered may be added. Patients may thus be protected from malicious gadgets that could transmit incorrect data on their behalf. With this system implementation each patient can exercise control over the access to his or her personal data by authorizing access to some healthcare organizations while denying access to others.

On one side, where patients can share their own data to the physicians and a group of healthcare service providers in a network, physicians can also provide an access to the healthcare network of their own. Physicians and their own network of healthcare providers have access to the health and real-time vital records of the patients to know the current health status precisely. This access ensures better risk-assessment of patients' condition from the providers' side with more transparency and security. On one side, where the network of healthcare providers has access to the regular patients, during the event of an emergency, physician should make a request for access while providing healthcare services to a patient who does not belong to the own network. Here, the network of healthcare providers will have temporary access to the data that they need for the treatment. However, in this case still patient oversees what data needs to be shared with the other network of healthcare providers (Fig. 11.2).

11.5 Conclusion

Medi-Block enables a seamless communication between patients and a network of healthcare providers with the enhanced security. Here, the primary role of a block chain technology is to provide enhanced security features to the existing framework of sensor based IoT enabled healthcare network access. The technology helps in reaching the goals of customer-dominant logic [19] as patients control data regarding their vitals. On the other end, patients provide the access to their healthcare data to the group of physicians and other allied network actors who provides external support to the patients. The technology also plays a pivotal role in order to achieve the service dominant logic. As the technology is helping physicians to provide better care with the current available resources, healthcare network actors can create better value from the value propositions. In short, the blockchain technology used for the personalized healthcare creates a secure ecosystem for the network actors to provide better care and patients receive better treatment.

References

1. Shukla, R.G., Agarwal, A., Shekhar, V.: Leveraging blockchain technology for Indian healthcare system: an assessment using value-focused thinking approach. J. High Technol. Manage. Res. **32** (2021). https://doi.org/10.1016/j.hitech.2021.100415
2. Chakraborty, S., Bhatt, V., Chakravorty, T.: Is telemedicine best alternative to reaching last mile: Investigation in the context of rural India. Ind. J. Public Health Res. Dev. **9** (2018). https://doi.org/10.5958/0976-5506.2018.01341.4
3. de Blok, C., Meijboom, B., Luijkx, K., et al.: Interfaces in service modularity: a typology developed in modular health care provision. J. Oper. Manag. **32** (2014). https://doi.org/10.1016/j.jom.2014.03.001
4. Chauhan, A., Jakhar, S.K., Jabbour, C.J.C.: Implications for sustainable healthcare operations in embracing telemedicine services during a pandemic. Technol. Forecast. Soc. Chang. **176** (2022). https://doi.org/10.1016/j.techfore.2021.121462
5. Sims, J.M.: Communities of practice: telemedicine and online medical communities. Technol. Forecast. Soc. Chang. **126**, 53–63 (2018). https://doi.org/10.1016/j.techfore.2016.08.030
6. Prahalad, C.K., Ramaswamy, V.: Co-opting customer competence. Harv. Bus. Rev. (2000)
7. Joiner, K., Lusch, R.: Evolving to a new service-dominant logic for health care. Innov. Entrepr. Health **25** (2016). https://doi.org/10.2147/ieh.s93473
8. Talukder, M.S., Sorwar, G., Bao, Y., et al.: Predicting antecedents of wearable healthcare technology acceptance by elderly: a combined SEM-Neural Network approach. Technol. Forecast. Soc. Chang. **150**, 119793 (2020). https://doi.org/10.1016/j.techfore.2019.119793
9. McColl-Kennedy, J.R., Vargo, S.L., Dagger, T.S., et al.: Health care customer value cocreation practice styles. J. Serv. Res. **15**, 370–389 (2012). https://doi.org/10.1177/1094670512442806
10. Bhatt, V., Chakraborty, S.: Enhancing service engagement and collaborative decision making through wearable device adoption (2021)
11. Bhatt, V., Chakraborty, S.: Improving service engagement in healthcare through internet of things based healthcare systems. J. Sci. Technol. Policy Manage. (2021). https://doi.org/10.1108/JSTPM-03-2021-0040
12. Thompson, S., Whitaker, J., Kohli, R., Jones, C.: Chronic disease management: how IT and analytics create healthcare value through the temporal displacement of care. MIS Q. **44**, 227–256 (2020). https://doi.org/10.25300/misq/2020/15085

13. Khuntia, J., Yim, D., Tanniru, M., Lim, S.: Patient empowerment and engagement with a health infomediary. Health Policy Technol. **6**, 40–50 (2017). https://doi.org/10.1016/j.hlpt.2016.11.003
14. Ghose, A., Sinha, P., Bhaumik, C., et al.: UbiHeld: ubiquitous healthcare monitoring system for elderly and chronic patients. dl.acm.org
15. Milošević, M., Shrove, M.T., Jovanov, E.: Applications of smartphones for ubiquitous health monitoring and wellbeing management. JITA—J. Inf. Technol. Appl. (Banja Luka)—APEIRON (2011). https://doi.org/10.7251/jit1101007m
16. Plageras, A.P., Psannis, K.E., Ishibashi, Y., Kim, B.G.: IoT-based surveillance system for ubiquitous healthcare. In: IECON Proceedings (Industrial Electronics Conference) (2016)
17. Vargo, S.L., Lusch, R.F.: Service-dominant logic 2025. Int. J. Res. Mark. **34**, 46–67 (2017). https://doi.org/10.1016/j.ijresmar.2016.11.001
18. Vargo, S.L., Lusch, R.F.: Service-dominant logic: continuing the evolution. J. Acad. Mark. Sci. **36**, 1–10 (2008). https://doi.org/10.1007/s11747-007-0069-6
19. Heinonen, K., Strandvik, T., Mickelsson, K.J., et al.: A customer-dominant logic of service. J. Serv. Manage. **21**,(2010). https://doi.org/10.1108/09564231011066088
20. Chakraborty, S., Bhatt, V., Chakravorty, T.: Impact of digital technology adoption on care service orchestration, agility and responsiveness. Int. J. Sci. Technol. Res. **9**, 4581–4586 (2020)
21. Chakraborty, S., Bhatt, V., Chakravorty, T., Chakraborty, K.: Analysis of digital technologies as antecedent to care service transparency and orchestration. Technol. Soc. **65**, 101568 (2021). https://doi.org/10.1016/j.techsoc.2021.101568
22. Bhatt, V., Chakraborty, S.: Real-time healthcare monitoring using smart systems: a step towards healthcare service orchestration Smart systems for futuristic healthcare. In: Proceedings—International Conference on Artificial Intelligence and Smart Systems, ICAIS 2021. Institute of Electrical and Electronics Engineers Inc., pp. 772–777 (2021)
23. Tran, C.D., Nguyen, T.T.: Health vs. privacy? the risk-risk tradeoff in using COVID-19 contact-tracing apps. Technol. Soc. **67** (2021). https://doi.org/10.1016/j.techsoc.2021.101755
24. Kim, D., Park, K., Park, Y., Ahn, J.H.: Willingness to provide personal information: perspective of privacy calculus in IoT services. Comput. Hum. Behav. **92**, 273–281 (2019). https://doi.org/10.1016/j.chb.2018.11.022
25. Xu, Z.: An empirical study of patients' privacy concerns for health informatics as a service. Technol. Forecast. Soc. Chang. **143**, 297–306 (2019). https://doi.org/10.1016/j.techfore.2019.01.018
26. Dhagarra, D., Goswami, M., Kumar, G.: Impact of trust and privacy concerns on technology acceptance in healthcare: an Indian perspective. Int. J. Med. Infor. **141** (2020). https://doi.org/10.1016/j.ijmedinf.2020.104164
27. Baudier, P., Kondrateva, G., Ammi, C., et al.: Digital transformation of healthcare during the COVID-19 pandemic: patients' teleconsultation acceptance and trusting beliefs. Technovation **102547** (2022). https://doi.org/10.1016/j.technovation.2022.102547
28. Caceres, R.C., Paparoidamis, N.G.: Service quality, relationship satisfaction, trust, commitment and business-to-business loyalty. Eur. J. Mark. **41**, 836–867 (2007). https://doi.org/10.1108/03090560710752429
29. Li, X., Hess, T.J., Valacich, J.S.: Why do we trust new technology? a study of initial trust formation with organizational information systems. J. Strateg. Inf. Syst. (2008). https://doi.org/10.1016/j.jsis.2008.01.001
30. Azbeg, K., Ouchetto, O., Jai Andaloussi, S.: BlockMedCare: a healthcare system based on IoT, Blockchain and IPFS for data management security. Egypt. Inf. J. (2022). https://doi.org/10.1016/j.eij.2022.02.004
31. Bhatt, V., Chakraborty, S.: Intrinsic antecedents to mHealth adoption intention: an SEM-ANN approach. Int. J. Electron. Govern. Res. **18** (2022). https://doi.org/10.4018/IJEGR.298139
32. Cunha, J., Duarte, R., Guimarães, T., et al.: Blockchain analytics in healthcare: an overview. Procedia Comput. Sci. **201**, 708–713 (2022). https://doi.org/10.1016/j.procs.2022.03.095
33. Nakamoto, S.: Bitcoin: a peer-to-peer electronic cash system. Consulted **1–9** (2008). https://doi.org/10.1007/s10838-008-9062-0stem.Consulted

34. Sharma, L., Olson, J., Guha, A., McDougal, L.: How blockchain will transform the healthcare ecosystem. Bus. Horiz. **64**, 673–682 (2021). https://doi.org/10.1016/j.bushor.2021.02.019
35. Cerchione, R., Centobelli, P., Riccio, E., et al.: Blockchain's coming to hospital to digitalize healthcare services: designing a distributed electronic health record ecosystem. Technovation (2022). https://doi.org/10.1016/j.technovation.2022.102480
36. Spanò, R., Massaro, M., Iacuzzi, S.: Blockchain for value creation in the healthcare sector. Technovation (2021). https://doi.org/10.1016/j.technovation.2021.102440
37. Chondrogiannis, E., Andronikou, V., Karanastasis, E., et al.: Using blockchain and semantic web technologies for the implementation of smart contracts between individuals and health insurance organizations. Blockchain: Res. Appl. **3**, 100049 (2022). https://doi.org/10.1016/j.bcra.2021.100049
38. Sadeghib, R.J.K., Prybutok, V.R., Sauser, B.: Theoretical and practical applications of blockchain in healthcare information management. Inf. Manage. **59** (2022). https://doi.org/10.1016/j.im.2022.103649
39. Abu-elezz, I., Hassan, A., Nazeemudeen, A., et al.: The benefits and threats of blockchain technology in healthcare: a scoping review. Int. J. Med. Inf. **142** (2020)
40. Shao, Z., Zhang, L., Brown, S.A., Zhao, T.: Understanding users' trust transfer mechanism in a blockchain-enabled platform: a mixed methods study. Decis. Support Syst. **155** (2022). https://doi.org/10.1016/j.dss.2021.113716
41. Muheidat, F., Patel, D., Tammisetty, S., et al.: Emerging concepts using blockchain and big data. In: Procedia Computer Science. Elsevier B.V., pp. 15–22 (2021)
42. Shukla, S., Thakur, S., Hussain, S., et al.: Identification and authentication in healthcare internet-of-things using integrated fog computing based blockchain model. Internet of Things (Netherlands) **15** (2021). https://doi.org/10.1016/j.iot.2021.100422
43. Adere, E.M.: Blockchain in healthcare and IoT: a systematic literature review. Array **14** (2022)
44. Cheng, A., Guan, Q., Su, Y., et al.: Integration of machine learning and blockchain technology in the healthcare field: a literature review and implications for cancer care. Asia Pac. J. Oncol. Nurs. **8**, 720–724 (2021)
45. Tariq, N., Qamar, A., Asim, M., Khan, F.A.: Blockchain and smart healthcare security: a survey. In: Procedia Computer Science. Elsevier B.V., pp. 615–620 (2020)
46. Varaprasada Rao, K., Panda, S.K. (2023). A design model of copyright protection system based on distributed ledger technology. In: Satapathy, S.C., Lin, J.C.W., Wee, L.K., Bhateja, V., Rajesh, T.M. (eds.) Computer Communication, Networking and IoT. Lecture Notes in Networks and Systems, vol. 459. Springer, Singapore. https://doi.org/10.1007/978-981-19-1976-3_17
47. Varaprasada Rao, K., Panda, S.K.: Secure electronic voting (E-voting) system based on blockchain on various platforms. In: Satapathy, S.C., Lin, J.C.W., Wee, L.K., Bhateja, V., Rajesh, T.M. (eds.) Computer Communication, Networking and IoT. Lecture Notes in Networks and Systems, vol. 459. Springer, Singapore (2023) https://doi.org/10.1007/978-981-19-1976-3_18
48. Rahman, M.S., Islam, M.A., Uddin, M.A., Stea, G.: A survey of blockchain-based IoT eHealthcare: applications, research issues, and challenges. Internet of Things **19**, 100551 (2022). https://doi.org/10.1016/j.iot.2022.100551
49. Heinonen, K., Strandvik, T.: Customer-dominant logic: foundations and implications. J. Serv. Mark. **29** (2015). https://doi.org/10.1108/JSM-02-2015-0096
50. Seppänen, K., Huiskonen, J., Koivuniemi, J., Karppinen, H.: Revealing customer dominant logic in healthcare services (2017)
51. Cheung, F.Y.M., To, W.M.: A customer-dominant logic on service recovery and customer satisfaction. Manag. Decis. **54** (2016). https://doi.org/10.1108/MD-03-2016-0165
52. Lusch, R.F., Vargo, S.L., O'Brien, M.: Competing through service: insights from service-dominant logic. J. Retail. **83** (2007). https://doi.org/10.1016/j.jretai.2006.10.002
53. Panda, S.K., Mohammad, G.B., Nandan Mohanty, S., Sahoo, S.: Smart contract-based land registry system to reduce frauds and time delay. Secur. Priv. **e172** (2021). https://doi.org/10.1002/spy2.172

54. Panda, S.K., Satapathy, S.C.: Drug traceability and transparency in medical supply chain using blockchain for easing the process and creating trust between stakeholders and consumers. Pers. Ubiquit. Comput. (2021). https://doi.org/10.1007/s00779-021-01588-3
55. Niveditha, V.R., Sekaran, K., Amandeep Singh, K., Panda, S.K.: Effective prediction of bitcoin price using wolf search algorithm and bidirectional LSTM on internet of things data. Int. J. Syst. Syst. Eng. **11**(3–4), 224–236
56. Sathya, A.R., Panda, S.K., Hanumanthakari, S.: Enabling smart education system using blockchain technology. In: Panda, S.K., Jena, A.K., Swain, S.K., Satapathy, S.C. (eds.) Blockchain Technology: Applications and Challenges. Intelligent Systems Reference Library, vol. 203. Springer, Cham (2021). https://doi.org/10.1007/978-3-030-69395-4_10
57. Lokre, S.S., Naman, V., Priya, S., Panda, S.K.: Gun tracking system using blockchain technology. In: Panda, S.K., Jena, A.K., Swain, S.K., Satapathy, S.C. (eds.) Blockchain Technology: Applications and Challenges. Intelligent Systems Reference Library, vol. 203. Springer, Cham (2021). https://doi.org/10.1007/978-3-030-69395-4_16
58. Panda, S.K., Daliyet, S.P., Lokre, S.S., Naman, V.: Distributed ledger technology in the construction industry using corda. New Adv. Soc. Artif. Intell. Ind. Internet Things Paradigm. https://doi.org/10.1002/9781119884392.ch2
59. Panda, S.K., Satapathy, S.C.: An investigation into smart contract deployment on Ethereum platform using Web3.js and solidity using blockchain. In: Bhateja, V., Satapathy, S.C., Travieso-González, C.M., Aradhya, V.N.M. (eds.) Data Engineering and Intelligent Computing. Advances in Intelligent Systems and Computing, vol. 1. Springer, Singapore (2021). https://doi.org/10.1007/978-981-16-0171-2_52
60. Panda, S.K., Rao, D.C., Satapathy, S.C.: An investigation into the usability of blockchain technology in internet of things. In: Bhateja, V., Satapathy, S.C., Travieso-González, C.M., Aradhya, V.N.M. (eds.) Data Engineering and Intelligent Computing. Advances in Intelligent Systems and Computing, vol. 1. Springer, Singapore (2021). https://doi.org/10.1007/978-981-16-0171-2_53
61. Panda, S.K., Dash, S.P., Jena, A.K.: Optimization of block query response using evolutionary algorithm. In: Bhateja, V., Satapathy, S.C., Travieso-González, C.M., Aradhya, V.N.M. (eds.) Data Engineering and Intelligent Computing. Advances in Intelligent Systems and Computing, vol. 1. Springer, Singapore (2021). https://doi.org/10.1007/978-981-16-0171-2_54
62. Nanda, S.K., Panda, S.K., Das, M., Satapathy, S.C.: Automating vehicle insurance process using smart contract and Ethereum. In: Chakravarthy, V.V.S.S.S., Flores-Fuentes, W., Bhateja, V., Biswal, B. (eds.) Advances in Micro-Electronics, Embedded Systems and IoT. Lecture Notes in Electrical Engineering, vol. 838. Springer, Singapore (2022). https://doi.org/10.1007/978-981-16-8550-7_23
63. Panda, S.K., Elngar, A.A., Balas, V.E., & Kayed, M. (eds.) Bitcoin and Blockchain: History and Current Applications, 1st edn. CRC Press (2020). https://doi.org/10.1201/9781003032588
64. Blockchain Technology: Applications and Challenges. In: Panda, S.K., Jena, A.K., Swain, S.K., Satapathy, S.C. (eds.). Springer, Intelligent Systems Reference Library. https://doi.org/10.1007/978-3-030-69395-4

Chapter 12
Application of Blockchain Technology in Human Resource Management

Musarrat Shaheen, Sode Raghavendra, and Swati Alok

Blockchain will impact not only IT but every function. HR leaders who fail to do sufficient scenario planning and experiment with the technology accordingly risk significant long-term disintermediation.

—Gartner Senior, Director Analyst Matthias Graf [39]

There are some current challenges, but once the network effect kicks in, the growth of blockchain technology could be exponential.

—Jeff Mike, Vice President, and HR Research Leader for New York City-based consulting firm Bersin by Deloitte [41]

Abstract The application of blockchain in finance and other related areas is quite evident, but the role of blockchain in human resource management is not well discussed so far. The objective of this chapter is to examine the pertinent part of blockchain technology in different areas of human resource management, ranging from the recruitment and selection process for certificate verification and skill mapping to payroll processing and employee data protection. The chapter concludes with the challenges behind the implementation of blockchain technology in the area of human resource management.

Keywords Blockchain and HR processes · Cryptocurrency and employee compensation · Distributed ledger and employee records

M. Shaheen (✉) · S. Raghavendra
Department of OB & HRM, IBS Hyderabad, IFHE (Deemed-to-be-university), Hyderabad, India
e-mail: drmusarrat.shaheen@gmail.com

S. Raghavendra
e-mail: raghavendra.s@ibsindia.org

Department of OB & HRM, IFHE (Deemed-to-Be University), Hyderabad, India

S. Alok
Department of OB & HRM, BITS Pilani, Hyderabad, India
e-mail: swathi@hyderabad.bits-pilani.ac.in

S. K. Panda et al. (eds.), *Recent Advances in Blockchain Technology*,
Intelligent Systems Reference Library 237, https://doi.org/10.1007/978-3-031-22835-3_12

12.1 Introduction

The progression of civilization is the result of three eras—the agricultural era, the Industrial era, and the Digital era. Our globe is changing technologically, especially how business is done and, the value being added to it, the way work is generated and communicated in the organizations. Digital technology is becoming infrastructural, just like electricity. It narrows the lines between the physical, digital, and biological systems. Digital technology has both optimistic and pessimistic scenarios for our living conditions, and it has the potential to create new jobs, alter work and workplaces, and displace jobs. Various technologies have their impression in almost every aspect of the business such as automation which reduces the financial professional's burden by reducing the repetitive text which is used to detect errors and mismatches on insurance forms; and machine learning is used to detect scams with better precision and can also be utilized to predict machine breakdowns, deep learning algorithms automates quality control operations that are done and on a routine basis and so on. Technology falsification of customer expectations, product and service improvement, inter-organizational interactions, designing of works and jobs, and many more. This change and technological advancement made companies look forward to a big picture focusing on new strategies and policies, project plans, investments, development plans, etc.

Moreover, technology has a novel approach to managing the workforce and improving the organization's performance, creating many opportunities for HRM and impacting employee relations. Overall, technology dramatically affects the labor market as it allows work to be done at any time. For example, cloud technology will enable access to their organization's server and allows to work from any location. it also allows the organization to monitor employee behavior working in remote areas. Technology has improved the employee relations. Nevertheless, these digital labor platforms are dependent on HRM activities for workforce planning, balancing the supply and demand of contingent workers, and for job design to have the autonomy to work, for compensation, and giving rewards to workers to retain them and encourage the desired behavior, for performance management and many more. HRM acts as an instrument for leading digital platform ecosystems as it allows connecting independent entities within the ecosystem and ensuring the proper supply of workers according to the demand. At the same time, HR faces challenges in every aspect of managing workers. In terms of recruitment and selection, it is essential to recruit the workers in parallel, i.e., avoiding the mismatches of demand and supply of workers, as this may cause disruptions in the ecosystem causing either workers or clients to leave the ecosystem. But this doesn't mean that anyone can access the online platform; HR managers should fulfill the purpose of deciding which workers should be allowed to access the online marketplace and which do not, depending on the information from client ratings. Since gig workers are temporarily working and are not contractually obligated to the organization, they take no repercussions. They can move on to another gig, which is a big challenge for HR. In terms of training and development, Gig worker's status is ambiguous; it becomes challenging to train these gig workers. Justifying the training and development investments becomes difficult. Organizations must use technology to learn the availability of gig workers to avoid poor service to the clients, which is a huge task. In terms of performance management, the gig economy is different compared to standard employment; hence it has become a challenge for HR managers. There is a need for gig workers to feel included and as a part of the workforce. Always open discussions should be encouraged, along with feedback and recognition. In terms of compensation and benefits, Gig compensation is assumed to be a difficult or rather complex decision as the professional relationship varies between employer and employee. The jobs should be viewed as a set of tasks and skills instead of a job position as gig workers prefer to work in the short term and remain flexible. Different Onboarding strategy for gig workers needs to be established by leveraging automation throughout the process. Overall, every step of HR will be affected while working with the gig economy.

Hence, HR needs to inculcate technology in their work by automating HR-related decision-making algorithms in different areas like workforce planning, selection, appraisal, and compensation, analyzing the worker's data by using data process techniques, including data extraction, sorting, and cleaning that can support decision making. HR must try to use various algorithms to understand the relevant metrics like absenteeism, performance, or traits. There are times when HR managers have to take some prescriptions on whether or not to follow a particular decision which can be made easy by simulations and scenario-based techniques. Online digital platforms have moved up to a stage where investors are being attracted and retained, thus creating more gig economies and increased gig workers. This quick growth becomes a rapid problem when not managed properly as it becomes complex. This has made it challenging for HR to safeguard gig workers' professional safety and well-being. There exists a complexity between the market and corporate logic; market logic treats gig workers as a small part of business and to be easily negotiable, whereas corporate sense expects gig workers to scale up the platform creating a challenge for HR to have control over gig workers and supporting them simultaneously and giving them autonomy. With the gig economy, there have been dramatic changes in working conditions in terms of the inability of individuals to be able to influence the working environment, missing the institutional connectedness, etc. it becomes a challenge for HR practitioners with this flexible approach to working and they are required to think how the organization can be benefited along with ensuring the employee well-being and security. HR decision-makers are using data mining approaches, regression techniques, and machine learning algorithms for forecasting, like recruitment and selection, workforce planning and performance management.

Thus, managing employees in the gig economy, especially generation Z who prefers to work from any place, any time, needs use of transformative technology in HR practices. Various technologies such as Artificial Intelligence (AI), machine learning, big data, Internet of Things (IoT), mobile technology and biometrics are being introduced and inculcated into organization's human resource management approaches for performing various activities. The introduction of Electronic HRIS reduced the cost of HRM functions specially employee performance appraisal and evaluation of job applicants. Advanced technologies are altering international HR practices by putting in place e-recruitment, e training and e-competence management, thereby enhancing HRM service quality globally. Robot technologies eliminate repetitive and routine activities precisely without error, thereby offering HR practitioners to engage in strategy roles and freedom to use their skills effectively. Intelligent "beings" such as robot, AI revolutionized traditional HR functions whereas deep learning algorithms, smart objects, and IoT fostered more productive coordination and cooperation, especially across the international border. The next section mentions the prominent advanced technologies that were adopted by HR practitioners in various industries.

12.1.1 Use of Technology in Human Resource Management

HR was primarily a paper-based job which was a slow process for storing employee information and sharing the organizational data. It was in the 1980's when fax machines started using by organizations that reduced their time for sharing information. In 1987, PeopleSoft, a purpose-built HR management system was released, a key development in the world of ERP. In 1990, the ERP was scaled up by World Wide Web, and HR has started using it for recruitment among the talent pool. In 1996, a bundle of application was released for creating intranets. Further, in 1998, Application Tracking Systems came into existence that are been used by HR as a database for job applicants which makes life easier for hiring managers. Later in 2000s intranets were considered as a boon for internal communications that helped HR for sharing all kinds of documents such as contact information, training manuals, scheduling, health insurance etc. which are later been used by employees as self-serve and access them anytime from a centralized location. Within 5 years of intranet emergence 50% of organizations have started using employee self-service HR technology. 2010 was a first step towards chatbots and HR portals which made two-way communications possible in easier manner and in less time. Purpose built HR portals is another technological development and has become a starting point for automation. Thought live chats have improved the HR efficiency, it remained a tough task in handling the enquiries, time management, locating the required information etc. There emerged AI, in 2020, the advances in AI made machines that can respond to enquiries in no time and on a much greater scale. Ever since last 40 years' globe has experienced a wind of change brought by technology and every bit of it is being used in HR making it a necessity for the organizations at large. The prominent technologies that are used by HR practitioners worldwide are briefed in chronological order.

12.1.2 Database Management—Enterprise Resource Planning (ERP)

ERP can be understood as binding together different computer systems of a large organization that helps to integrate the flow of information and make every department to be on the same page. In the age of increasing competition among enterprises, HR managers have been very keen in attracting the best talent at the same time reducing the personnel costs in order to improve the competitiveness of enterprises. The personnel cost analysis helps managers to analyze the past, present and future personnel costs with which corporate cost analysis can be done through an integrated ERP environment. Based on the production needs managers are using HR systems in ERP so that they can develop organizational and staffing plans. HRM has been inculcated in the ERP system which in turn helps in recruitment management, Payroll accounting and Job management.

12.1.3 Data Mining—Descriptive and Predictive

Data mining and descriptive analytics techniques are automating the decision-making process of human resource functions. Personnel recruitment, Talent management and employee turnover analysis are some of the HR functions which immensely has used data mining techniques. Decision trees are used to explore data for predictions. For talent management, classification, association, clustering and prediction are the data mining techniques being used. For understanding the problems of turnover decision trees, logistic regression and neural networks are some of the techniques used from data mining. Predictive analysis is being adopted when the data sources are in line with the strategic business outcomes. The important elements of HRM such as recruitment and attrition are predicted using predictive scores. For example, by mining data on the relationship among antecedents like previous employers, previous work experience, university attended companies target certain universities for future hiring. Similarly, using historical turnover data of past several years, companies are able to predict which highly skilled employees may show high probability of leaving, thereby targeting them for retention and incentive programs.

12.1.4 Design Thinking and Human Resource Management

Design thinking is an approach for creative problem-solving technique that will change HR in its core. It helps the HR team to organize work, deliver value and find solutions. It is a human-centric approach that helps in all stages of the design process. Design thinking is not correlative; it gives freedom to check with customers or clients' changing needs in the middle of the project to make adjustments and meet their expectations. Design thinking has given a completely new dimension to developing strategies. It has been found that there is a need for alignment in individuals and businesses, Design thinking plays an important role in this alignment and meets all kinds of HR challenges especially related to the workforce. It allows HR to reinvent themselves, focus on employee experiences, wide innovation, and create a result-oriented mindset.

12.1.5 Internet of Things (IoT) and Human Resource Management

IoT, the interrelated computing devices, helps to operate business across borders with more coordination and cooperation. The organization is a combination of employees and the work which is carried out by them, in order to manage both the employees and work, HR needs to access huge amounts of data and get involved in complex decision-making tasks, emails etc. Such quantifiable data can be obtained from various devices that are driven by the technology of IoT. This will make the HR task easier to develop expertise work culture by evaluating employee productivity, their communication patterns and collaborations. IoT technology keeps managers, HR and employees connected with internet enabled devices, thus enhancing the work experience. IoT also helps HR to create flexible workspaces by allowing them to track the alertness of employees on the job and keeping track of their attendance. Strategy and decision-making needs huge amounts of data and IoT helps to accumulate such huge data related to people and procedures.

12.1.6 Artificial Intelligence and Human Resource

Artificial Intelligence (AI) is one of the technologies that allow a computer to execute tasks that generally require human ability, and adaptive decision-making. AI has applications, which are inbuilt with knowledge and decision-making abilities hence increasing the HRM decision accuracy either in support of human intelligence, or in place of it. Current AI based HRM technologies are supporting recruitment to selection, retention, talent search etc. The tricky task like HR planning which includes assigning the right candidate for the right job is being handled by AI in a much easier way. It makes the perfect recruitment decisions by analyzing future employee needs. AI algorithms help to improve job candidate identification, that is who is most suited for the job and provide effective communication of the job opening. It also makes a simpler interview process with asynchronous video interviews. AI Algorithms allows for background checking of the job applicants and makes it easier to develop compensation packages without any bias of race, gender and sexual orientation. AI guides HR managers to assess training efficiency and decisions on employee competency, including emotional and intellectual abilities and experiences level, in order to assign the right candidate with the corresponding talents to the correct positions. AI helps to complete the required courses by identifying the training needs depending on the employee skill gap. Automation helps HR in handling the HR payroll activities and AI systems helps in identifying the salary parameters with respect to employee jobs.

12.1.7 Blockchain (Distributive Ledger) and Human Resource Management

Blockchain is the one among the advanced technologies which is termed as decentralized and distributed ledger technology (DLT) [27, 20] which stores the source of digital assets. This technology is booming as it allows not only information, but everything of value-money, titles, deeds, identity, and votes to be exchanged securely, transparently and more importantly without risk of tampering. As it creates trust through consensus and ensures that the parties authorized to access agree that any changes made to it are valid, it provides greater benefits to various HR functions. HR process or functions that takes longer time, highly labor intensive, expensive due to need for large authentic data collection or that requires third party verification such as identifying and verifying potential candidates, managing international HRM activities of cross border payments & mobility, data heavy process like VAT administration, payroll could be managed and benefited through use of Block chain technology.

Thus, the adoption of technology helps i.e. ERP [28, 2], data mining, and design thinking helps in streamlining its process, reducing administrative burdens, and compliance costs, competing for talent globally, providing real-time metrics to spot trends/patterns of workforce behavior, and above all play a significant strategic role in organizations growth. Such technological advancement has made HR viewed as a portal rather than a person. However, IoT, AI, and blockchain three transformational technologies that have emerged in the same generation are transformational technologies. IoT, AI, and blockchain, any one standalone has the power to alter complete HR practices but together, their transformative impact will be unprecedented. As it said consider these three as interconnected organic processes, similar to the human nervous system where IoT senses thereby converting the world of things to a world of data, AI and machine learning analysis part of the brain which thinks by exploring data and making decisions which were made initially by human and finally Blockchain, memory part of nervous system which creates a secure transaction.

The details about one of the most important transformative technologies i.e., the use of blockchain technology in various Human Resource Processes are explained in the next section.

12.2 Blockchain Technology and Human Resource Processes

Human capital is one of the most valuable assets for any organization. However, we frequently come across instances where an employee falsified their credentials, including their education, training, promotions, awards, etc., which costs both the employer and the employee [40]. According to a career builder poll, 33% of employees lied about their qualifications [10] and 30% of employers claimed they lacked the necessary technological resources to find the right hire [35]. A fake certificate racket was busted in India which led to the termination of 1832 employees [31]. Technology integration with HRM greatly aided organizations in resolving these problems [29], but other persistent problems such as biased hiring, training evaluation, and performance reviews are challenging to resolve.

Andrew Spencer, consultant, and researcher of a UK-based HR technology firm suggested that blockchain in HR can be used in the four pertinent areas—recruits' credential verification, compensation and payments, job matching platforms, and employees' identity management. Blockchain technology has the potential to address such issues and some researchers emphasized that integrating technology with HRM functions will lead to efficiency and effectiveness [29]. Since the induction of mainframe computers, ERP systems, and online-based systems in the late twentieth century, HRM has evolved to incorporate technology. More recent advancements include the use of mobile devices, social media, artificial intelligence, IoT, ML, and Blockchain Technology [7, 19]. Blockchain technology is a game-changing breakthrough that has impacted businesses [8] and its applications are safe and process-efficient [32]. Blockchain is a decentralized, shared, secure, fair, cost-effective, traceable, and asset management system for enterprises' digital ledger. Blockchain entails the sender recording the transaction, the blockchain members validating the transaction, and the sender and recipient approving the transaction. It has the ability to bring all the stakeholders together on one platform, increasing the transparency of business transactions. This chapter describes the potential application of blockchain technology in human resource processes such as recruitment, training effectiveness, payroll and performance evaluation.

12.2.1 Blockchain Technology in Recruitment and Selection

Lussier [16] and Kavanagh [12] described technology integration in HRM. They also discussed emerging trends of social media, mobile gadgets, technological impact on deliverables and problems faced by HR managers in pursuit of technological integration of HR functions. Even though technological integration of HR applications improved efficiency and effectiveness of HR processes, there was a need to make the HR system more secure, safe, and transparent [1, 17]. Blockchain based recruitment management system is a way forward to make recruitment and selection more

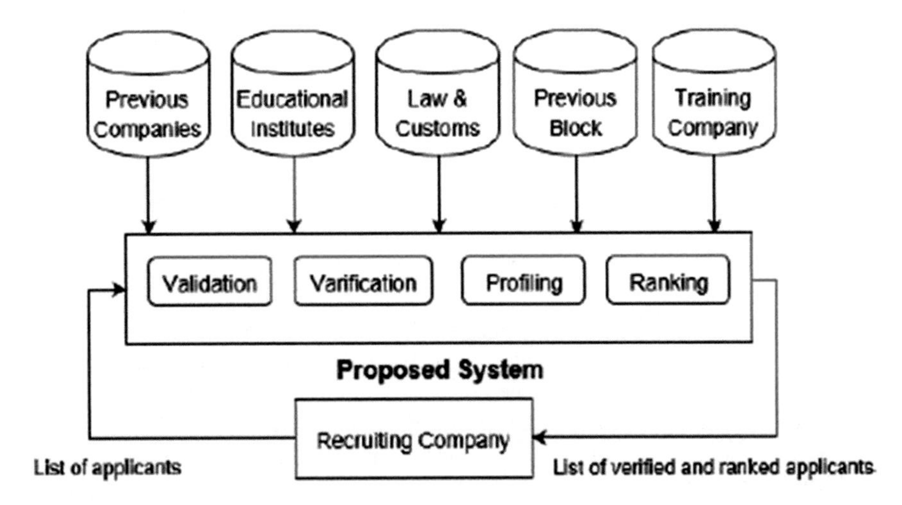

Fig. 12.1 Blockchain-based recruitment process (adopted from [17])

secure and transparent. Md Mehedi Hasan Onik [17] developed and implemented a blockchain based recruitment management system. The premise of the framework is depicted in Fig. 12.1. The framework helps to develop a blockchain recruitment management system that collects, verifies, and ranks the applicants. Organization gets list of applicants pertaining to the industry. The applicant's credentials such as previous employment, educational institute, training firm, any criminal records or violation of law etc. are databases validated and verified. The next step is applicants profiling wherein applicant's information is profiled against the organizational vacancy. The mismatched applicant's data is discarded and other applicants are ranked. Based on the rank applicants are shortlisted and the organization takes a call to proceed further with the selection process.

12.2.2 Blockchain Technology in Training and Development

For employees to advance in their careers and to provide organizations a competitive advantage training and development is considered a critical process. Learning organizations achieve a competitive advantage to withstand the changing environmental dynamics by engaging employees in the training process [9]. In pursuit of competitive advantage, organizations build innovation and creativity as core competencies [5] to deliver products and services. However, the effort, time, and expenses put into designing and evaluating the training effectiveness have an impact on the company's value addition [38]. Kirkpatrick's [14] model of training evaluation, cost–benefit analysis, and other methods help organizations to determine training effectiveness. Still, organizations are constantly under pressure to lower the cost of training [30] and are questioned regarding the underlying assumptions for evaluation [13]. E-learning was a significant advancement that brought a paradigm shift in the training process by reducing the costs and time constraints associated and improved the efficiency of the training process but faced backlash due to the social isolation problem [9, 18]. Instant feedback and isolation learning obstacles were addressed by more advanced training methods such as blended learning that became more effective and efficient. However, fast-growing new industrial technologies (Industrial revolution 4.0 or 5.0) such as IoT, automation, and blockchain technology alter how training effectiveness is assessed [32]. Blockchain technology integration with HR processes and other business operations is becoming more popular among practitioners and academicians [8]. The zero-change characteristic of blockchain technology makes it more practical to use it for training effectiveness [11]. The framework for evaluating the effectiveness of training using blockchain technology was suggested [11], and emphasis was placed on the need for organizations to measure employee learning using blockchain blocks that contained information on their training program performance.

The information included feedback from the instructor, peers, supervisors, and others, and it informs the participants of their skills and competencies that enable them to perform better. The development of such blockchain-based training effectiveness systems will assist in the development of learning organizations. Jain [11] through their qualitative research found that blockchain technology compromises the chain of distributed data storage of employees in term of training and other information that is transparent as it facilitates employees' training and development needs, training evaluation, outcomes, etc., and the data is traceable, trustworthy, and temper proof. As employee information is dispersed, companies may access it even if employees switch from one company to another. Also, the employees advance in their careers and other additional information keeps on accumulating as a block. Employee training and development records, certificates, and future learning needs may all be easily tracked using blockchain technology.

12.2.3 Blockchain Technology for Payroll

Nearly 90% of the time for human resources managers is spent processing payroll, including payments for employees from other countries [4]. According to Gartner's report [8], the capacity to adapt and perform well has led to a 32 percent growth in the contingent workforce or gig workers. The implementation of ERP systems or the outsourcing of payroll administration gave technological integration for payroll management a boost. Organizations were able to reduce the expenses by means of outsourcing and ERP systems, however, there exist security and confidentiality concerns [32]. By validating employee identities, calculating work hours, and paying in real-time for employees and gig workers, blockchain technology overcomes worries of confidentiality and security. Further, the adoption of blockchain technology allows to pay overseas workers in the domestic currency where they are working and also takes care of tax compliance [8, 32, 23]. All transactions in the blockchain are digitally signed, kept in a shared ledger, and shared among all the members in the network. The ledger is updated as each payment is made and validated by members as in when the transactions take place. Payroll management is made easier, flexible, and real-time for both domestic and international payments by the use of blockchain technology.

12.2.4 Blockchain Technology for Performance Management

Employee performance management requires data confidentiality and security. Performance evaluation records are updated periodically to aid employees career advancement in the organization. As a result, businesses want technology that is more secure, offers real-time feedback, supports professional growth, and fosters effective performance management systems [3, 22]. A new performance management system based on blockchain is proposed [6, 21, 33] that is objective, error-free, and collects performance information from all the stakeholders. A block is established as soon as a new employee joins the organization and begins working all stakeholders give feedback on his or her performance, giving a multi-rating view about the performance evaluation. Integration of blockchain technology with a performance management system minimizes paperwork, raters bias, and HR's work while ensuring the security and confidentiality of employees' data [34, 15, 24].

12.2.5 Employment Records Management and Blockchain

One of the tedious works where HR managers spend a relatively higher number of hours is creating and updating the employees' records, work history, and other HRIS[1] enrolments. The information which is recorded pertains to creating a new account for each employee, their joining information, their termination, if any, details, training records, performance scores, and notifications to legal entities such as tax or immigration data. A blockchain-enabled distributed ledger in HRIS (BCHRIS) can be used to verify a candidate's job data. Further information can be added into the blocks for administration purposes related to pay and benefits such as. health care and medical insurance and compensations. Blockchain technology not only can save time but also prevent repeating data input and administration processes.

Other than record maintenance, blockchain also ensures keeping attendance data secure. ID2020, a non-governmental organization that provided digital IDs, is aggressively using blockchain technology to verify, store disseminates information such

[1] A human resources information system (HRIS) is a software solution that maintains, manages, and processes detailed employee information and human resources-related policies and procedures. As an interactive system of information management, the HRIS standardizes human resources (HR) tasks and processes while facilitating accurate record keeping and reporting.

as biometric data. In a similar way, human resource departments can use deploy blockchain to improve time tracking and payroll systems.

12.2.6 *Employees' Compensation—Blockchain Enabled Cryptocurrency*

The primary usage of blockchain—the cryptocurrency, can't be ignored. Remunerating employees in the form of cryptocurrency holds more meaning in today's context, where workers are global citizens. In developed nations, implementing crypto-based payroll may not seem necessary. Here, compensating in standard, fiat currency makes more sense. Whereas, when a company hire on a global scale, compensating workers in cryptocurrencies such as Bitcoin, Ethereum, and others are more beneficial. Workers can use this money with much ease to buy goods and services. Global market and few economies are quiet uncertain, where the currency can be devalued, banking systems are rigid and sometimes untrustworthy, and the government has certain restrictions. In these nations crypto-based payroll systems backed by blockchain technology will attract more workers on a global level.

12.2.7 *Managing Contractual/Temporary Workers Through Blockchain*

One of the unique features of blockchain- the smart contracts can be extensively used to manage and handle contractual or temporary workers. The smart contract develops an enforceable and immutable obligations and rights for all the parties in a network. These immutable smart contracts can be used to generate payments automatically once the worker has completed the assigned tasks and work. Thus, Smart Contracts can make the cash flow smooth for the companies with regards temporary workers.

12.3 Advantages of Blockchain in Human Resource Management

Although blockchain technology is popular and commonly known for cryptocurrency, scholars have opined that it will change the functions of all areas of business. It started with disruption in the finance, real estate, and energy sectors, but in a recent article, Forbes stated that all industries should be prepared for the "game-changing" possibilities of the blockchain (Ahmed 2019). In the domain of human resource management as well, now organizations are moving towards blockchain and started leveraging its benefits at all the stages of employees' life cycle from recruitment, onboarding, training, and appraising to retirement. It is said that HR managers who "fail to do sufficient scenario planning and experiment with the technology accordingly risk significant long-term disintermediation," [39, 25, 26]. It is expected that blockchain will change the way large amounts of sensitive employee data are handled and deployed through processes of human resource management.

As an employee's records are confidential in nature, the first and the foremost thing which comes to one's mind is how secure is blockchain technology and whether one can trust it or not. Leaks of data such as the bank details, Government Identity cards, performance records, reimbursements, clients' expenses and others will create a management level fiasco and will lead to legal complications. Hence, when it comes to use of a technology, then the most important thing is to verify it. The first and the foremost thing to weigh is whether the technology is trustworthy? Is data secure when one uses the technology? Who all can have access to data? When it comes to the usage of blockchain technology as well, one has similar queries as it uses shared and distributed ledgers that are accessible not by one but by many members. Due to this, blockchain is sometimes called a "trustless" network. But this is not correct, as the reason is not that network members don't have trust on each other, rather it is not required, as by nature of its configuration blockchain technology creates an environment of trust due to the advanced level of security, enhanced transparency, and immutability. No other technology provides such secure and trustworthy features. Other than Trust blockchain has many more other benefits, some of which are explained below.

12.3.1 Enhanced Security

The human resource department are responsible for managing confidential employees' personal records while undergoing high volume financial transactions pertaining to pay, remunerations, provident funds, healthcare, bank records, performance data and other expense reimbursements. Data is quite sensitive, important, crucial here. Leak of data is a major concern here. Using blockchain for handling this large chipmunk of data and financial transactions will be beneficial, as blockchain technology ensures security and monitors how these sensitive and critical information are viewed by network members. It prevents unauthorized activity and fraud by creating records that are difficult to alter and by encrypting data. It also anonymizes personal data and permissions are used to prevent access. Further, information is stored across a network of computers and not on a single server, hence it is difficult to view and hack data.

12.3.2 Higher Transparency

Blockchain uses distributed ledger records where identical data and transactions are stored and recorded in multiple locations. To have smooth functioning, it is essential that all the HR executives who are working on employees' data should be able to view and access identical and similar information. For instance, the training manager can view similar information about the new joiners which the onboarding team is viewing. Here comes the issue of transparency. Blockchain allows this transparency. All network members have permission to access and view similar information at the same time which ensures full transparency. All transactions are immutably recorded, and are time- and date-stamped. This enables members to view the entire history of a transaction and virtually eliminates any opportunity for fraud.

12.3.3 Automation

Smart contracts are another feature of blockchain which further enhance the efficiency and the speed with which the data is processed. Smart contracts allow us to fulfil the pre-specified conditions. Once the conditions are met, the next step related to the transaction or process is automatically initiated. Human intervention or reliance on third party for verifying the terms of the contracts are automatically eliminated through these smart contracts. For instance, once the employees have furnished all the KYC documents for employment, the next round of scheduling the interview is automatically initiated.

Chillakuri and Attili [6] have listed several features of blockchain, such as decentralized networks, open distributed ledger, real-time updates, transition validity, automation through smart contracts and others, which leads to expected outcomes and importance of the same in the area of human resource management. Expected outcomes are the benefits of blockchain which is related to HR. Benefits such as the single source of truth, transparency, traceability, limited cost, effective collaboration, faster settlements, fraud prevention and analytics support. These outcomes are useful in validation of certificates and employment history, skill mapping, payroll processing, data protection of employees and performance management (Fig. 12.2).

12.4 Challenges Behind Implementation of Blockchain Technology

Adopting blockchain has many advantages. It facilitates smooth function with multiple members on common networks. But, since blockchain is a new technology and is still evolving, one needs to look for the challenges behind adoption of

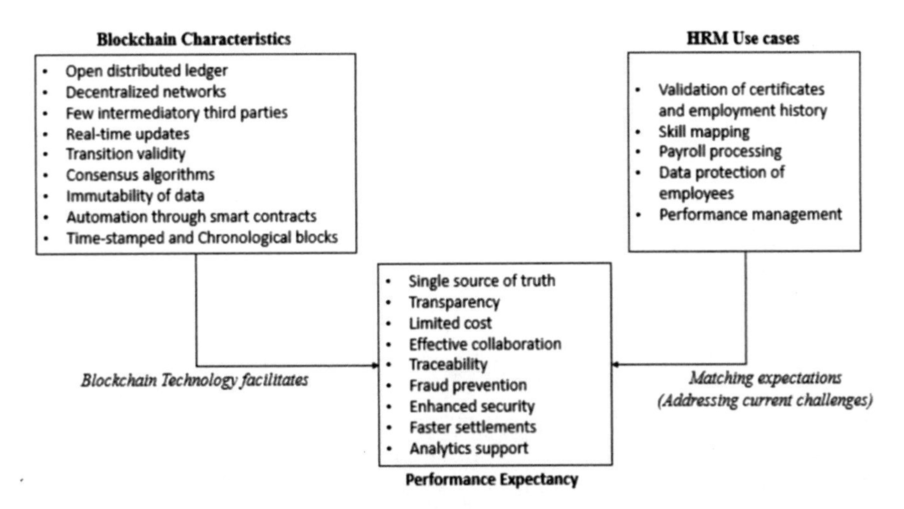

Fig. 12.2 Blockchain and HRM. *Source* Chillakuri and Attili, 2021

Fig. 12.3 Challenges behind adoption of blockchain technology

blockchain technology. Whether big or small medium enterprise, one need to review these challenges to leverage benefits of blockchain (Fig. 12.3).

12.4.1 *Inefficient Technological Design*

In spite of one of the major advantages of blockchain technology is security, but there is a major loophole. There is a coding flaw with regard to decentralized application (dApps) development. Though Ethereum has tried to cover up all the limitations of Bitcoin, it still holds several loopholes and one of them is dApps. Ethereum gave access to developers to download and install dApps based on their system. Users can utilize these loopholes and can hack into the system quickly. Employees' records especially their transactions and bank details are more crucial. The dApps flaw makes these data vulnerable. These loophole with regards to dApps needs to be fixed to make blockchain more secure.

12.4.2 Low Scalability

Scalability is another challenge behind the implementation of blockchain. The problem relates to mass integration. As it is known that Ethereum and Bitcoin are now having the highest number of members on the network which leads to long transitions and time to process. As a result, the transactions get delayed and cost higher than usual. Large scale companies will require more members while handling big employee data. Slowing down the process will delay the whole process [36, 37].

12.4.3 No Regulation

Regulations and legal compliance are yet to be standardized. Organizations are aggressively opting cryptocurrency to handle their transactions, but government regulations are yet to be framed properly. There are no specific rules are there when it comes to the blockchain.

12.4.4 Lack of Adequate Skilled Workers

As we know blockchain is an emerging software and still evolving. Skilled professionals who can work on blockchain is still not available. The shortage of talent will lead to hiring experts and which in turn increase the cost. Like any other technological innovation, blockchain technology will continue to evolve.

12.5 Companies Using Blockchain in HRM

IBM Garage

IBM is perhaps the first company to have its operation conducted on a distributed ledger of blockchain. It uses Hyperledger Fabric, an open-source blockchain platform, which was developed by the Linux Foundation. IBM Garage enables human resource professionals from diverse industries to monitor applicants' work history and past performance while simultaneously allowing applicants to share and upload their resumes with employers of their interest across the network.

Persol Career Company

It is Japan's one of the largest employment service firms. The company provides manpower services ranging from mid-level recruitment to different management support services. The company uses IBM blockchain to have a more rigorous and efficient way of validating and verifying job applicants' educational qualifications and work experiences.

AWS Blockchain

After IBM another big firm which has adopted blockshian is Amazon Web Services (AWS). AWS blockchain provide on-demand cloud computing to enterprises, individuals and governments on a paid basis. AWS blockchain started in the year 2019, and provides four different solutions to its customers and partners. It offers a ledger database that records all application's data changes. These records are immutable, fast and cryptographically verifiable.

Beowulf: Streamlining Workplace Communication

Beowulf to simplify organizational internal communication in various business settings provides multiple solutions such as corporate communication, ready-to-use software development kits (SDKs) for corporate communication, and other features. The has several renowned customers and clients such as Asia Life Insurance Group), the Vietnam University of Science, and the U.S.-based cybersecurity firm OPSWAT.

The BeSure Network

The company uses blockchain to validate safety protocol at the workplace. The blockchain pulls safety compliance data from various sources such as from factory floor supervisors, employees, and regulatory bodies. The main objective is to reduce cases of workers who are unverifiable and are working in hazardous environments. Smart contracts and automated data entry are used to make it easier. The system is gaining popularity in UK.

ETCH—Instant Payroll

ETCH provides a blockchain-based payroll solution which facilitates payment at the individuals' leisure time. The payments are credited as ECH tokens which are stored in digital wallet. Employees can utilize this digital money at different locations worldwide. ECH issue a card which can used to withdraw this digital pay as cash.

12.6 Conclusion—Future Ahead

Though blockchain has secured its place worldwide, it still needs more time to get mature in the area of human resource management. Gartner[2] [8] has predicted that blockchain technology will create a business value of $3.1 trillion by 2030. The significant amount of returns will in general be resulted from efficiency improvements in existing operating models and business processes. But the real value will be from the way it brings a paradigm shift in how businesses, partners, societies, customers, and individuals are interacting, creating, and exchanging value. To be more precise, blockchain technology has created an environment where participants/members of a network, who may or may not know each other, exchange information and value in a digital environment. That is blockchain entrusted trust in an untrusted environment thereby eliminating the requirement of a trusted central authority. The usage of blockchain in the area of Human Resource can be categorised under three waves. The first wave pertains to the usage of blockchain for job applicant's education and other verification and real-time employees' payments and remunerations. The second wave focus on the labour markets where blockchain enhances the visibility of work, and the match with workers. This in turn resulted in an increase of market trust. The third wave relates to smart contracts for managing temporary and gig workers.

Blockchain technology is one of the safest technologies that aids in streamlining human resource management processes from hiring, and training to compensate workers. It can identify prospective pitfalls and inefficiency in various HR procedures which require time, money and manpower. As noted, blockchain ledger provides a reliable instrument for verifying the past record of job applicants. It also helps in saving hiring cost of related to logistics and communication for the candidates' verification.

Thus, it is understandable that managing HRM using advanced technologies leads to efficiency, speed, quality of service and scalability in various HR functions, but at the same time implication and managing technology requires HR practitioners to look strategically at issues pertaining to scope of job replacement, decision making and learning opportunities for their employees as it effects both workers and working process of employees. Employees may experience change in tasks, require different qualifications, upgradation of skill and may impact their well-being. Developing favorable and socio-technology interventions such as, supportive transformational leaders, flexible structure, ethical work culture, upskilling employees and redesigning job roles; change management training would be the need of the hour to realize the positive outcomes of these technologies.

[2] Gartner, Inc is a technological research and consulting firm based in Stamford, Connecticut that conducts research on technology and shares this research both through private consulting as well as executive programs and conferences.

References

1. Analoui, F: The Changing Patterns of Human Resource Management. Routledge (2017)
2. Blockchain technology: applications and challenges. In: Panda, S.K., Jena, A.K., Swain, S.K., Satapathy, S.C.: (eds.). Springer, Intelligent Systems Reference Library. https://doi.org/10.1007/978-3-030-69395-4
3. Buckingham, M.: Reinventing performance management. Harv. Bus. Rev. **93**(4), 40–50 (2015)
4. Burke, I: It's not all paperwork: how does HR really spend their time, retrieved from www.business.com/articles/its-no-all-paperwork-how-does-hr-really-spend-their-time. Accessed 31July 2022
5. Caputo, F.C: Innovation through digital revolution: the role of soft skills and Big Data in increasing firm performance. Manage. Decis. **57**(8), 2023–2051 (2019). https://doi.org/10.1108/MD-07-2018-0833
6. Chillakuri, B., Attili, V.P: Role of blockchain in HR's response to new-normal. Int. J. Organ. Anal. (2021). https://doi.org/10.1108/IJOA-08-2020-2363
7. Crosby, M.P.: Blockchain Technology: Beyond Bitcoin. Appl. Innov. **2**, 6–10 (2016)
8. Gartner: Future of Work trends Post-COVID 19, retrieved from https://www.gartner.com/smarterwithgartner/9-future-of-work-trends-post-covid-19 (2020)
9. Gil, A.G.-A: The training demand in organizational changes processes in the Spanish wine sector. Eur. J. Train. Develop. **39**(5), 315–331 (2015). https://doi.org/10.1108/EJTD-09-2014-0067
10. Hayes, L: Nearly three in four employers affected by bad hiring. Career Builder Surve (2017). Retrieved from https://press.careerbuilder.com/2017-12-07-Nearly-Three-in-Four-Employers-Affected-by-a-Bad-Hire-According-to-a-Recent-CareerBuilder-Survey
11. Jain, G.S.: Enhancing training effectiveness for organizations through blockchain-enabled training effectiveness measurement (BETEM). J. Organ. Change **34**(2), 439–459 (2021)
12. Kavanagh, M: Human Resource Information Systems: Basics, Applications, and Future Directions. Sage Publications (2017)
13. Kraiger, K: Decision-based evaluation. In: Kraiger, K. (ed.) Creating, Implementing, and Maintaining Effective Training and Development: State-of-the-Art Lessons for Practice. Mahwah, NJ: Jossey-Bass (2002)
14. Krickpatrick, D.: Evaluating Training Programs: The Four levels. Berrett- Koehler Publisher, San Francisco, CA (2006)
15. Lokre, S.S., Naman, V., Priya, S., Panda, S.K.: Gun tracking system using blockchain technology. In: Panda, S.K., Jena, A.K., Swain, S.K., Satapathy, S.C. (eds.) Blockchain Technology: Applications and Challenges. Intelligent Systems Reference Library, vol. 203. Springer, Cham (2021). https://doi.org/10.1007/978-3-030-69395-4_16
16. Lussier, R: Human Resource Management: Functions, Applications, and Skill Development. Sage Publications (2018)
17. Md Mehedi Hasan Onik, M.H.-S: A Recruitment and Human Resource Management Technique Using Blockchain Technology for Industry 4.0. Smart Cities Symposium (SCS -2018), pp. 11–16. Manama, Bahrain: IET (2018)
18. Miller, L.: ASTD 2012 state of the industry report: organizations continue to invest in workplace learning. Train. Develop. Mag. **66**, 42–48 (2012)
19. Miraz, M.H.: Application of blockchain technology beyond cryptocurrency. Ann. Emerg. Technol. Comput. **2**(1), 1–6 (2018)
20. Nanda, S.K., Panda, S.K., Das, M., Satapathy, S.C.: Automating vehicle insurance process using smart contract and Ethereum. In: Chakravarthy, V.V.S.S.S., Flores-Fuentes, W., Bhateja, V., Biswal, B. (eds.) Advances in Micro-Electronics, Embedded Systems and IoT. Lecture Notes in Electrical Engineering, vol. 838. Springer, Singapore (2022). https://doi.org/10.1007/978-981-16-8550-7_23
21. Niveditha, V.R., Karthik Sekaran, K., Singh, A., Panda, S.K.: Effective prediction of bitcoin price using wolf search algorithm and bidirectional LSTM on internet of things data. Int. J. Syst. Syst. Eng. **11**(3–4), 224–236

22. Panda, S.K., Satapathy, S.C.: Drug traceability and transparency in medical supply chain using blockchain for easing the process and creating trust between stakeholders and consumers. Pers. Ubiquit. Comput. (2021). https://doi.org/10.1007/s00779-021-01588-3
23. Panda, S.K., Mohammad, G.B., Nandan Mohanty, S., Sahoo, S.: Smart contract-based land registry system to reduce frauds and time delay. Secur. Privacy. e172 (2021). https://doi.org/10.1002/spy2.172
24. Panda, S.K., Daliyet, S.P., Lokre, S.S., Naman, V.: Distributed ledger technology in the construction industry using corda. Distributed Ledger Technology in the Construction Industry Using Corda. https://doi.org/10.1002/9781119884392.ch2
25. Panda, S.K., Satapathy, S.C.: An investigation into smart contract deployment on Ethereum platform using web3.js and solidity using blockchain. In: Bhateja, V., Satapathy, S.C., Travieso-González, C.M., Aradhya, V.N.M. (eds.) Data Engineering and Intelligent Computing. Advances in Intelligent Systems and Computing, vol. 1. Springer, Singapore (2021). https://doi.org/10.1007/978-981-16-0171-2_52
26. Panda, S.K., Rao, D.C., Satapathy, S.C.: An investigation into the usability of blockchain technology in internet of things. In: Bhateja, V., Satapathy, S.C., Travieso-González, C.M., Aradhya, V.N.M. (eds.) Data Engineering and Intelligent Computing. Advances in Intelligent Systems and Computing, vol. 1. Springer, Singapore (2021). https://doi.org/10.1007/978-981-16-0171-2_53
27. Panda, S.K., Dash, S.P., Jena, A.K.: Optimization of block query response using evolutionary algorithm. In: Bhateja, V., Satapathy, S.C., Travieso-González, C.M., Aradhya, V.N.M. (eds.) Data Engineering and Intelligent Computing. Advances in Intelligent Systems and Computing, vol. 1. Springer, Singapore (2021). https://doi.org/10.1007/978-981-16-0171-2_54
28. Panda, S.K., Elngar, A.A., Balas, V.E., Kayed, M. (eds.): Bitcoin and Blockchain: History and Current Applications, 1st edn. CRC Press (2020). https://doi.org/10.1201/9781003032588
29. Papadopoulos, T.B: The use of digital technologies by small and medium enterprises during COVID-19: implications for theory and practice. Int. J. Inform. Manage. **55** (2020). https://doi.org/10.1016/j.ijinfomgt.2020.102192
30. Pineda, P: Evaluation of training in organizations: a proposal for an integrated model. J. Eur. Indus. Train. **34**(7), 673–693. https://doi.org/10.1108/03090591011070789
31. Press Trust of India, Economic Times (PTI, E. T): Employees with fake certificates will be sacked: Govt, retrieved from https://economictimes.indiatimes.com/news/politics-and-nation/employees-with-fake-caste-certificates-will-be-sacked-govt/articleshow/59159793.cms (2017)
32. Price Waterhouse Coopers (PwC): How blockchain technology could impact HR and the world of work. Accessed from https://www.pwc.ch/en/insights/hr/how-blockchain-can-impact-hr-and-the-world-of-work.html#:~:text=%22Blockchain%20could%20shake%20up%20the,transparency%20to%20name%20a%20few.%22
33. Sathya, A.R., Panda, S.K., Hanumanthakari, S.: Enabling smart education system using blockchain technology. In: Panda, S.K., Jena, A.K., Swain, S.K., Satapathy, S.C. (eds.) Blockchain Technology: Applications and Challenges. Intelligent Systems Reference Library, vol 203. Springer, Cham (2021). https://doi.org/10.1007/978-3-030-69395-4_10
34. Sekhar, C: Enhance employee performance management experience with blockchain (2017). Retrieved from https://www.linkedin.com/pulse/enhance-employee-performance-management-experience-blockchain-aknr/
35. Seymour, L: Poor recruitment practices put Australian businesses at risk (2017). Retrieved on July 27, 2022, from https://atcevent.com/sourcing/poor-recruitment-practices-put-australian-businesses-risk/
36. Varaprasada Rao, K., Panda, S.K.: A design model of copyright protection system based on distributed ledger technology. In: Satapathy, S.C., Lin, J.CW., Wee, L.K., Bhateja, V., Rajesh, T.M. (eds.) Computer Communication, Networking and IoT. Lecture Notes in Networks and Systems, vol. 459. Springer, Singapore (2023). https://doi.org/10.1007/978-981-19-1976-3_17
37. Varaprasada Rao, K., Panda, S.K.: Secure electronic voting (E-voting) system based on blockchain on various platforms. In: Satapathy, S.C., Lin, J.CW., Wee, L.K., Bhateja, V., Rajesh,

T.M. (eds.) Computer Communication, Networking and IoT. Lecture Notes in Networks and Systems, vol. 459. Springer, Singapore (2023). https://doi.org/10.1007/978-981-19-1976-3_18
38. Ward, P., Williams, A.M., Hancock, P.A: Simulation for performance and training. In: Ericsson, K.A., Charness, N., Feltovich, P.J., Hoffman, R.R. (eds.) The Cambridge Handbook of Expertise and Expert Performance, pp. 243–262. Cambridge University Press (2006)
39. Wiles, J.: 5 Ways Blockchain Will Affect HR. Accessed from https://www.gartner.com/smarterwithgartner/5-ways-blockchain-will-affect-hr. 27 Aug 2019
40. Zhou, C.T.: An uncertain search model for recruitment problems with enterprise performance. J. Intell. Manuf. **28**(3), 695–704 (2017)
41. Zielinski, D.: Is HR Ready for Blockchain? Accessed from https://www.shrm.org/hr-today/news/hr-magazine/0318/pages/is-hr-ready-for-blockchain.aspx. 23 Feb 2018

Chapter 13
P2P-The Key Behind Regulatory Framework of DeFi Services

H. S. Shalini, K. Ravichandran, and P. V. Raveendra

Abstract Decentralized Finance (DeFi) is an emerging financial technology that functions on the similar mechanism underlying cryptocurrencies. In DeFi the transactions will happen over a secure distributed ledger system which maintains anonymity of the person who carries out that transaction. DeFi completely removes the control that banks have on the customers transactions thereby making it safe and secured. Peer-to-Peer (P2P) is one amongst the premises on which DeFi operates. This chapter makes an attempt to study the role of P2P in DeFi Financial Transactions (DFFT).

Keywords Cryptocurrency · Decentralized Finance (DeFi) · Decentralized Finance Application (d App) · Peer-to-Peer(P2P) financial transactions · Total Value Locked (TVL)

13.1 Introduction

Though Decentralized Finance is not a new concept, it gained its momentum from June 2021 with a TLV of $13 billion. This is because of its characteristic feature that it eliminates the transaction fees that is levied by banks and other financial institutions while using their services [1, 2]. Due to lack of security and increasing rate of scams in banking sector, the consumer lost his trust in banks which is the major hook that binds the banker and the customer together. This paves the way to change the entire workflow model on which the banking transaction happens. Bank customers are also unhappy due to tedious and time-consuming documentation and approval formalities that is followed in banks. This creates the need for a cost-sensitive and hassle-free borrowing which consumer prefer while carrying out banking transactions.

H. S. Shalini (✉) · K. Ravichandran
MBA Department, Acharya Bangalore B-School, Bangalore, India
e-mail: shalini.hs@abbs.edu.in

K. Ravichandran
e-mail: drravi@abbs.edu.in

P. V. Raveendra
Department of MBA, MS Ramaiah Institute of Technology, Bangalore, India

S. K. Panda et al. (eds.), *Recent Advances in Blockchain Technology*,
Intelligent Systems Reference Library 237, https://doi.org/10.1007/978-3-031-22835-3_13

In decentralized finance the consumers will hold electronic money in their digital wallets instead of depositing the same in a bank. Moreover, the consumer need not approach a banker to carry out his transaction. An internet connection is sufficient for the consumer to transfer funds in few seconds or a minute. De-Fi works on Peer-to-Peer (P2P) [3] mechanism in which a consumer can enter his loan specifications and get a suitable lender to finance his loan.

13.2 Current Status of De-Fi in the World

World Economic Forum in collaboration with the Wharton Blockchain and Digital Asset Project developed a policy maker tool kit in June 2021. This forum was consisted of Ms. Sumedha Deshmukh, who is the platform curator, Block-chain and digital assets World Economic forum; Ms. Sheila Warren, Deputy head, Center for the fourth Industrial Revolution, Member of the Executive Committee-World Economic Forum and Mr. Kevin Werbach, Professor of legal studies and Business Ethics and Director, Block Chain Digital Asset Project, Wharton School University of Pennsylvania. In this tool kit it is stated that the De-Fi originated along with the Blockchain its technology which was first invented by Satoshi Nakamoto, a pseudonym for a person, or a set of people. De-Fi gained its significance from the year 2021 as the total locked in value of digital assets was $13 billion compared to its previous year TLV of $ 670 million.

13.3 Centralized Versus Decentralized Finance

Before discussing about advantages and disadvantages of DeFi it is important to distinguish between centralized and decentralized finance. In centralized finance the customer's money will be with the intermediaries who work for money and they charge fee for each and every service they offer. For instance, if a customer purchase grocery and makes payment through credit card then the transaction details will be sent by the merchant to the customer's bank through credit card network. The credit card network will complete the transaction and requests the payment from customer's bank. The customer's bank approves the payment and sends the approval to network if there is sufficient credit limit available to the customer. Likewise, all other financial transactions such as application for loan and or issue of cheque book involves cost. The customer may not access the financial services while travelling. Keeping these set of backdrops in mind decentralized finance seems to be the solution for easy and continuous access to financial services [4, 5].

In decentralized finance the Peer-to-peer mechanism is involved wherein a customer enters his requirements in the De-Fi software application which will be matched with the specifications of the peers who are ready to lend money to that consumer. The programming language used here is algorithms which again keeps the consumer details confidential. Then the customer needs to agree to the terms and conditions of the lender. Thus, entire transaction will be recorded in the blockchain, the loan is disbursed once there is consensus of opinion between the lender and the borrower. After disbursal of loan, the lender can start collecting EMI payments in agreed upon intervals.

13.4 What is DeFi?

DeFi is currently an evolving trend which consists of a category of blockchain based decentralized applications (Dapps) that is into providing financial services. DeFi consists of a variety of business models, new technologies and organizational layouts which are innovative and these models will replace the traditional ones. DeFi consists of a set of protocols (a set of policies and procedures) which defines a specific software. They help in creating and managing easy interface [6]. DeFi services also comprises of DeFi protocols to create financial services and related functions which can be used to manage financial risk and to hedge changes in interest rates. DeFi has a friendly user interface.

13.5 Distinctive Characteristics of DeFi

DeFi has some distinguishing features compared to traditional financial services and auxiliary services. Any DeFi protocol, model or service has four characteristic features which are unique to DeFi. They are as follows.

13.5.1 Financial Products or Services

Unlike other information services which indirectly supports transfer of money DeFi will process the transactions directly facilitating transfer of value among the parties involved in transaction.

13.5.2 Permissionless Operation and Settlement

DeFi operations are built on permissionless block chains. Till to date most of the transactions were based on Ethereum blockchain but with the advancement in technology and growing network the new block chain mechanisms such as Tezos, Solana, Avalanche, Binance smart chain and Polkadot. The service functionality of DeFi is defined by a set of smart contracts which have different governance structures that are managed by communities or firms that establish conditions for protocol.

13.5.3 Unalterable Custodial Design

The assets that are managed by DeFi services cannot be altered other than the account owner carrying out the transaction. The tokens issued under DeFi transactions cannot be altered as they are subjected to the explicit logic of smart contract accompanied by concerned DeFi protocols.

13.5.4 Composed and Programmable Database

DeFi services are enabled by Application programming interface (API) which enables service composability which is same as the financial services provided by banks and other financial institutions. The term composability stands for combining different programmatic components to create financial products and services including multiple DeFi services and protocols. For instance, a loan derivative subject to an insurance contract could be used for stablecoin blockchain [7, 8].

13.6 Conceptual Overview of De-Fi Architecture

Figure 13.1 shows the conceptual framework of De-Fi wherein different set of activities are piled up as stack of coins. The entire system is dependent on base layer which is blockchain system making it possible for the users to store, modify and exchange the asset ownership information in a secured manner. This replaces the conventional execution and settlement part of traditional financial services which makes the transaction instant so that the user need not wait for long time for settlement. This process allows for the creation of digital financial products and services which will be implemented in De-Fi system applications. Apart from this there will be additional layers of applications which functions as aggregators permitting users to switch from one De-Fi service to the other. At this level financial assets can be transferred freely based on contractual logic and they may restrict other users to provide collateral for the same. Oracle services can also be used for implementing information and content to the blockchain into De-Fi transactions. Oracle services will supply reliable data which can be recorded outside the settlement layer.

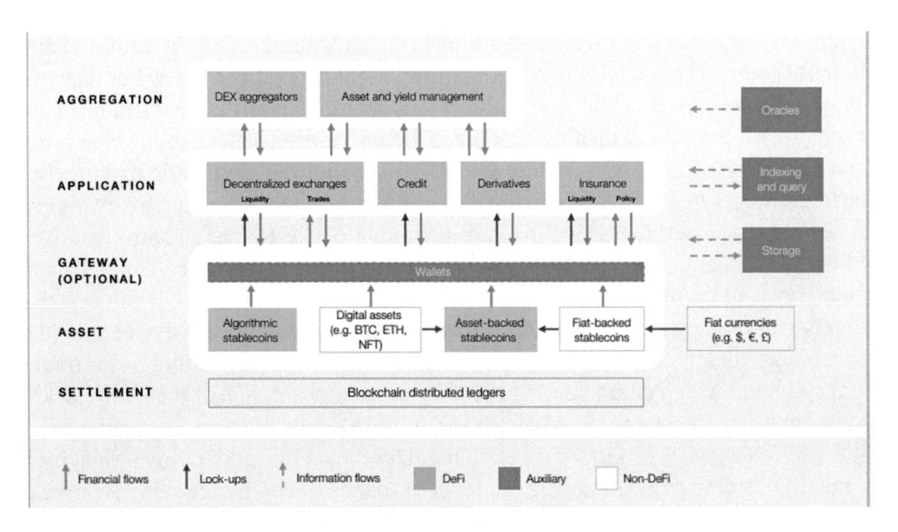

Fig. 13.1 Depicting conceptual overview of DeFi. *Source* WEF_DeFi_Policy_Maker_Toolkit_2021.pdf

13.7 Decentralized Governance

The term governance refers the way in which an economy is controlled through decisions by resolving the conflicts that arises out of decisions made. In DeFi, governance moderates the activity between the applications and the payment settlement layer including decisions on changes in interest rates, requirement of collateral etc. This model will probe us ask new questions like who is accountable for the decisions made, how are the decisions made and how does the performance management team work?

The concept of governance emerged way back during invention of Bitcoins. Governance is applicable from non-profit companies to public and private companies which are operating under the purview of a government in a country. If an investor has a stock in a company for example Apple, then it gives the investor a right to vote in the company and also rights in the profits of the company. The voting rights here refers to the vote to Board of Directors. It gives the investor the right to attend shareholders meetings. The Board of Directors will vote on the managers like CEO or the president of the company. The management runs the day today operation of the company. Managers are put in place because they are good in running day today operations of the job and complain about things using many sources. As the history of Apple says there is not enough dividends paid to shareholders, the investors have the rights to demand for voting rights of the company. There are many types of shares in fact some shares come with the voting rights others won't. In case of Biocoins there is no Bitcoin company and what determines the operation of a Bitcoin is the Code underlying the Bitcoin. The developers of the Bitcoin decide on the code of Bitcoin and they may take out the part of the block through Segwit. If the changes are not acceptable then they need to search for fork which is acceptable code. This might increase the size of blockchain or bring in some kind of structural change in the Bitcoin. There could be equal number of chains that can offer. For example, Tezos which is the SDC broker who can talk about proposal issues. If any change happens in Tezos chain then we can make the necessary changes in the chain itself and we can bring in the governance in the blockchain itself. But while making these changes the change makers should make sure that the change is made for everyone's sake not for one or two traders. In Tezos not a single investor has a control on tokens of the blockchain or else we cannot call it as decentralized. Our code gives us the ability to vote through tokens which can be delegated to others.

There are many incentive systems that are used by many DeFi service providers to promote their services in the market starting from imparting liquidity for trading and collateral which could be used in credit transactions. Some of them are Lock up yields pays the interest received on a bond or the dividend of share for trading fees to immobilize digital assets to serves a collateral for the service underlying the payment; Liquidity mining pays interest in the form of tokens issued by the service provider. These tokens are the governance tokens, which are completely regulated; Liquidity fees pays market makers a percentage of total transaction as commission though not spontaneously but for a purpose; Yield farming moves funds among liquidity mining and lock up yields; Airdrops pays wallet addresses having tokens which promotes awareness of digital assets. These mechanisms may not be integral part of DeFi but are strongly associated with DeFi transactions (Fig. 13.2).

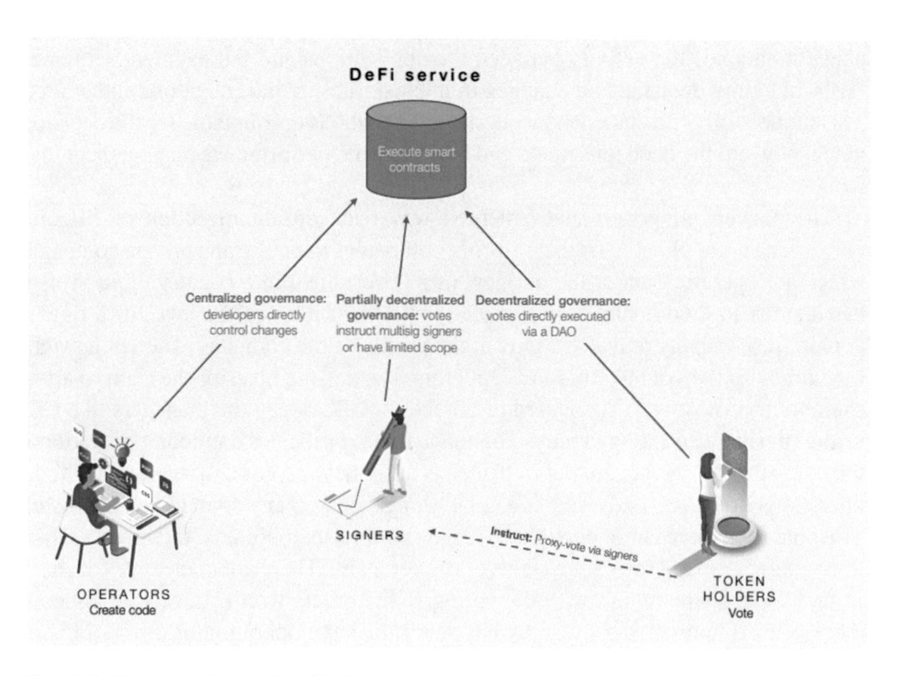

Fig. 13.2 Depicting forms of DeFi Governance. *Source* world economic forum proceedings

13.8 Types of Services Offered Under DeFi

Because of the involvement of coding programs and composability the possible services that could be provided under SEBI is endless. Some of the categories of services provided by DeFi are as follows.

13.8.1 Exchanges

By the exchanges our mind refers to stock or commodity exchanges where the stocks or commodities are traded under the regulatory framework. But in case of trading transactions wherein the underlying assets are stable coins which allows the customer to trade on digital asset for another we cannot expect a centralized exchange mechanism. This calls for a Decentralized exchange (DEX) protocol as the transaction does not take into account the collateral underlying the trade. Under DEX protocols an important category that can be traced out is Automated Market Makers (AMM) in which the algorithm will continuously keeps on pricing the underlying assets which are traded imparting liquidity to the orders instead of comparing the order with the order book.

13.8.2 Credits

Credits are nothing but interest-bearing instruments created and traded by financial institutions which should be repaid at the time of maturity. The quantum of credit depends on the relationship between the lender and the borrower which could be bilateral or it could be a common pool of capital. Deciding on the credit terms and conditions is quite complex and credit instruments can be self-securitized and traded in the market. The hallmark of DeFi is that it allows flash loans which will allow the borrower to avail credit facility without having collateral. This credit is for a very short term wherein the borrower needs to repay the same within a single block [9].

13.8.3 Stable Coins

The price of stable coins is linked to a standard asset such as US dollar. There are two forms of stable coins. Asset based stable coins will use smart contracts for dealing with the collateral in terms of its aggregation or liquidation. They are considered as digital assets. Algorithmic stable coins belong to the second category which maintain the contraction through tokens and expansion through dynamic process.

13.8.4 Derivatives

Derivatives are financial instruments which are used to hedge price risk. Derivatives are risk management instruments which give the user to trade in synthetic assets. The value of derivatives will depend on the value of the underlying asset. There are four types of derivative contracts namely forwards, futures, options and Swaps. Amongst four only futures and options are the most frequently traded derivative contracts in the market. DeFi provides the opportunity to trade in synthetic assets wherein the underlying asset can be a share, debenture, commodity, Swap or any other digital asset. It involves an NFT non fungible token which is uniquely associated with an art or real estate asset which might be tied to an activity of a business related to a crowdfunding business. The value of the derivative could also be tied to the outcome of a sport event or happening of a political campaign turning the derivatives into a predictive model [10].

13.8.5 Insurance

It is a hedging tool very similar to derivatives as in both the cases the payoff depends on happening of an event. In insurance the payoff depends on happening of an unforeseen event which will pay off the buyer of insurance contract with a premium. Insurance acts as a medium of pooling the risk by investors who want to hedge their risk and DeFi provides this hedge through smart contracts [11].

13.8.6 Asset Management

This involves asset management companies which will pool investments by a set of individual investors and invest the same in the form of mutual funds for a small amount of fees called asset management fee. DApp will facilitate the asset management life cycle through transparency and efficiency.

13.9 Risks Associated with DeFi

Risks associated with DeFi can be broadly categorized into financial risk, operational risk, technical risk, legal compliance risk and emergent risk. Financial risk arises due to depletion of funds due to non-performance from any of the parties involved in a credit transaction which can also be termed at default risk. Financial risk can be subclassified into Market risk, Counterparty risk and liquidity risk. Market risk is the decline in price of the underlying asset due to unavoidable reasons such as creation of a complex underlying instrument which makes the trade very complex and also the regulatory frameworks pose lot of restrictions so that the underlying loses its liquidity. Counterparty risk is the risk arises due to default by any of the parties involved in a cash or credit transaction. This arises due to failing to repay a loan along with interest or failing to settle a transaction by providing a proper collateral.

Liquidity risk is the risk that arises due to non-availability of market to buy or sell an underlying asset. Lack of liquidity will lead to inefficiencies in the market as the underlying asset may not perform as per the expectation. Under DeFi liquidation process differs from the regulated transactions as in the case of regulated transactions there will be intermediaries such as a bank, clearing corporations and stock or commodity exchanges which frames the regulatory mechanism for trading which is not found in DeFi transactions. DeFi services due to the speed at which they operate will compensate and incentivize the service providers by imparting liquidity to the underlying assets may it be under-collateralized loans DeFi liquidity risks can be managed by proper governance strategy and designing a incentive structure carefully. Flash loans paly a vital role in managing this unique set of liquidity risk imparting artificial liquidity for a short-term period of time which takes care of liquidity and counterparty risk [12].

Technical risk is the failure of a transaction due to unforeseen events like digital assets hacks. As per the report submitted by Ciphertrace it is estimated that half of digital asset hacks which happened in the year 2020 were through Decentralized Finance services rather than traditional financial services. While Bitcoin and Ethereum which were known to be the largest public block chain networks were able to avoid majority of these hacks. The degree to which the DeFi transactions are interconnected will act as the major cause for malicious attacks. Transaction risks is another dimension of transaction failures in DeFi system. This will happen due to double spending which makes the transaction very expensive and the settlement layer is attacked which in turn affects application layer. Smart contract risks deal with the codes which will not get executed as it was intended to be. As it is a well-known fact that all software has significant threat of getting affected by bugs. The DAO a decentralized crowdfunding platform was the first viable DeFi service in the year 2016 when the ether was trading at $150 million was blocked up in the smart contract with an objective to funds decentralized way of funding transactions. To take care of these losses the miners introduced a new crypto known as Ethereum classic which reversed that entire transaction related to Ethereum classic through which the traders were able to prevent losses [13, 14].

Miner risk refers to the case in which the transaction processing mechanism behave maliciously to some transactions, in every case although the nature of threat posed is different the level of severity depends on analogous level of transaction. It is the miners who will decide on the frequency and the order of transactions to execute them. The miners will not execute the trade in free order, the miners will look for low transaction orders compare to high transactions orders.

Operational risk is also found in DeFi transactions even though the entire DeFi service is automated. Operational risk can be found in areas wherein human operators play a major role in DeFi services. Higher the decentralization of service lower is the operational risk found. Daily maintenance and upgradation is not possible in decentralized services. Code forks acts as an avenue for the groups which wishes to alter the code of block chain thereby bringing in the desired change. A major problem for block-chain based systems is the key management as these platforms identify users and their transactions underlying the assets through cryptographic keys which are specially designed to carry out a transaction. Governance will have an impact on block chain-based services which raise complex potential risks associated with block-chain technology. When participation rates are low, we can follow "one token-one vote" policy in order to increase the level of order placement and execution. In a recent study it is found that the DeFi transaction tokens are not evenly distributed and they are concentrated making the system complicated which lacks transparency.

Redressal of disputes is the final risk which arises due to irreversibility of transactions. The smart contract after execution could not be changed or reversed just because of enforcement from a government entity or an individual player.

Another category of risk is legal risk which arises due to ono-compliance with the legal regulations set by the local government of that geographical area. Money laundering is an example of how DeFi can be used to convert black money into white money. DeFi structure is not favorable of money laundering activity as the likelihood of those events will complicate enforcement. In legal compliance financial crime is one of the factors involving breach of anti-money laundering or countering the financing or anti-social activist groups restrictions, financial sanctions and related legal regimes. DeFi developers should also consider the fraud which are created through scams and manipulating the financial markets through misappropriation of the transactions by the investors. In some cases called rug pulls the scam makers will drag the investors into a legitimate DeFi service and after the transaction they will disappear. Sometimes the users may evade the regulatory framework through technical glitches by obfuscating activity.

Apart from the above-mentioned regular risks there could be emergent risks which are results of multiple event failures. As an example, sometimes a bank or a financial institution may be too big to fail. In this case a single default on a mortgage loan will create a cascading effect which leads to multiple impact. As DeFi operates at international level DeFi components might produce risks that cannot be allocated with any individual service. In traditional markets brokers can influence the performance of financial markets as they have control over the transactions manually but in a DeFi system where all transactions are automated it takes a few minutes to have the cascading effect of transactions wherein there will be a huge volatility and investors will face losses due to liquidations of collaterals in their positions. Quantifying and managing such risks are difficult. In DeFi the traditional value at risk breaks down as the investors prefer liquidating their positions only when it is profitable for them unlike in case of a traditional transaction system wherein the liquidation of collateral will happen even though if it is not profitable. These kinds of losses can be better estimated using Monte Carlo Simulation by creating the worst-case scenarios and the maximum losses which can be incurred over a period of time.

13.10 Financial Regulation Related to DeFi

Before we discuss the regulatory framework governing DeFi it is necessary to understand the basic objective of any DeFi service which should be protecting the basic rights of the investors, notice and prevent doubtful activities, capital mobilization, a safe and sound digital platform. DeFi operates under a wide variety of activities such as stocks, derivatives, cybercrime and insurance. Keeping these aspects in mind the following recommendations can be made:

- stoicism deals with decision not to allow new regulations in DeFi
- timely issuance of security warnings to the end users
- eliminate regulations which are not required from DeFi context
- granting license with limited regulatory framework making the transaction easy and smooth
- to prohibit some of the unnecessary measures which are prohibitive in nature
- design and develop new frameworks often with public voting or in consultation with the end users

13.11 Policy Tools for DeFi

In this section we will look at different frameworks followed by policy makers to impart regulatory framework to DeFi.

a. **Decision Tree Analytics**: Decision tree analytics is a series of decisions and chance points which are involved in a DeFi transaction where the policy makers can make a set of decisions mentioned below:

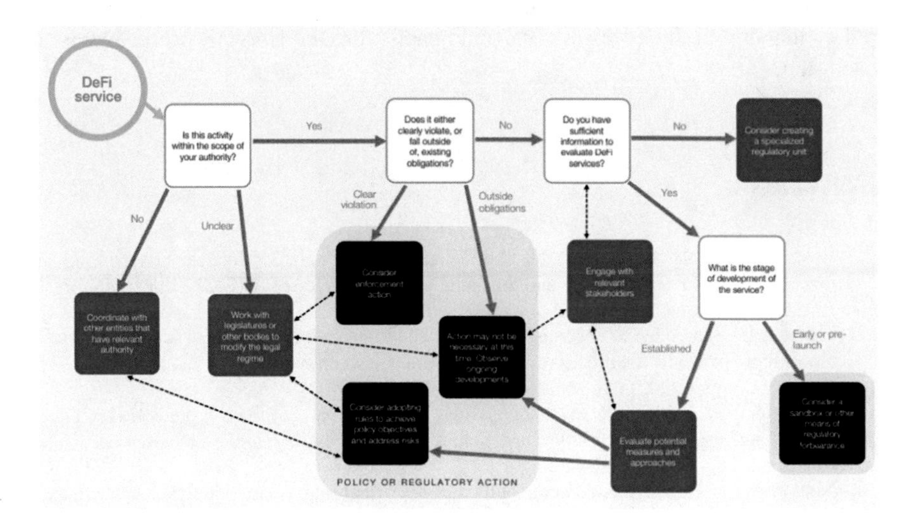

Source Decentralized Finance (DeFi) Policy-Maker Toolkit

b. **DeFi Life cycle-based regulation**: Just like a product or a human being, DeFi services also have a life cycle. The life cycle starts from Centralized system to decentralize system. There are mainly 4 main phases in DeFi service life cycle namely 1. Development 2. Publication 3. Deployment and 4. Operation.
c. **Transition Process**: Compared to the year 2017 wherein initial coin offerings (ICO's) was the means through which the digital assets were addressed. Specialized regulatory units are the desks targeted towards staffing and selecting the people who are well versed with technology, who can interact with industry people and guide others. Disclosure is the hallmark of any financial transaction. Therefore, the policymakers should make sure that the timely disclosures are made at the right time. Regulatory sandboxes will also act as regulatory forbearance program tools which act as carve outs can be defined by interactive activities and regulatory authority. There are a variety of cases which needs immediate attention and which raises red flags and other cases might address the grey areas. In such cases it is necessary for policymakers to intervene and narrow down the uncertainty and incentivize the risky activities.

13.12 Conclusion

The contents mentioned in this chapter are just the Starting points for the policy makers who can understand the risks involved in a typical DeFi service transactions and who can exploit the opportunities posed by DeFi businesses and services. Digital assets concept was first introduced in the year 2009 and smart contract platforms started in 2015. Therefore, any development in DeFi services can be considered in a rapidly evolving space. Together it can be concluded that DeFi financial services are a distinct and rapidly developing financial services. As mentioned in this chapter the policy makers should evaluate costs and benefits of DeFi services while evaluating each step.

References

1. Panda, S.K., Mohammad, G.B., Nandan Mohanty, S., Sahoo, S.: Smart contract-based land registry system to reduce frauds and time delay. Secur. Privacy e172 (2021). https://doi.org/10.1002/spy2.172
2. Panda, S.K., Satapathy, S.C.: Drug traceability and transparency in medical supply chain using blockchain for easing the process and creating trust between stakeholders and consumers. Pers. Ubiquit. Comput. (2021). https://doi.org/10.1007/s00779-021-01588-3
3. Niveditha, V.R., Sekaran, K., Amandeep Singh, K., Panda, S.K.: Effective prediction of bitcoin price using wolf search algorithm and bidirectional LSTM on internet of things data. Int. J. Syst. Syst. Eng. **11**(3–4), 224–236
4. Sathya, A.R., Panda, S.K., Hanumanthakari, S.: Enabling smart education system using blockchain technology. In: Panda, S.K., Jena. A.K., Swain, S.K., Satapathy. S.C. (eds)

Blockchain Technology: Applications and Challenges. Intelligent Systems Reference Library, vol. 203. Springer, Cham (2021). https://doi.org/10.1007/978-3-030-69395-4_10

5. Lokre, S.S., Naman, V., Priya, S., Panda, S.K.: Gun tracking system using blockchain technology. In: Panda, S.K., Jena, A.K., Swain, S.K., Satapathy, S.C. (eds) Blockchain Technology: Applications and Challenges. Intelligent Systems Reference Library, vol 203. Springer, Cham (2021). https://doi.org/10.1007/978-3-030-69395-4_16
6. Panda, S.K., Daliyet, S.P., Lokre, S.S., Naman, V.: Distributed ledger technology in the construction industry using Corda. The New Advanced Society: Artificial Intelligence and Industrial Internet of Things Paradigm. https://doi.org/10.1002/9781119884392.ch2
7. Panda, S.K., Satapathy, S.C.: An investigation into smart contract deployment on Ethereum platform using Web3.js and solidity using blockchain. In: Bhateja, V., Satapathy, S.C., Travieso-González, C.M., Aradhya, V.N.M. (eds) Data Engineering and Intelligent Computing. Advances in Intelligent Systems and Computing, vol. 1. Springer, Singapore (2021). https://doi.org/10.1007/978-981-16-0171-2_52
8. Panda, S.K., Rao, D.C., Satapathy, S.C.: An investigation into the usability of blockchain technology in internet of things. In: Bhateja, V., Satapathy, S.C., Travieso-González, C.M., Aradhya, V.N.M. (eds) Data Engineering and Intelligent Computing. Advances in Intelligent Systems and Computing, vol 1. Springer, Singapore (2021). https://doi.org/10.1007/978-981-16-0171-2_53
9. Panda, S.K., Dash, S.P., Jena, A.K.: Optimization of block query response using evolutionary algorithm. In: Bhateja, V., Satapathy, S.C., Travieso-González, C.M., Aradhya, V.N.M. (eds.) Data Engineering and Intelligent Computing. Advances in Intelligent Systems and Computing, vol. 1. Springer, Singapore (2021). https://doi.org/10.1007/978-981-16-0171-2_54
10. Nanda, S.K., Panda, S.K., Das, M., Satapathy, S.C.: Automating vehicle insurance process using smart contract and Ethereum. In: Chakravarthy, V.V.S.S.S., Flores-Fuentes, W., Bhateja, V., Biswal, B. (eds.) Advances in Micro-Electronics, Embedded Systems and IoT. Lecture Notes in Electrical Engineering, vol. 838. Springer, Singapore (2022). https://doi.org/10.1007/978-981-16-8550-7_23
11. Panda, S.K., Elngar, A.A., Balas, V.E., Kayed, M. (Eds.).: Bitcoin and Blockchain: History and Current Applications (1st ed.). CRC Press (2020). https://doi.org/10.1201/9781003032588
12. Blockchain technology: applications and challenges. In: Panda, S.K., Jena, A.K., Swain, S.K., Satapathy, S.C. (eds.). Springer, Intelligent Systems Reference Library. https://doi.org/10.1007/978-3-030-69395-4.
13. Varaprasada Rao, K., Panda, S.K.: A design model of copyright protection system based on distributed ledger technology. In: Satapathy, S.C., Lin, J.CW., Wee, L.K., Bhateja, V., Rajesh, T.M. (eds.) Computer Communication, Networking and IoT. Lecture Notes in Networks and Systems, vol. 459. Springer, Singapore (2023). https://doi.org/10.1007/978-981-19-1976-3_17
14. Varaprasada Rao, K., Panda, S.K.: Secure electronic voting (e-voting) system based on blockchain on various platforms. In: Satapathy, S.C., Lin, J.CW., Wee, L.K., Bhateja, V., Rajesh, T.M. (eds.) Computer Communication, Networking and IoT. Lecture Notes in Networks and Systems, vol. 459. Springer, Singapore (2023). https://doi.org/10.1007/978-981-19-1976-3_18

Chapter 14
An Overview of Blockchain Technology and Its Adoption in Industry

Riju Chaudhary, Devyanshi Bansal, and Sumit Kaur Bhatia

Abstract The development of a cryptocurrency cannot occur without the use of a vital piece of technology known as blockchain. It is a distributed ledger that has progressed to the point where they are now even more applicable and beneficial. This is due to the passage of time and the development of many sectors. These days, blockchain technology is being employed in certain ways throughout all the industries. In this study, each and every significant attribute and weakness is discussed and analyzed. We have also conducted a literature analysis on blockchain-related topics and highlighted some of the most current sectors in which blockchain technology has found the greatest utility.

Keywords Blockchain · Cryptocurrency · 51% attack · Healthcare · Supply chain management · Smart contract · COVID-19

14.1 Introduction

The ever-expanding world has led to advancements in technology, particularly in blockchain technology, which has led to important new innovations. The power to build anything in the digital world that cannot be replicated is very valuable, and blockchain has the tools to protect Bitcoin, which, in the current economic environment, is a great accomplishment in cryptography.

A distributed database or ledger that is shared across the nodes of a network connection is one example of what blockchain may be characterized as. A blockchain

R. Chaudhary · S. K. Bhatia (✉)
Department of Mathematics, Amity Institute of Applied Sciences, Amity University Uttar Pradesh, Noida, India
e-mail: sumit2212@gmail.com

R. Chaudhary
e-mail: rchaudhary@amity.edu; riju.chaudhary@gmail.com

D. Bansal
Amity School of Engineering and Technology, Amity University Uttar Pradesh, Noida, India
e-mail: Devyanshi.bansal@s.amity.edu; bansaldevyanshi049@gmail.com

S. K. Panda et al. (eds.), *Recent Advances in Blockchain Technology*,
Intelligent Systems Reference Library 237, https://doi.org/10.1007/978-3-031-22835-3_14

may be thought of as an electronic database that holds information in a format that is digital. The revolution that is brought by a blockchain is that it ensures the accuracy and safety of a track of data and establishes confidence without the requirement of a third party that can be relied upon. Let's begin at the very beginning: what precisely is meant by the term 'blockchain technology'?

14.1.1 What is Blockchain?

"Adaptive collective action is superior to bureaucracy," as stated by Hendrith Vanlon Smith Jr., CEO of Mayflower-Plymouth, defines blockchain as a way for customers to get the required transparency in their transactions. Customers have full access to a product's production history, making it impossible for manufacturers to conceal any information from them.

Blockchain is a decentralized, unchangeable record that makes it easier to keep track of resources and verify transactions in a network of businesses. Every unit in the computer network has access to this data. It saves information in digital representation since it is a database. The blockchains' blocks are where they organize a bunch of pieces of information. Each of these blocks has storage capacity, and when they are full, they are sealed and connected to previously filled blocks to form the blockchain. The data is permanently kept on the blockchain, and once it has been saved, it cannot be modified. It is a peer-to-peer network that is used for transaction values and there is no requirement of any third party for settling the transactions. Blockchain technology is now being used in a variety of industries, including cryptocurrencies in the banking industry such as Bitcoin, Ethereum, and Zcash (Zerocash).

As a result of this, we now have a better understanding of how blockchain technology works. With that in mind, let's look at one of the most crucial areas of application for blockchain technology: Bitcoin.

Bitcoin is a sort of electronic money that works irrespective of a central bank, keeping a track of trades and generating fresh bits of cryptocurrency, through the computer solving of algebraic puzzles. Furthermore, this blockchain network employs a hash-based Proof-of-Work (PoW) distributed consensus mechanism. There must be a question at this point, and that is what does this PoW mean? So let's simply have a look at it.

PoW is defined as a method that necessitates a considerable but manageable quantity of efforts to discourage wasteful or harmful uses of computer resources, such as generating phishing emails or conducting interruption of service attacks. PoW is commonly employed in cryptocurrency mining to validate deals and generate new coins. Because of PoW, Bitcoin and other cryptocurrency transactions are99 conducted securely, eliminating the requirement for a reliable intermediate group. PoW at a level demands enormous quantity of power, which only grows as more miners enter the system.

14.1.2 History of Blockchain

In 1992, the Merkel tree was invented, which increased the efficiency of blockchain by allowing several records to be gathered in a node. These trees produced a safe chain of building blocks. The blocks were linked together using the same method of connecting one block at a time, and the chain's most recent block included all the information on the previous blocks' value. But ultimately, this innovation was underutilized.

The research scientists Stuart Haber and W. Scott Stornetta first proposed the blockchain technology in 1991. To prevent backdating or tampering with digital documents, they fundamentally tried to offer a technologically workable alternative for it. They created a system of encrypted blocks with security that would be used to store time-stamped data.

A highly well-known cryptocurrency called bitcoin was used in year 2008 by Satoshi Nakamoto to introduce blockchain, a technology that is now widely used. It timestamps and verifies each trade via a peer-to-peer connection [1]. Without the need for a centralized authority, it might be controlled independently. These developments were so advantageous that they established blockchains as the foundation of cryptocurrency.

In 2013, Ethereum pioneered the novel idea of blockchain by presenting it as "a decentralized network that executes smart contracts." It was noted that blockchain "facilitates programmers to establish trades, keep registrations of liabilities, move money in line with commands issued a very long time ago (such as a will or an upcoming contract), and many more things that have not been conceived yet, all without a mediator or collateral risk."

Ethereum is a cryptocurrency-transfer system that enables you to send bitcoin to anyone for a nominal charge. In contrast, Bitcoin is only a money that people can use to buy things. In addition to this, it operates programs that everybody may use and that nobody can stop. An extensive study of the miners in the Bitcoin and Crypto Ethereum blockchains is presented by the authors to demonstrate the computational variety of possibilities and to facilitate the creation of accounts that would make it possible to identify aberrant behaviors and stop 51% of attacks. Also, the analytical research of the researchers has revealed that in recent years, there have been an increasing accumulation of hash rate knowledge in a relatively small number of miners, which poses a true danger to existing blockchains. In addition, there is a mining tendency among main hackers, which enables the discovery of aberrant behavior. This trend enables the detection of prime hackers.

In recent years, blockchain technology has not only advanced but also made efforts to expand into other markets, including those in the healthcare, government, Sc, entertainment, and many more. It is anticipated that smart contracts enabled by blockchain would transform several conventional sectors, including the financial sector, healthcare, the energy sector, and others [2].

14.2 The Fundamental Components of a Blockchain

It may be challenging to conduct an accurate study of the many pieces that make up a blockchain. It is common practice to classify nodes, distributed ledgers, immutability, decentralization as components of a blockchain; however, this classification is not totally true. We are going to make use of this chance to talk about the four primary aspects of a blockchain, and then we are going to go on to discussing the additional characters and contributors afterwards.

a. **The technique of distributed ledgers**

The distributed database and its irreversible log of events are accessible to all network users. Expenses are validated just individually with this digital ledger, reducing the unnecessary repetition that is usual commercial stations. This digital ledger records interactions that include the transfer of commodities or information across members of the network. Members in the network control and reach an agreement on modifications to the ledger information. There is no central body or third-party broker engaged, such as a commercial bank or agency. Every entry in the public ledger contains a timestamp and a unique cryptographic identity, giving the ledger an independently audited and unchangeable account of all network interactions.

b. **Unchangeable data—immutability**

No participant is allowed to modify or otherwise alter a transaction after it has been entered into the shared ledger and recorded there. If an error is discovered in the history of a transaction, it is necessary to enter a new transaction in order to rectify the situation; both of these procedures may then be seen. It is impossible to change, which means that it provides an exceptionally high degree of protection.

c. **Node**

The devices known as nodes are responsible for storing these enormous volumes of data. Nodes might be small computers like laptops or large servers like data centers. Each of the nodes that make up a blockchain network is connected to the others. It might be either a Full Node or a Partial Node depending on the circumstances. Full Node: This kind of node stores an unaltered copy of all transactions on the blockchain. Transactions may be validated, accepted, or rejected by it depending on its discretion. Because it does not keep a complete copy of the blockchain ledger, it is often referred to as a "Lightweight Node." Partial Nodes fall under this category. The whole history of a blockchain network is stored among its nodes. After confirming the authenticity of the associated data and signatures, the nodes add a new block to the blockchain network. This block is subject to a second round of verification using the hash code. It is possible for nodes to exist both online and offline. In addition to this, nodes do quality assurance checks on the block of transactions.

d. **Decentralization**

Decentralization means a design of the network in which more than one authority acts as a centralized location for a subset of the members. Because some individuals are located behind a centralized location, the absence of that hub will prohibit those participants from being able to communicate with one another [3].

The term "decentralization" is used to describe the process by which power and decision-making authority are moved from a centralized entity (a person, organization, or set of these entities) to a distributed system in the context of blockchain technology. Mostly in manner of a distributed ledger, each participant in the network has an identical copy of the data that is being tracked. If the ledger of a member is in any way changed or corrupted, it will be discarded by the rest of the other users in the network.

e. **Proof of stake**

Proof of stake, sometimes known as PoS, is an alternative to proof of work that is more energy efficient. Validators take the role of miners in a proof-of-stake system. The validators will begin the process of confirming the blocks as soon as they have secured a certain number of coins as a stake. After the block is attached to the blockchain, the validators will each get a payout according to the amount of stake they own. The PoS makes far better use of available resources than the PoW does in this regard. Ethereum, one of the most widely used blockchain protocols, is now dependent on PoW, although it has plans to switch to PoS sometime in the early part of 2018 [4].

14.3 Recognizing the Immutability of Blockchain

A popular word used in many publications and research papers is immutability. The capacity of a blockchain ledger to stay intact, unmodified, and everlasting is referred to as immutability. It is a blockchain tool that allows for a rapid, accurate, and cost-effective approach. Before delving into the specifics of blockchain immutability, it is critical to grasp a fundamental principle known as cryptographic hashing or hash value.

The authors investigated the immutability of blockchain node storage using a distributed hash architecture, resulting in immutable or irrevocable blockchain node memory. They employed thought mapping as their approach. As a conclusion, the distributed hash model simply keeps a fraction of the block data at every node, and the transaction data is considered a commodity [5].

The cryptographic theory aids in the execution of trade information contained within every chunk of blockchain. The hash value is an alphanumeric string that is unique for every block since it is created independently by every other block. The hash function also acts as a digital signature to every block, ensuring that nobody can interfere with the computer or change the information which has already been put into the block. Every new block incorporates the meta-data from the previous block's hash value in this hashing mechanism, making the blockchain procedure indestructible and making it nearly impossible to change the data once it is included in the node. Presently, programming languages have an array of hash functions which when employed, generate a checksum signature. Irrespective of the size of the input, the span of the resulting string is usually 64 characters. From the above information it is well understood that an immutable ledger on blockchain provides users with a strong encryption and transparency. Thus, while cybercriminals can try to attack the information at any moment, it is nearly unfeasible due to the chain formation. It also ensures that data continues to flow easily and securely, as well as protects data from corruption.

Everyone is aware that the blockchain industry is expanding at a breakneck pace, but unfortunately, this also means that there are more risks. There are a variety of methods that may be used to compromise a blockchain. The performance of these assaults gets more and more difficult over time as more processing power is connected to the system, making it more secure. In this part of the chapter, we will discuss the most recent and widespread security flaws involving blockchain technology.

14.4 Evolution of Blockchain from Industry 1.0 to 4.0

Since the first Bitcoin application was released in 2008, a number of trust-based worldwide solutions have been created to circumvent the bad intent that may arise in essential apps. The use of blockchain technology offers a solution to the problems caused by fraud in many different industries, including banking, finance, and the government. Because fraudulent operations were carried out on such a broad scale in order to manipulate the company functioning, this was a significant cause for worry. The capabilities of a blockchain network are more than enough to solve the problems that have been outlined above. Because blockchain eliminated the need for third-party services and attached a digital signature to each and every operation, it may be considered a trustworthy and safe network [6] (Fig. 14.1).

- **Blockchain 1.0: cryptocurrency**

The initial iteration of the blockchain technology, also known as Blockchain 1.0, emerged from the digital ledger technology. It does this by providing a dispersed

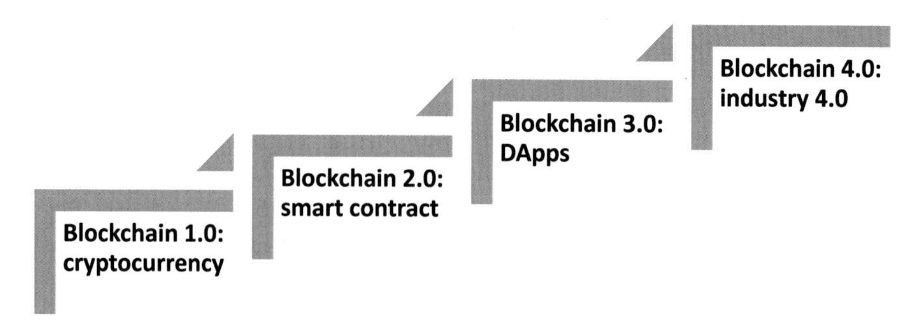

Fig. 14.1 Evolution of blockchain

network to all of the participants, which helps to overcome the issue of double spending. Mining is a technique that is engaged in blockchain version 1.0. Its purpose is to mine or verify data transfers, or to alleviate the problems associated with centralized systems' lack of security [7].

- **Blockchain 2.0: Smart Contact**

The term "blockchain 2.0" refers to a new technology that is the second generation of the original blockchain and integrates numerous of its previous innovations. The new fundamental ideas are embodied in "Smart Contracts." The Ethereum Blockchain, which has its core task as the facilitation of the use of Smart Contracts, is currently the most well-known initiative in this sector [8].

- **Blockchain 3.0: DApps**

Following the release of version 2.0, a new version was released, which included DApps, also known as decentralized application software. A decentralized application (DApp) functions similarly to a traditional app in that it may have a frontend written in any language that communicates with its backend and that the backend code is executed on a distributed peer-to-peer network. in addition, it makes use of other consensus mechanisms, like as proof of work and proof of stake, both of which are useful when it comes to the implementation of smart contracts in blockchain [6].

- **Blockchain 4.0: Industry 4.0**

The Blockchain 4.0 is an additional forthcoming step that will be beneficial to the advancement of Blockchain technology. It intends to do this by delivering Blockchain Technology as a platform that can be used by businesses to build and operate apps, therefore bringing the technology into the mainstream. Blockchain may be used with other successful technologies, such as artificial intelligence, if this option is realized. 4.0 makes it feasible for several platforms to be coherently integrated into a single system in a smooth manner, allowing for the fulfilment of the requirements of businesses and industries. Unibright is the introduction platform that will be put forth to bring forward Blockchain 4.0 utilities. It permits an integration of multiple different blockchain business models. One such example of this would be the SEELE Platform, which enables integration in the blockchain realm by facilitating cross-communication across multiple protocols across a variety of services in a harmonious manner. Transactional speeds of up to one million per second are presently not attainable with any of the generations that are in use; however, the fourth generation has the potential to make this practical.

14.5 Blockchain Security Breaches

Blockchain is now becoming a dominant technology, however this has raised the number of security vulnerabilities. There are many researchers who have examined

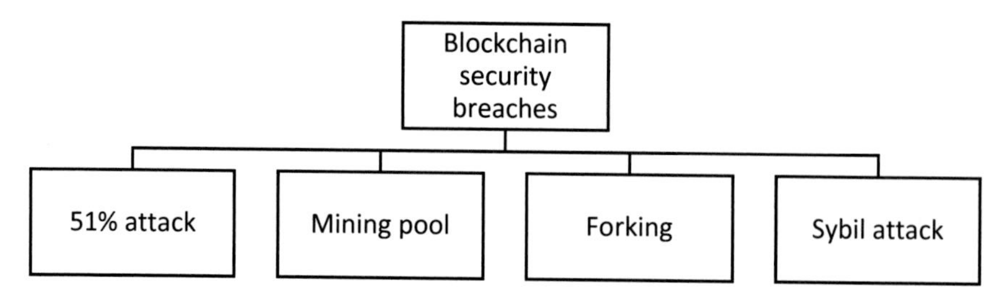

Fig. 14.2 Blockchain security breaches

about, but a comprehensive examination of all concerns in one location is essential. So, the various threats are (Fig. 14.2):

a. **51% Attack**

51% assault is one in which a team of hacker's controls more than 50% of the site's processing bandwidth or computer power. The intruders would indeed be able to block fresh transactions from receiving verification, halting transfers among selected or all consumers. They would also be able to rectify transactions made while in control of the network, allowing them to double-spend bitcoin (Double spending occurs when a customer uses the similar cryptocurrency for several payments. It is comparatively easier with POW-based blockchains since attackers may simply manipulate the interval among both the commencement and verification of two operations. They would very probably be unable to modify existing blocks. Successful 51% assaults are only possible on smaller cryptos with lesser mining systems.

For a long time, the 51% attack was thought to be inherently unknowable due to the enormous offensive cost. But, in latest years, the attempt has occurred on a regular basis, losing several cryptocurrencies millions of dollars. The authors have addressed the five most sophisticated defense tactics for preventing the attack, as well as their key weaknesses. They find that in most circumstances, security measures struggle to supply effective defense against the 51% approach because the flaws are acquired from network configuration [9].

Different techniques which can prevent 51% attack:

- Delayed Proof-of-Work (DPoW)—It is a blended agreement mechanism that allows a blockchain to benefit from the security provided by secondary blockchain hashing ability. There is a set of notary nodes that assist in the transfer of information from the first ledger to the next. It disregards the largest chain rule, and its notarial nodes contribute to the security procedures. It also does not identify the susceptibility ahead of time.
- PirlGuard—It is based on the Horizon punishment system. This security protocol substantially discourages intruders from trying malicious gazing, providing somewhat that boosts the security and the network infrastructure. It uses a compensation mechanism for attaching nodes. Master nodes degrade the system while also identifying vulnerabilities as soon as feasible.

- **Threats on Mining pools**

A mining pool is a collaborative group of cryptocurrency miners that pool their computing capabilities over web. Workers of the group attempt to locate a block and, if they succeed, then they obtain a benefit in the form of the corresponding currency.

According to the authors of this review, an assailant needs just a small proportion of the assets of a susceptible mining pool, making this attack method incredibly inexpensive to a somewhat less powerful competitive mining pool [8].

There are three ways to mine a pool:

- Proportional mining pool—In this form of pool, miners that contribute to the pool's computing resources get equities when the pool successfully finds a block. Following that, miners are rewarded in accordance with the number of equities they own.
- Pay-per-share pools—These pools give reward instantly irrespective of when the block is discovered. A miner who contributes to this form of pool can trade stakes at any moment for an equitable reward.
- Peer-to-peer pool—They incorporate a distinct blockchain relating to the pool overall, which is meant to prevent both the pool's administrators and the pool as a whole from collapsing caused by a single centralized fault.

There are 2 kinds of mining pool attacks:

- Internal assaults happen when a miner intentionally takes more than the required rewards, which disrupts the usual functioning of the pool and forces it to neglect subsequent processing activities.
- When a worker uses more hash power than necessary to attack the group, this is an example of an external assault since it results in double expenditure [10].

The primary drawbacks of mining pools are that disruptions can occur at any moment, followed by an undesirable and convoluted reward structure since the awards are divided, reducing the miners' revenue. There are two sides to every story, and just as everything has a downside, it also has a benefit. As a result of pooling, mining expenditures have fallen, and the likelihood of creating more money over time has increased.

c. **Forking**

Because blockchain is the best cryptographic solution and is used in almost all businesses, it is subject to several massive assaults. The most typical attack following the 51% strike is the forking attack. Before we go into the details of forking, let's first define Most Trusted Chain (MTC). The forking attack is a type of risky conduct that seeks to substitute the MTC by creating an alternate chain in order to obtain profits.

Three basic features of forking attack:

- Simpleness—In a forking attack, the assailant's side-chain nodes are also validated by processing capacity just like the main-chain blocks. Simply mining blocks and picking the right time to broadcast them is all that the hacker requires to do. One of most noticeable characteristics of a forking assault is its simplicity, that enables

it to be conducted with no need for embedded processing assistance. It implies that the intruder faces a very limited field.
- Uncertainty—The pause time that leads to the fork architecture is a relatively regular occurrence in a public blockchain because of the intricate nature of system circumstances. To obtain a superior impact, the beginning of a forking attack might be concealed in large forks.
- Predictability of achievement—The assault could have an extremely strong accuracy chance if the assailant has sufficient computer power and a purposeful publication strategy.

There are 3 basic ways to protect against the forking attack: decreasing fork, limiting collaborative mining, and selfish mining.

Each form of blockchain has a separate forking process depending on its design and use case. Forking comes in three varieties:

- Soft fork—It occurs whenever the blockchain technology is modified in a reverse manner.
- Hard fork—It occurs when the bitcoin technology is updated in a manner that is not available digitally.
- Temporary fork—When multiple miners extract a new block at the exact moment, this occurs.

Because blockchain is used in so many industries, including cryptocurrencies, a successful forking assault would result in massive economic damage. Since blocks employ hash functions to join as a network, the forking attack is one of the few ways to exploit the infrastructure.

d. **Sybil Attack**

An assault known as a Sybil attack is one in which the attacker appears to be a large number of different persons all at once. Through the creation of several false names, it exerts its influence on the network and ultimately controls the whole network. These several identities, when seen from a single perspective, seem to be ordinary users. However, behind the scenes, there is a single entity known as an unidentified assailant that commands all of these bogus entities at the same time [11].

Because there isn't a reliable identity management system in place, classic attacks may be launched against blockchain. An attacker can undermine a permissionless blockchain using a technique known as the Sybil attack, in which they create a large number of pseudonymous identities (also known as fake user accounts) and force genuine entities into a minority position. This can have a significant impact on the blockchain. These virtual nodes are able to simulate the behavior of real nodes, which enables them to exert an outsized amount of impact on the network. This might result in a chain reaction of assaults including DoS, DDoS, and others [12].

14.6 Blockchain for Different Industries

Nevertheless, not all industrial sectors have acquired the same amount of attention from the blockchain, and for certain applications, blockchain research has already come to an end. Although not in all industries and fields, blockchain has the potential to influence the future.

To concentrate a little more on industry verticals that have a great potential of reaching the consumer and significant influence on the market, scholars, researchers and several blockchain practitioners have proposed several sectors that should but not have got enough blockchain attention [13] (Fig. 14.3).

14.6.1 Security and Privacy in Blockchain

Nowadays, there are over a thousand different blockchains: some are simply copies of Bitcoin, while others differ dramatically in structure and offer distinct performance and security assurances. Security and privacy are at the heart of blockchain technologies, and the power to either create or ruin these systems. The concept is that there are two distinct kinds of keys which contribute to the increased safety of the blockchain, which are essential to the technology. These two categories of keys: Public and private key.

The usage of public and private keys in the case of confidentiality is perhaps the most significant aspect of blockchain. Everyone seems to be familiar that ledger is an asymmetric cryptography used to safeguard transaction records. Both keys are arbitrary strings of integers that are related to cryptography. It is almost hard to predict other users' keys, increasing blockchain security and protecting it from hackers. Because it includes no personal information, the public key may be exchanged with anyone.

Based on these keys, we are also capable of determining the two different kinds of blockchains, which are the Public and the Private Blockchain. The public blockchain, also termed as a permissionless blockchain, allows anybody to read and verify logs. On the other side, private blockchains, also referred as permission blockchains, are restricted and typically only allow links between one firm or partnership.

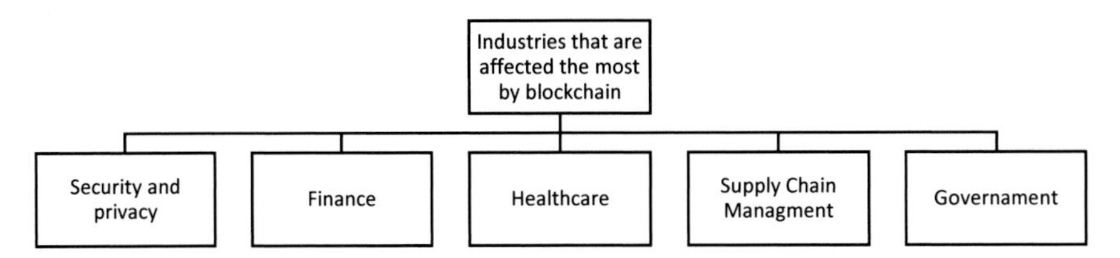

Fig. 14.3 Industries that are affected the most by blockchain

14.6.2 Blockchain in Healthcare

The blockchain is indeed a relatively young technology with the potential to change several sectors, including the medical industry. The extent to which it can realize its potential in the healthcare industry is highly dependent on the rate, at which connected cutting-edge technologies are incorporated into the ecosystem. Tracking of systems, medical insurance, the movement of medicines, and clinical studies are all part of this. By using monitoring of devices, hospitals can chart their services within the context of a blockchain, and this capability extends even across the whole of the life cycle. The use of blockchain technology has the potential to significantly enhance patient history management, in particular monitoring the process of insurance settlement. As a result, treatment options may be accelerated with efficient data preservation [14].

A blockchain is a digital ledger that keeps data and is used to strengthen the trust in the procedures that are used to exchange healthcare data. This is accomplished via the use of computerized monitoring of the origination of the data and responsible identity authentication.

Two potential solutions for efficient and secure permission management might be made available via the use of blockchain technology.

The risk of an error being made by a human being is reduced thanks to the use of smart contracts and cryptographic keys, which also help to cut down on the amount of time that elapses between the collection of patient data and the fulfilment of operations like healthcare billing and fee processing.

Blockchain technology's advantages in medical data/information handling:

It is essential to examine and maintain the security of the data, since any loop might result in the loss of data. The following table provides an explanation of how blockchain technology is used to efficiently manage the administration of healthcare records or information (Fig. 14.4):

Information clarity, accountability, data integrity, inspection, data authenticity, customizable entry, confidence, anonymity, and safety are all major concerns for changing healthcare database systems. Furthermore, a major number of present healthcare systems used for data management are centralized, which poses the danger of data loss in the event of a major disaster. Blockchain is a new and decentralized innovation with the possibility to drastically revolutionize, restructure, and alter the way information is managed in the medical industry [15].

So here are some opportunities that can be utilized in healthcare sector using blockchain (Fig. 14.5):

a. Increased tracking of drugs—What occurs in this case is that blockchain prohibits medication changes, making drug testing more transparent and accountable.
b. Integrative healthcare and clinical testing—Comprehensive utilization of healthcare databases is offered, coupled by the prohibition of repeating diagnostic tests, and blockchain also assists new physicians in quickly learning about past patients' histories.

S. No.	Characteristics	Benefits of Blockchain
1.	Data Reliability	The management of all health records across several locations is standardized. It makes it possible to securely store client health records.
2.	Data Consistency	Every piece of information kept on a blockchain has a predefined code.
3.	Data Certainty	Blockchain eliminates the possibility of security breaches here. The saved data is protected from harm that might be done by hackers and any information handling errors.
4.	Managing expenses for health data	Pharmaceutical firms may now simply avail patients records without having to visit several different places.
5.	Exchanging of information in world wellness	In this, healthcare organizations are provided with characteristics for worldwide connectivity and traceability.
6.	Enhanced monitoring of patient records	It maintains medical organizations compliant with legislative laws and guidelines and makes it simple for examiners to monitor the transactions. Even so, excessive data duplication is avoided.

Fig. 14.4 Administration of health records using blockchain

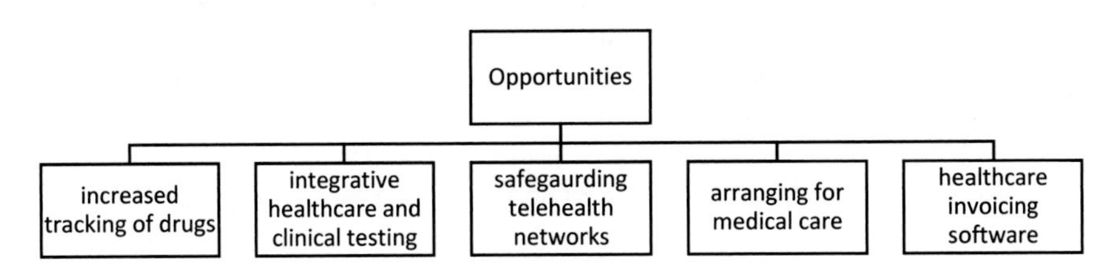

Fig. 14.5 Opportunities of blockchain in healthcare sector

c. Safeguarding telehealth networks—Telehealth is defined by the Health Resources and Services Management as the utilization of digital data and telecommunications technology to assist widely prescribed medical services, patient and expert wellness learning, global health, and healthcare management. The effectiveness of telehealth systems is primarily determined by how well they manage vulnerabilities. Blockchain can assist telehealth systems by providing enough confidence, confidentiality, and online privacy.

d. Arranging for medical healthcare policy—Blockchain technology provides unparalleled clarity by logging all purchases made on it in a decentralized, tamper-proof, identifiable, irreversible, and encrypted manner. It can help to streamline the health insurance process and improve supplier listing reliability by gathering smart contracts, agreement data, and payments in an organized way, therefore addressing the incompatibility challenge and making the administrative task easier.

e. Healthcare invoicing software—One of the major contributing factors of unintended billing mistakes is the extensive encoding required in the healthcare accounting system. Blockchain is a potential innovation that can make payments more convenient and safer. Blockchain can reduce such constraints by preserving all information in an irreversible way, allowing health insurers to settle insurance claims faster while using less assets, effort, and money.

A Blockchain is utilized to keep record of drug duties and to supervise the delivery of healthcare services. It keeps healthcare income reports up to date while reducing data translation effort and money. Each chunk of patient medical records is hashed using blockchain based technology.

14.6.3 Blockchain and Government

The implementation of an e-government system has substantially increased both the effectiveness and the openness of day-to-day governmental activities. The vast majority of the current electronic services offered by the government are delivered in a centralized fashion and significantly depend on human beings to manage. The highly centralized information technology infrastructure is more susceptible to assaults from the outside. Additionally, it is not too difficult for malicious individuals who are already within the system to undermine the data's integrity. To find solutions to these problems, researchers have come up with the idea of enhancing the e-Government system with the help of the blockchain technology and the decentralized autonomous organization (DAO) [16].

The capability of blockchain technology to capture payments on distributed ledgers presents significant options for authorities to make it more transparent, reduce the likelihood of fraudulent activity, and build confidence in the public sector [17].

At this moment, Blockchain is being utilized in a variety of settings; nevertheless, there are still a great deal of potential adoption of Blockchain technology that have not yet been uncovered or put into practice. The decentralized nature of blockchain, together with the transparency of the network and the authenticity of the information that is kept there, are the distinguishing features of this technology. Because of these qualities, another potential use for Blockchain technology is in the process of allocating public funding among various projects. In most cases, when money is allotted to a project, there is no awareness as to how these resources are being spent, and a significant portion of it is never displayed in records owing to the fact that it is corrupted. In order to address this issue, a solution that makes use of Blockchain technology to guarantee transparency has been presented [18].

14.6.4 How Blockchain Has Helped Government in COVID-19?

Because COVID-19 has created such a substantial disruption, the government is obligated to take prompt action in response to the situation. Within a month of the virus's full-scale spread, the World Health Organization (WHO) was forced to proclaim that COVID-19 had reached pandemic proportions because of the intensity of the outbreak. Consequently, in order to combat this COVID-19 dilemma, we need a strategy that is supported by technology. The various attributes of blockchain technology, such as decentralization, transparency, and immutability, have the potential to contribute to the reduction of this pandemic through the early detection of epidemics, the timely delivery of medication, and the protection of user privacy during treatment [19].

As there was a significant increase in the demand for medical supplies and equipment such as ventilators, face masks, hand sanitizer, personal protective equipment (PPE), and testing kits and probes for polymerase chain reaction (PCR). If there hadn't been adequate management of the supply chain, these goods never would have made it to the location where they were required in a timely way. As a result, the use of blockchain technology in supply chain management proved quite effective in assisting the government.

The ability to readily exchange data is one of the characteristics of blockchain that was covered above. This feature proved to be incredibly valuable during a pandemic since it made it possible for the information of patients to be easily transmitted to any healthcare facility. The use of smart contracts has made contactless delivery simpler and more efficient. The government received significant assistance from all these factors in its battle against the epidemic and in its efforts to restore normalcy to the situation. Nonetheless, the government was also accountable for bearing the loss. More than 200 nations, including the United States of America, the European Union, South Africa, and Russia, are dependent on the supply of generic medications that India provides. India is responsible for more than 20% of the worldwide supply of generic medicines. Because to COVID-19, India has placed limits on the shipment of twenty-six bulk pharmaceuticals, also known as active pharmaceutical ingredients (APIs). These 26 APIs collectively account for around ten percent of India's total exports [20, 21].

14.6.5 Supply Chain Management and Blockchain

The administration of a Supply chain (Sc) involves taking care of all aspects of a product or service's manufacturing process, beginning with the purchasing of its raw materials, and continuing all the way through the distribution of the finished item to the end user. Sc executives may utilize the data provided by blockchain solutions for

Sc, to better control the disruptions that are occurring now and to develop resilience for the future.

Members in a blockchain supply may find it easier to capture important data such as price, date, location, quality, certification, and other pertinent data in order to facilitate more efficient Sc management. The accessibility of these details within blockchain has the capacity to boost the reliability of the resource Sc, minimize cost that is caused as a result of counterfeiting. the grey market, significantly improve awareness and regulation regarding outsourced product manufacturing, and possibly strengthen an organization's leadership in the market [22].

In 2020, the worldwide national emergency caused by the Covid-19 pandemic emphasized the necessity for SC digitization. Across the whole SC network, the digitization phenomenon has been producing new realities and relationship patterns. Blockchain technology is amongst the most prominent revolutionary technologies for SC development [23]. The attempts to incorporate blockchain technology in Sc often confront a large number of hurdles and obstacles, which may pose a significant risk to the sustainability of such efforts. As a result, it is very necessary to conduct an in-depth investigation of the difficulties associated with the use of blockchain technology [24].

Benefits of blockchain technology in SC management.

- The opacity of BCT is rising [25] allowing for tasks such as mapping, capturing, and monitoring [26].
- Blockchain technology offers a security assurance [27] method as well as data security for information exchange.
- It also improves effectiveness as well as system-wide performance [27, 28].
- It expedites overseas shipment and paperwork procedures and allows faster response to SC irregularities (Rejeb et al., 2021).
- Adoption of Bctdecreases fraudulent activity and eliminates risk in acquiring, commodities, and trades. Additionally, reduces operational risk in general. It even bans the exchange of bogus or fraudulent assets [25, 26, 29, 30].

14.7 Smart Contracts and Blockchain

Briefly said, smart contracts are computer programs that are recorded on a blockchain and are activated automatically when certain criteria are satisfied. They are often used to streamline the implementation of an arrangement so that all parties involved may be instantly confident of the results. This eliminates the need for any intermediary and prevents any wastage because of their participation. They also have the ability to automate a process, which will cause the next step to be triggered when the circumstances are satisfied.

Without the assistance of a dependable third party, smart contracts enable, carry out, and maintain agreements between unreliable parties [31].

Network automation and the capacity to transform paper contracts into digital ones were made possible by smart contracts. By allowing automatic transactions without the intervention of a central authority, smart contracts allowed users to formalize their agreements and trust relationships [32].

Smart contracts are replicated to each node of the blockchain network to avoid contract manipulation. With this Human error might be decreased to lower the likelihood of conflicts involving such contracts by allowing the execution of operations by machines and using the capabilities offered by blockchain platforms [2].

Despite the development over the years, Smart contracts continue to encounter several hurdles. For example, in 2016, the Decentralized Autonomous Organization (DAO) Smart contract was exploited to steal about 2 million Ether due to its vulnerability. Aside from the security issue, smart contracts have a number of other obstacles, including privacy, legal, and performance concerns [2].

Blockchain-based smart contracts have emerged as a primary area of academic interest because to its unique qualities, which include the decentralized storing of transaction data, the independent implementation of contract algorithms, and the decentralized creation of trust. It has the ability to completely transform the operational framework of almost any company, which would ultimately lead to improved levels of service. It also has manufacturing implications, such as cryptocurrency systems, as well as applications in logistics, agriculture, real estate, energy trading, and other field [33].

14.8 Conclusion

There is no shadow of a doubt that the advantages provided by Blockchain technology will quickly entice companies and organizations all over the globe to increase their level of investment in it. As was said in the preceding sections, blockchain technology offers a multitude of benefits; yet there are a vast number of sectors that do not make significant use of blockchain technology. Blockchain technology may allow for the expansion of certain sectors over time, but since blockchain itself is getting mature, it will take some time for certain industries to become more advanced. Therefore, industries may profit from the use of blockchain technology, and they can expand with the help of the new technology over time. The insurance business, real estate market, transportation industry, cloud storage industry, food industry, and agriculture industry are all still expanding, and we will see significant growth in these areas over the next several years due to the adoption of blockchain. Last but not the least there is a significant amount of growth for development in blockchain as the industry moves from the 4.0 to the 5.0 level.

References

1. Popovski, L., Soussou, G., Webb, P.B.: A Brief History of Blockchain, p. 3 (2018)
2. Khan, S.N., Loukil, F., Ghedira-Guegan, C., Benkhelifa, E., Bani-Hani, A.: Blockchain smart contracts: applications, challenges, and future trends. Peer-Peer Netw. Appl. **14**(5), 2901–2925 (2021)
3. Yaga, D., Mell, P., Roby, N., Scarfone, K.: Blockchain technology overview. National Institute of Standards and Technology, Gaithersburg, MD, NIST IR 8202, Oct 2018. https://doi.org/10.6028/NIST.IR.8202
4. Veeramani, K., Jaganathan, S.: A quick synopsis of blockchain technology. Int. J. Blockchains Cryptocurrencies **1**, 54 (2019). https://doi.org/10.1504/IJBC.2019.101852
5. Rahardja, U., Hidayanto, A.N., Lutfiani, N., Febiani, D.A., Aini, Q.: Immutability of distributed hash model on blockchain node storage. Sci. J. Inform. **8**(1), 137–143 (2021)
6. Tanwar, S.: Blockchain revolution from 1.0 to 5.0: technological perspective. In: Tanwar, S. (ed.) Blockchain Technology: From Theory to Practice, pp. 43–61. Springer Nature, Singapore (2022). https://doi.org/10.1007/978-981-19-1488-1_2
7. Tanwar, S.: Blockchain Technology: From Theory to Practice. Springer Nature (2022)
8. Aggarwal, S., Kumar, N.: Chapter Fifteen—Blockchain 2.0: smart contracts☆☆working model. In: Aggarwal, S., Kumar, N., Raj, P. (eds.) Advances in Computers, vol. 121, pp. 301–322. Elsevier (2021). https://doi.org/10.1016/bs.adcom.2020.08.015
9. Aponte-Novoa, F.A., Orozco, A.L.S., Villanueva-Polanco, R., Wightman, P.: The 51% attack on blockchains: a mining behavior study. IEEE Access **9**, 140549–140564 (2021)
10. Gupta, N.: A deep dive into security and privacy issues of blockchain technologies. In: Handbook of Research on Blockchain Technology, pp. 95–112. Elsevier (2020)
11. Islam, M.R., Rahman, M.M., Mahmud, M., Rahman, M.A., Mohamad, M.H.S.: A review on blockchain security issues and challenges. In: 2021 IEEE 12th Control and System Graduate Research Colloquium (ICSGRC), pp. 227–232 (2021)
12. Swathi, P., Modi, C., Patel, D.: Preventing sybil attack in blockchain using distributed behavior monitoring of miners. In: 2019 10th International Conference on Computing, Communication and Networking Technologies (ICCCNT). IEEE (2019)
13. Zeadally, S., Abdo, J.B.: Blockchain: trends and future opportunities. Internet Technol. Lett. **2**(6), e130 (2019)
14. Haleem, A., Javaid, M., Singh, R.P., Suman, R., Rab, S.: Blockchain technology applications in healthcare: an overview. Int. J. Intell. Netw. **2**, 130–139 (2021). https://doi.org/10.1016/j.ijin.2021.09.005
15. Yaqoob, I., Salah, K., Jayaraman, R., Al-Hammadi, Y.: Blockchain for healthcare data management: opportunities, challenges, and future recommendations. Neural Comput. Appl. 1–16 (2021)
16. Diallo N., et al.: eGov-DAO: a better government using blockchain based decentralized autonomous organization. In: 2018 International Conference on eDemocracy & eGovernment (ICEDEG), Apr 2018, pp. 166–171. https://doi.org/10.1109/ICEDEG.2018.8372356
17. Batubara, F.R., Ubacht, J., Janssen, M.: Challenges of blockchain technology adoption for e-government. In: Proceedings of the 19th Annual International Conference on Digital Government Research: Governance in the Data Age. ACM Other conferences. (2018). https://doi.org/10.1145/3209281.3209317. Accessed 28 June 2022
18. Mohite, A., Acharya, A.: Blockchain for government fund tracking using Hyperledger. In: 2018 International Conference on Computational Techniques, Electronics and Mechanical Systems (CTEMS), Dec 2018, pp. 231–234. https://doi.org/10.1109/CTEMS.2018.8769200
19. Sharma, A., Bahl, S., Bagha, A.K., Javaid, M., Shukla, D.K., Haleem, A.: Blockchain technology and its applications to combat COVID-19 pandemic. Res. Biomed. Eng. **38**(1), 173–180 (2022). https://doi.org/10.1007/s42600-020-00106-3
20. Guerin, P.J., Singh-Phulgenda, S., Strub-Wourgaft, N.: The consequence of COVID-19 on the global supply of medical products: why Indian generics matter for the world? F1000Research **9** (2020)

21. Kalla, A., et al.: The role of blockchain to fight against COVID-19. IEEE Eng. Manag. Rev. **48**(3), 85–96 (2020)
22. Apte, S., Petrovsky, N.: Will blockchain technology revolutionize excipient supply chain management? J. Excip. Food Chem. **7**(3), 910 (2016)
23. Tokkozhina, U., Martins, A.L., Ferreira, J.C.: Uncovering dimensions of the impact of blockchain technology in supply chain management. Oper. Manag. Res. 1–27 (2022)
24. Almutairi, K., et al.: Blockchain technology application challenges in renewable energy supply chain management. Environ. Sci. Pollut. Res. (2022). https://doi.org/10.1007/s11356-021-18311-7
25. Min, H.: Blockchain technology for enhancing supply chain resilience. Bus. Horiz. **62**(1), 35–45 (2019). https://doi.org/10.1016/j.bushor.2018.08.012
26. Kamble, S.S., Gunasekaran, A., Sharma, R.: Modeling the blockchain enabled traceability in agriculture supply chain. Int. J. Inf. Manag. **52**, 101967 (2020). https://doi.org/10.1016/j.ijinfomgt.2019.05.023
27. Zhong, Z., Carr, T.R.: Application of mixed kernels function (MKF) based support vector regression model (SVR) for CO2–reservoir oil minimum miscibility pressure prediction. Fuel **184**, 590–603 (2016)
28. Sternberg, H.S., Hofmann, E., Roeck, D.: The struggle is real: insights from a supply chain blockchain case. J. Bus. Logist. **42**(1), 71–87 (2021)
29. Choi, T.-M.: Supply chain financing using blockchain: impacts on supply chains selling fashionable products. Ann. Oper. Res. 1–23 (2020)
30. Wan, P.K., Huang, L., Holtskog, H.: Blockchain-enabled information sharing within a supply chain: a systematic literature review. IEEE Access **8**, 49645–49656 (2020)
31. Buterin, V.: A next-generation smart contract and decentralized application platform. White Pap. **3**(37), 2-1 (2014)
32. Singh, A., Parizi, R.M., Zhang, Q., Choo, K.-K.R., Dehghantanha, A.: Blockchain smart contracts formalization: approaches and challenges to address vulnerabilities. Comput. Secur. **88**, 101654 (2020). https://doi.org/10.1016/j.cose.2019.101654
33. Hewa, T.M., et al.: Survey on blockchain-based smart contracts: technical aspects and future research. IEEE Access **9**, 87643–87662 (2021)
34. Hölbl, M., Kompara, M., Kamišalić, A., Nemec Zlatolas, L.: A systematic review of the use of blockchain in healthcare. Symmetry **10**(10), 470 (2018)
35. Dimitrov, D.V.: Blockchain applications for healthcare data management. Healthc. Inform. Res. **25**(1), 51–56 (2019)

Chapter 15
A Smart Contract-Based Framework for Value Addition in Retail Market

Subhasish Mohapatra and Roneeta Purkayastha

Abstract Recent advancement of smart contract is considered as one of the greatest innovations of the society. Blockchain is applied to many sectors like banking, healthcare, retail etc. It is providing benefits for the different applications by offering high-end security. In today's world, smart contract plays a very vital role in many commercial sections due to its immutability property. Smart contract is bringing real transformation, so it is expected that in future many industries will leverage the advantages of blockchain to make their technology more vibrant. In this research work, the authors try to introduce smart contract in retail sector. Product management and tracking is one of the major issues in the retail sector, in that case to make the transaction more rapid and reliable, the authors propose a smart contract-based solution in retail management. In this chapter, the authors suggest an Ethereum based retail commodity management and tracking. It identifies user and supplier and track node to node delivery status of product without the help of any third-party service. Moreover, the impact of smart contract in transaction in blockchain will bring revolution in retail sector. So, authors anticipate that smart contract improves the life span of retail industry. Though it enhances time and security management but to establish quality tracking in retail sector, we need a precise model. As we all know that during the pandemic when people relied on web-based transaction at that point of time block chain played a major role in peer-to-peer communication. It not only saves the history of product delivery but also boost consumer confidence.

Keywords Blockchain · Retail sector · Smart contract · Ethereum · Immutability

15.1 Introduction

Retail industries are flourishing now-a-days. However, smart contract proposes a limit less opportunities for safe selling. Digitization is the new norm for many sectors. After the recent pandemic all of the major work sectors are trying to embrace block

S. Mohapatra (✉) · R. Purkayastha
Department of Computer Science and Engineering, Adamas University, Kolkata, India
e-mail: mohapatra.subhasish@gmail.com

S. K. Panda et al. (eds.), *Recent Advances in Blockchain Technology*,
Intelligent Systems Reference Library 237, https://doi.org/10.1007/978-3-031-22835-3_15

chain technology. To mitigate cyber fraud in retail sector and to streamline their business process they are trying to maintain smart contract-based identity management. So, that all the information is immutable. In that case, customer gets a high-end satisfaction. However, retail industry faces some real time challenges like tracking of supply chain, safe selling of product. These issues could be resolved in smart contract. Gradually, retail industry is trying to integrate smart contract with their legacy system. So, the researcher thinks it is the right time for superlative digitization. Blockchain in reality is helping management to resolve many internal issues. Now the question is why we need smart contract in retail industry. To highlight some major research finding of various research work, authors go through many papers and, find one major area i.e., inventory management and identity management yet to be selectively traversed. It further mitigates cyber fraud and attack. Retail industry is adding major revenue to growth of any nation but at the same time researcher try to deal with some internal issues like ineffective internal communication, spurious goods and usage of advance technology like block chain etc. Digital transformation influences potential mapping of any industry [1–4]. This transaction changes the consumer behavior. Indeed, block chain delivers a substantial change for rapid advancement. Smart contract guarantees fraud tracking. As we know that it uses incorruptible digital ledger so it ensures authenticity between buyer and seller. Smart contract prevents fraud in entire supply chain. So, trade conflict of high-end products like diamond, saffron, sandal wood, gold etc. can be counterfeited. All the transaction information could be entered into the ledger for making the supply chain more informative. Researcher tries to put invoices on smart contract so that transaction must have to be verified by all nodes to minimize fraud. It ensures integrity of retail information for quality assessment of product [5–10]. As every retail industry has a sharp attention on effective supply chain management, Smart retail application verifies user identity as well as seller identity. This can construct a vibrant consumer experience for example if any retail industry goes with this mindset, then only it will be able to create better brand loyalty among buyer. To scale up their business there are many retail companies who can use smart contract application. For example, Amazon, Alibaba, Walmart are using smart contract to expand their business process automation. Amazon grab the benefit of block chain to provide high end experience to their user presently all advertising business in Amazon is done by block chain. To trace product in retailing Alibaba, developed a smart contract solution that recorded entire logistic details. Indeed, smart contract is the only panacea for cross border e-commerce transaction. Walmart is the major food chain use smart contract to improve their business. In a true sense block chain provides massive benefits for tracking of ownership and improving robustness in retail industry. In this profound research work authors give evolution of smart contract in Sect. 15.2. Sections 15.3 and 15.4 outline genesis of block chain and Ethereum. Sections 15.4 and 15.5 highlight the concept of block chain in retail sector. Section 15.6 give proposed Blockchain model and Smart Contract for Retail sector.

15.2 Related Study

This section draws a reader attention regarding the basic structure of block chain and smart contract. As we know block chain is a peer-to-peer distributed ledger without any central authority. It can abolish brokerage service and establish trust worthy transaction between peer to peer. In this section authors go through some potential use case of other researcher to find a specific use case in retail domain. Cyber hacking and fraud are proliferating day by day. To lessen these issues authors, suggest an Ethereum based smart contract solution to retail domain. It helps to furnish integrity of transaction. The global transformation of retail industry needs tamper proof transaction that is only possible through smart contract. Retail domain is one of the intense areas where ample of scope is present to inculcate smart contract. Moreover, smart contract solution magnifies plethora of retail applications to lessen issues like brokerage service, goods and service tax reduction etc. Cryptographic hash is integrated with every block in block chain. Nonce is the first block in block chain all previous block stores digital information electronically gradually smart contract is gaining popularity in retail industry. Block chain application in retail industry streamlines administrative work force [11–15]. For example, by the use of smart contract, payment system is becoming hassle free. It can be digitized and tracked in real time tax over head in payment system. So, it's relevance is included in protection of fraudulent activity in payment system. However, it can foster high end customer trust in transaction of digital or physical asset. In the last couple of years retail industry has been aiming to improve customer experience. As side of block chain, it propagates a transitive relationship between customer experience with retail world. Smart contract is bringing innovation constantly. In this case researcher's work explain basic smart contract model in Fig. 15.1. Figure 15.1 explains techno integration of smart contract to retail industry. Usually there are three basic pillars in retail application. The first one is producer whose role is to produce supplies to numerous retailers. The retailer act is to provide end product to consumer. Smart contract manages to restore an address delivery delay. It usually gives opaque information to all consumers [16, 17]. So, drafting of smart contract bridges a security lapse in supply chain. It incorporates smooth collaboration for digital contract management. The actionable insight of smart contract has some built in capabilities that facilitate resource optimization and metamorphose legacy retail contract management system to smart contract assisted compliance management system. The rapid growth of retail industry needs speedy supplier registration. Constantly this is bringing new innovation in retail domain. In that case, a blockchain enthusiast is trying to bring dynamic innovation for order fulfillment and facilitate quicker onboarding of supplier in smart retail management. There is a chance of theft and replacement in paper-based retail contract management system. But smart contract follows stringent cyber security norms as well as for consumer and supplier.

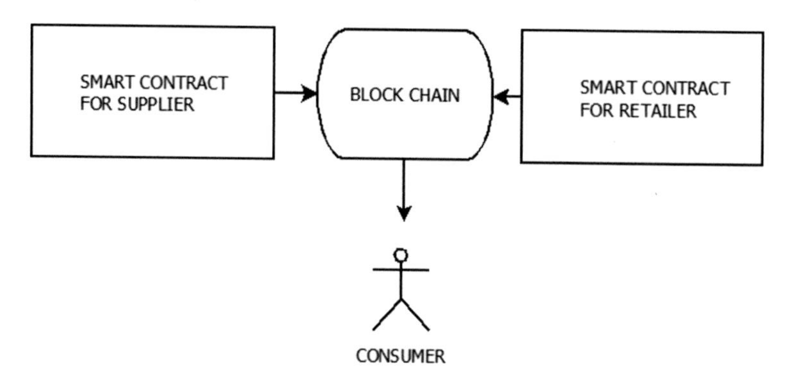

Fig. 15.1 Supplier to retail transaction model

15.3 Background of Blockchain Technology

15.3.1 Genesis of the Blockchain Technology

The first crypto-currency, Bitcoin was published by an individual in 2008 by the pseudonym Satoshi Nakamato [18]. The benefit of Bitcoin is that it does not require a central bank or any other intermediary for transaction. The Bitcoin is based on Blockchain technology, which is a distributed peer-to-peer database consisting of a network of nodes. The Blockchain technology is not only used in the field of crypto-currencies but it is being increasingly used in other areas such as Logistics, Retail Management, Supply Chain Management etc. [19–21].

Although the Bitcoin network shares similarities with conventional payment services like PayPal, it differs from the traditional currencies available in digital form through financial service provider like PayPal. Digital currencies which are dependent on a central node are prone to malicious attacks which can cause serious damage. Bitcoin works on a decentralized structure. If the type of blockchain is public, then anyone is allowed to become a user of the network. As the network is distributed in nature, a network attack is near to impossible [22]. The enabler technology of Bitcoin is the blockchain technology. The following section elaborates on the basic structure of a Blockchain and discusses its important features and properties.

15.3.2 Fundamental Structure of the Blockchain

The Blockchain consists of a chain of blocks, where each block has an identical structure. Each block comprises of a head and a body. Each block is linked to the previous block to form the chain structure. The exception to this rule is the Genesis block, which is the first block of the Blockchain network. A process, called mining, is used to combine all the transactions and store in the next block.

The header of a block contains various types of information such as reference to the previous block, timestamp information and the version of the block. The blocks are assigned a unique name by an algorithm for their unique identity. As the block becomes contained in the chain deeper, a greater number of participants have to be manipulated for any kind of change. This will reduce the chances of manipulation automatically. The most prominent characteristic of Blockchain is that the previous block cannot undergo any change without reconstructing the entire blockchain.

The vital part of each block is the body. Each block has a fixed size and hence the number of stored transaction information in each block is strongly limited. Some additional information is stored in the body of the block as well. If we consider the Bitcoin, this information is about the sender, the recipient and the amount in Bitcoin [23–25].

Each node on the Blockchain network carries a complete copy of the Blockchain. The more the number of nodes in the Blockchain, the greater number of copies of Blockchain data exist and the more secure the network becomes. In order to construct a strong and unique network, there must be an agreement on the Blockchain network. This can be achieved by consensus mechanisms. Consensus mechanism is the process to determine the validity of a Blockchain transaction. This decision is performed by a group of peers or nodes on the Blockchain network. These collections of rules safeguard the network from malicious attacks. There are different types of consensus mechanisms, which depend on the Blockchain and its application [26–30].

15.3.3 Important Features of the Blockchain Technology

Some of the important features of the Blockchain technology are discussed below:

1. **Distributed Ledger**: The distributed ledger can be accessed by all network participants. The transactions can be recorded only once in the shared ledger upon consensus from all network participants. This eliminates the duplication of effort which is typical in a traditional business network [31].
2. **Immutability**: The property of immutability means that a transaction cannot be altered after it has been stored in the shared ledger. If there is an erroneous transaction, a new transaction must be appended for reversing the error and both transactions become apparent thereafter [32].
3. **Smart Contracts**: A smart contract is a set of rules which is executed automatically and is stored on the Blockchain. It defines the conditions applicable for the problem domain and accelerates the transactions [33].
4. **Relative user anonymity**: Here anonymity refers to the user identities being hidden in the blockchain network. Only the digital addresses with their corresponding units are visible on the Blockchain. This is made possible due to the application of public key cryptography which permits the blockchain to be shared

worldwide while maintaining user anonymity. The anonymity is referred to as relative, as in Bitcoin, as the user's transaction history is possible to trace if Bitcoin address is known. Various algorithms have been used to offer different levels of anonymity to users [34, 35].

5. **Security of data**: Bitcoin units are stored in the blockchain itself. These Bitcoins can be accessed using private/public key pairs. The Blockchain data is stored and maintained on multiple devices typically. Even if one or more of the devices face network issues or are compromised, the data is protected [5, 9, 10].

15.4 Ethereum in Blockchain

Ethereum is a blockchain platform, which was introduced in the paper of Vitalik Buterin and handled different limitations of the scripting language of Bitcoin. Ethereum supports all types of computations, including loops which can be referred to as full Turing-completeness. Ethereum includes several other improvements over the Blockchain structure. One of them being supporting the state of the transaction. The platform is based on the decentralized, open-source, public and cryptographic Blockchain technology. Ethereum works on "Smart Contracts", a set of self-executing cryptographic rules that execute only when certain conditions are met. This eliminates the requirement of a third party for code execution on behalf of users, thus creating a decentralized system. The Ethereum state comprises of accounts. Each account consists of a 20-byte address and state transitions. The mapping between address and account state is considered to be a world state [36].

Ethereum leverages two types of accounts: externally owned (controlled by private keys) and contract accounts (controlled by their contract code). An Ethereum account is made of four components: nonce, ether balance, contract code hash and storage root. Nonce is used as a guarantee to indicate that each transaction can be processed only once. It indicates the number of transactions sent from a particular address. Ether balance is the number of Wei owned by the address (Wei indicates the smallest fraction of Ether, one Ether being equal to 10^{18} Wei). The Smart Contract execution and facilitation is performed by paying fees to miners in ETH units. The term Gas indicates the amount of fee charged for a transaction. The Gas is essential in motivating miners to process and verify transactions to gain monetary benefits. Contract code hash denotes the Keccak—256 hash of Ethereum Virtual Machine (EVM) code of the account, which executes upon receiving a message call. Storage root is the 256-bit hash of the root node of a Merkle tree representing the account content [37–39].

Ethereum is grabbing attention due to its decentralized block chain infrastructure. It constructs peer to peer network. It is quite secure with verified application code and that application code is known as smart contract. So, it establishes transaction between participants to initiate transaction between peer to peer. Here all the transaction record are immutable and run without central authority. Absolute ownership is granted to all user i.e., all user information is transparent to both consumer and supplier. In that scenario sender has to sign an agreement in transaction and keep track of ether amount spent in transaction from the account. Ether is a crypto currency which is meant for tracking transaction cost over the network. Ethereum virtual machine (EVM) is a flexible platform so it is the major advantage for researcher. In that case it is quite helpful for them to build their decentralized application in Ethereum. Solidity scripting language is used to build their application in Ethereum platform. In decentralized environment blockchain developer can deploy smart contract. Ethereum environment provide maturity quality experience for developer to deploy smart contract. The future of Ethereum is compatible with non-fungible token (NFT). In NFT ownership record can be transferred. In current scenario it is conceptualized as cryptographic asset with unique identification. Meta data present in NFT strongly differ from each other.

15.4.1 Smart Contract

A Smart Contract is a digital agreement that executes automatically and enables the transacting parties to exchange assets, properties, products or anything of value in a transparent and a hassle-free manner, thus eliminating the need for a third party. In a Blockchain network, a computer program verifies, executes and enforces a smart contract. When both the transacting parties involved in the smart contract agree to the contract terms and conditions, the program executes automatically. This eliminates the involvement of a third party as everything including verification and enforcement is done by the Blockchain network [40].

Nick Szabo had developed the notion of automated contracts already in 1990s. There was no secure platform on which Smart Contract could have executed safely, prior to Blockchain technology. The fundamental Blockchain characteristics discussed above enable Smart Contracts to run safely. Due to the predictive behavior of Smart Contracts, both the transacting parties can trust the Smart Contract which provide better security at lower costs. Hence, Smart Contracts are given much importance in the application of Blockchain technology in areas like Retail sector etc.

Smart Contract is nothing but an application code that stays at a precise location on the blockchain network known as contract address. In the meantime, application would call the smart contract, so the change of state can be triggered to initiate transaction. There are many scripting languages to write smart contract. Mostly, Solidity and Viper have inbuilt compilation process in EVM. It constructs bytecode and prepares the code for execution in blockchain. Signed data message is the transaction output and it will propagate one Ethereum account to next. It carries all vital credential information like sender and recipient information in addition to aggregate amount of Ether transmission. The specific Ether amount is known as validation cost that the network is willing to pay for the generation of byte code in smart contract. This further decides gas price limit. 1 Ether unit equals 10^9 Gwei. Ether is designed to serve two primary purposes. In first case, it avoids congestion in network by abolishing unnecessary transaction and secondly it incentivizes user in transaction mining precisely by verifying all nodes before transaction and listing down verified blocks in ledger. Every transaction is associated with gas limit and specific amount of fee. It is always safe to assign gas limit above threshold so that transaction will run smoothly and fees would not lose during transaction.

The product movement in retail supply chain needs to be monitored. So, to manage retail industry in supply chain authors are trying to develop a close look in this. Retail industry is grabbing this opportunity to be in forefront among all it contemporary. Underlying business process of retail industry is quite complex. Intersection of various factor involve with retail industry blockchain is designed to maximize their business process management. Database of transaction is created in blockchain as it links the block to previous block. The entire EVM ecosystem is decentralized in nature. Time stamp is associated with all transaction blocks. This makes the digital information immutable and irreversible. Impact of smart contract is quite prominent in logistics, tracking, product addition into EVM [41, 42].

15.5 Use Cases of Blockchain Technology in Retail Sector

The financial sector is one of the industry areas which has been leveraging the power of Blockchain technology to a great extent during the recent times. There are a variety of other industry sectors which are utilizing the capability of Blockchain technology, one of them being the Retail sector. Major retailers such as Amazon, Walmart and Alibaba have taken up Blockchain-based projects which facilitate the minimization of retail-based challenges. In this context, retailers can be benefitted from Blockchain technology by knowing how to store supplier data, track the source of origin of products and complete payments and contracts seamlessly. While most of the retail purchases move online, it becomes important for the retailers to upgrade their processes and prepare themselves in the way in which customers explore and purchase goods. In this section, we present some of the most prevalent Blockchain usage in the Retail sector.

i. **Tracking fake products**

Counterfeit products are one of the primary issues for the retailers in the luxury consumer products sector due to flexible policies in product return and high margins. In this context, Blockchain technology provides a technique to allocate a unique scannable code to every product. This code enables every customer to access the detailed history of the product from the source of origin to the final reseller and every movement down the track. During the recent times, IBM has established Trust Chain, a Blockchain-based technique that proves provenance of jewelry items by tracking every step of the supply chain from mine to shop.

ii. **Tracking supply chain**

Tracking of Supply chain is one of the most prevalent applications of block chain technology which enables the monitoring of goods across the supply chain. This efficient step only helps in verifying origin of products but it influences good health as well. Among the different application and scenarios of block chain, Walmart and Starbucks are developing blockchain technology to monitor the flow of products along the supply chain and also fetch real time supply chain details. More transparency is being brought about in supply chain by retailors across all sectors by adopting procurement practices which are more responsible in nature.

iii. **Revolution in payment system**

Although cryptocurrencies have started to being accepted as a method of payment, however it will take considerable time for digital wallets and cryptocurrency payments to gain acceptance in main stream retail space. Most of the retailers tend to offer traditional form of payment, due to the extra ordinary volatile nature of crypto currency prices. Some have found its implementation too difficult. The retailers who are willing to experiment with cryptocurrency payment will be benefitted by an alternative revenue source, thus gaining a wide section of customer segment. Blockchain technology provides retailers more flexibility and power in delivering discounts or coupon and redeeming them from their customers. Master card and American Express both are using Blockchain technology to authenticate discount coupons and renovate the system of loyalty rewards for customers.

iv. **Loyalty Reward System**

Loyalty programs play a vital role in engaging customers and rendering them competitive by giving them opportunity to earn grocery store points to airline miles. Those loyalty points are highly sensitive as they prone to misuse or abuse which can put the retailers in problem if not properly maintained. From the customers' point of view, many of them are dissatisfied due to difficulties in the sign-up process or it is too time consuming to receive the reward points. Blockchain technology addresses many of those issues by providing contemporary methods to protect and gather centralized data from loyalty program. Blockchain can not only allow reward points to be redeemed in a secure way but also simplifies the production and point sharing via retailers and programs.

v. **Data management, security, and sharing of customer Data**

The volume of customer data, which is to be stored and used is huge. Block chain applications streamline processes and provide a promising field in this problem area. Block chain applications help in fulfilling customer needs effectively and provide more control over data to consumers. Recommendation system is an advanced application of artificial intelligence which is capable of recognizing the customer needs and flash targeted deals to specific customers according to their needs.

15.6 Proposed Blockchain Model and Smart Contract for Retail Sector

Solidity programming language works on open source platform like Ethereum that allow users to develop a smart supplier-to-consumer business model. This research finding demonstrates execution of order transaction between consumer and supplier however the Fig. 15.2 highlights the smart contract framework for secure product addition in retail industry. It implements automated product execution in a decentralized fashion. Currently in this section authors try to integrate high-level overview of product mapping in retail chain and benefit of transaction settlement through Smart Contract (SC). There are many securities vulnerability concerned to retail domain So, we have developed a standard retail model for product addition into the logistic supply chain and proper tracking of invoice request into the initial phase of smart retail application. SC frame a trusted execution environment in retail domain. Here the efficiency of SC evaluates the secure hashing outcome to provide high-end performance. The code snippet in SC is self-verifiable, so it can handle transaction vulnerabilities before data synchronization in network. Once the code is written in SC, it is hard and quite challenging for hacker to modify.

In this section, the authors illustrate a use case of Jewellery retail sector using Blockchain technology. The concept of Smart Contract has been applied to build a

```
1   pragma solidity ^0.5.17;
2   contract retailmanagement{
3       address payable consumer;
4       address payable supplier;
5       uint public no_of_products = 0;
6       uint public no_of_invoice = 0;
7       uint public no_of_transactioncost = 0;
8        struct product{
9           uint productid;
10          uint invoiceid;
11          string productname;
12          string productsource;
13          uint transactioncost_per_unit;
14          uint securityDeposit;
15          uint timestamp;
16          bool vacant;
17          address payable supplier;
18          address payable currentconsumer;
19      }
```

Fig. 15.2 Partial solidity code for smart contract

secure and robust supplier-to-consumer model for Jewellery items transaction. The tool used for building the Smart Contract is Remix IDE. The different steps involved in building this system are as follows:

i. The "retailmanagement" Smart Contract is designed in Remix IDE as indicated in Fig. 15.2. The consumer and supplier are considered to be addresses in the Smart Contract. The following state variables are created to store the state of the Smart Contract with public modifier:

    - no_of_products,
    - no_of_invoice and
    - no_of_transactioncost

ii. The struct types created for the Smart Contract are:

    a. **product**: It contains the product item details of the jewellery item i.e. productid, invoiceid, productname etc. among others.
    b. **productAgreement**: It is the agreement details of the product such as productid, invoiceid, securityDeposit etc.
    c. **transactioncost**: It contains the cost details of the transaction which includes productid, productname, transactioncost_per_unit etc. among others.

iii. The relevant modifiers for the Solidity function in the Smart Contract have been declared like onlysupplier(uint _index) etc., which enable a function to execute only if the condition of the modifier is satisfied.

iv. The relevant functions created for the retail management Smart Contract are:

    a. **addproduct()**: This Solidity function is used for adding the product details to the Blockchain network. It increases the product count by one. Only the supplier account has access to this Solidity function.
    b. **signinvoice()**: This Solidity function is used for transferring the total transaction fee to the supplier account. The number of invoice is incremented by one.
    c. **invoiceCompleted() and invoiceTerminated()**: These Solidity functions trace the status of the transaction of Jewellery items.

v. After the Smart Contract has been developed, it is compiled using Solidity compiler within Remix IDE.

vi. The Smart Contract is deployed to the Blockchain network.

vii. As in Fig. 15.3, addproduct() Solidity function is invoked and the following inputs are given as follows:

    a. Product name: Platinum Jewellery
    b. Product address: Kolkata
    c. Transaction cost: 10^{18} Wei
    d. Security Deposit: 10^{18} Wei

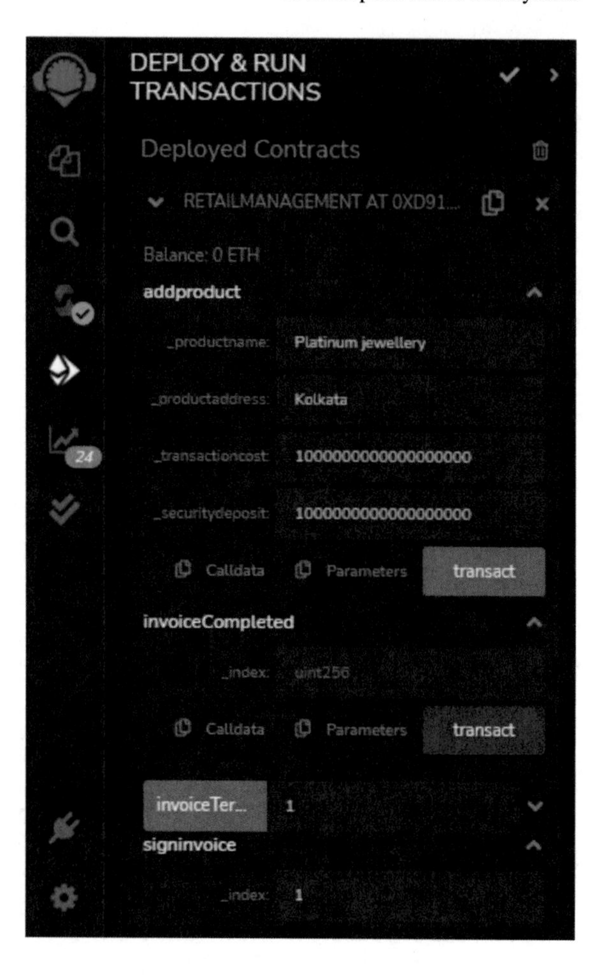

Fig. 15.3 Addproduct() screen in Remix IDE

On hitting the Transact button under Deploy and Run transactions, the success message is shown along with transaction hash generated in the console screen within Remix IDE as shown in Figs. 15.4 and 15.5. It shows the transaction cost and execution cost as well.

viii. Since we have added two Jewellery products, one for Gold jewellery and another for Platinum jewellery, the number of products is 2 as shown in Fig. 15.6.

ix. Since the transaction cost and security deposit together make up 2 ethers, 2 ethers is chosen as the Ether balance and the signinvoice() Solidity function is invoked generating a transaction hash as indicated in Fig. 15.7, the transaction cost and execution cost are generated in the console screen within the Remix IDE.

x. This completes the supplier-to-consumer transaction of Jewellery items successfully using Blockchain platform.

Fig. 15.4 Add product() for gold jewellery

Fig. 15.5 Add product() for platinum jewellery

The proposed model for Jewellery Retail sector shows how Blockchain can be applied in real-time transactions to perform the product mapping in retail chain and it establishes consumer trust in the supplier of products and builds brand value as well.

15.7 Conclusion

As far as smart contract technology is concerned, it is essentially executing all digital events which have been shared among participating agents. Each transaction in blockchain contains verified record. However, the advantage of block chain technology revolutionizes many sectors. So, wide range of institutions successfully regulate their digital asset with utmost privacy. In this research work authors highlight

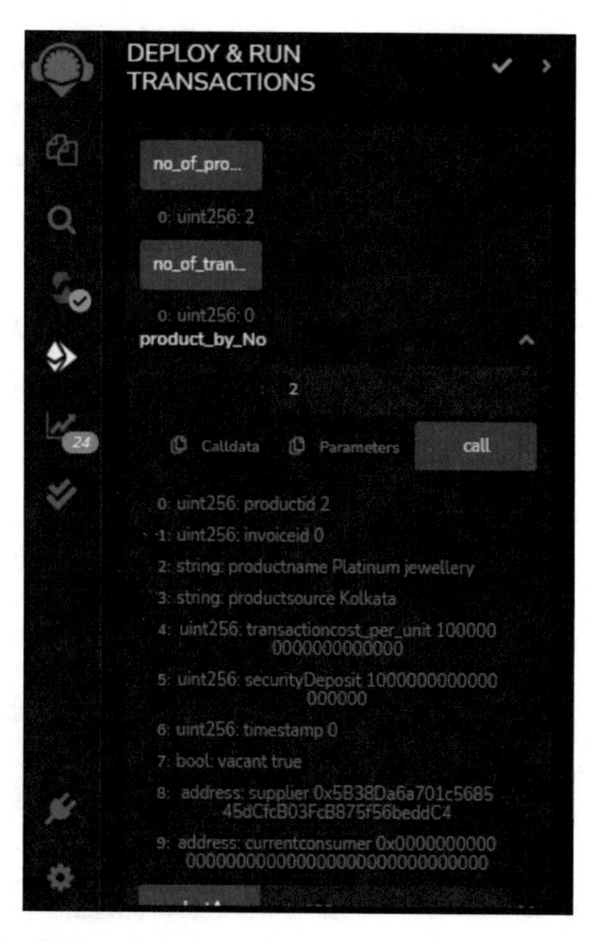

Fig. 15.6 Product details screen in Remix IDE

Fig. 15.7 Sign invoice() screen with transaction hash in Remix IDE

the benefit of smart contract in retail domain. Tracking and monitoring of data plays a vital role in retail shipment but still a lot of scope lies in this domain for further exploration. It upgrades quality tracking in retail shipment order processing. As the information stored in each node is visible to customer so customer can get a genuine product. Retail industry is gradually enhancing customer satisfaction, just in time inventory upgradation and systematic risk mitigation in supply chain. Globally smart contract is exploring a new dimension intrinsically. There are still some regulatory issues which need to be verified before it's adoption to any industry.

References

1. Aich, S., et al.: A review on benefits of IoT integrated blockchain based supply chain management implementations across different sectors with case study. In: 2019 21st International Conference on Advanced Communication Technology (ICACT). IEEE (2019)
2. Chen, J., et al.: A blockchain-driven supply chain finance application for auto retail industry. Entropy **22**(1), 95 (2020)
3. Xu, X., et al.: Coordination of a supply chain with an online platform considering green technology in the blockchain era. Int. J. Prod. Res. 1–18 (2021)
4. van Hoek, R.: Developing a framework for considering blockchain pilots in the supply chain–lessons from early industry adopters. Supply Chain Manag. Int. J. (2019)
5. Zhang, P., White, J., Schmidt, D.C., Lenz, G., Rosenbloom, S.T.: FHIRchain: applying blockchain to securely and scalably share clinical data. Comput. Struct. Biotechnol. J. **16**, 267–278 (2018) [CrossRef] [PubMed]
6. Hader, M., Elmhamedi, A., Abouabdellah, A.: Blockchain technology in supply chain management and loyalty programs: toward blockchain implementation in retail market. In: 2020 IEEE 13th International Colloquium of Logistics and Supply Chain Management (LOGISTIQUA). IEEE (2020)
7. Subramanian, N., Chaudhuri, A., Kayıkcı, Y.: Blockchain and Supply Chain Logistics: Evolutionary Case Studies. Springer Nature (2020)
8. Azaria, A., Ekblaw, A., Vieira, T., Lippman, A.: MedRec: using blockchain for medical data access and permission management. In: Proceedings of the 2nd International Conference on Open and Big Data (OBD 16), Vienna, Austria, 22–24 August 2016, pp. 25–30
9. Raja Santhi, A., Muthuswamy, P.: Influence of blockchain technology in manufacturing supply chain and logistics. Logistics **6**(1), 15 (2022)
10. Kuo, T.-T., Kim, H., Ohno-Machado, L.: Blockchain distributed ledger technologies for biomedical and health care applications. J. Am. Med. Inform. Assoc. **24**, 1211–1220 (2017). [CrossRef] [PubMed]
11. Angraal, S., Krumholz, H.M., Schulz, W.L.: Blockchain technology: applications in health care. Circ. Cardiovasc. Qual. Outcomes **10**, e003800 (2017). [CrossRef] [PubMed]
12. Yue, X., Wang, H., Jin, D., Li, M., Jiang, W.: Healthcare data gateways: found healthcare intelligence on blockchain with novel privacy risk control. J. Med. Syst. **40**, 218 (2016). [CrossRef] [PubMed]
13. Wu, X.-Y., Fan, Z.-P., Cao, B.-B.: An analysis of strategies for adopting blockchain technology in the fresh product supply chain. Int. J. Prod. Res. 1–18 (2021)
14. Ivan, D.: Moving toward a blockchain-based method for the secure storage of patient records. In: ONC/NIST Use of Blockchain for Healthcare and Research Workshop. ONC/NIST, Gaithersburg, MD, USA (2016)
15. Mohapatra, S., Parija, S.: A brief overview of blockchain algorithm and its impact upon cloud-connected environment. Bitcoin Blockchain 99–113 (2020)

16. Mohapatra, S., Roy, A.: A study on mediclaim processing in connected healthcare system. Trends Wirel. Commun. Inf. Secur. 363–374 (2021)
17. Das, T., Mohapatra, S., Roy, A.: Insurance policy claim verification model of unnatural death cases–an artificial intelligence based approach. In: International Conference on Innovations in Bio-Inspired Computing and Applications. Springer, Cham (2020)
18. Nakamoto, S.: Bitcoin: a peer-to-peer electronic cash system bitcoin: a peer-to-peer electronic cash system. Bitcoin.org. Disponible en (2009). https://bitcoin.org/en/bitcoin-paper
19. Cyran, M.A.: Blockchain as a foundation for sharing healthcare data. Blockchain Healthc. Today (2018). [CrossRef]
20. Shubbar, S.: Ultrasound medical imaging systems using telemedicine and blockchain for remote monitoring of responses to neoadjuvant chemotherapy in women's breast cancer: concept and implementation. Master's Thesis, Kent State University, Kent, OH, USA (2017)
21. Ianculescu, M., Stanciu, A., Bica, O., Neagu, G.: Innovative, adapted online services that can support the active, healthy and independent living of ageing people. A case study. Int. J. Econ. Manag. Syst. **2**, 321–329 (2017)
22. Kim, J.S., Shin, N.: The impact of blockchain technology application on supply chain partnership and performance. Sustainability **11**(21), 6181 (2019)
23. Zheng, Z., Xie, S., Dai, H., Chen, X., Wang, H.: An overview of blockchain technology: architecture, consensus, and future trends. In: Proceedings of the 2017 IEEE International Congress on Big Data (BigData Congress), Honolulu, HI, USA, 25–30 June 2017, pp. 557–564
24. Panda, S.K., Mohammad, G.B., Nandan Mohanty, S., Sahoo, S.: Smart contract-based land registry system to reduce frauds and time delay. Secur. Privacy e172 (2021). https://doi.org/10.1002/spy2.172
25. Panda, S.K., Satapathy, S.C.: Drug traceability and transparency in medical supply chain using blockchain for easing the process and creating trust between stakeholders and consumers. Pers. Ubiquit. Comput. (2021). https://doi.org/10.1007/s00779-021-01588-3
26. Mohapatra, S., Parija, S., Roy, A.: UML conceptual analysis of smart contract for health claim processing. Blockchain Technology: Applications and Challenges, pp. 151–168. Springer, Cham (2021)
27. Mohapatra, S., Parija, S.: A brief understanding of blockchain-based healthcare service model over a remotely cloud-connected environment. Evolutionary Computing and Mobile Sustainable Networks, pp. 949–955. Springer, Singapore (2021)
28. Mohapatra, S., Anand, K.: A brief model overview of personalized recommendation to citizens in the health-care industry. Recommender System with Machine Learning and Artificial Intelligence: Practical Tools and Applications in Medical, Agricultural and Other Industries, pp. 27–44 (2020)
29. Mahapatra, S., Sinha, D., Das, A.K.: A secure health management framework with anti-fraud healthcare insurance using blockchain. Pattern Recognition and Data Analysis with Applications. Springer, Singapore, pp. 491–505 (2022)
30. Dujak, D., Sajter, D.: Blockchain applications in supply chain. SMART Supply Network, pp. 21–46. Springer, Cham (2019)
31. Niveditha, V.R., Sekaran, K., Amandeep Singh, K., Panda, S.K.: Effective prediction of bitcoin price using wolf search algorithm and bidirectional LSTM on internet of things data. Int. J. Syst. Syst. Eng. **11**(3–4), 224–236 (2021)
32. Sathya, A.R., Panda, S.K., Hanumanthakari, S.: Enabling smart education system using blockchain technology. In: Panda, S.K., Jena, A.K., Swain, S.K., Satapathy, S.C. (eds.) Blockchain Technology: Applications and Challenges. Intelligent Systems Reference Library, vol. 203. Springer, Cham (2021). https://doi.org/10.1007/978-3-030-69395-4_10
33. Lokre, S.S., Naman, V., Priya, S., Panda, S.K.: Gun tracking system using blockchain technology. In: Panda, S.K., Jena, A.K., Swain, S.K., Satapathy, S.C. (eds.) Blockchain Technology: Applications and Challenges. Intelligent Systems Reference Library, vol. 203. Springer, Cham (2021). https://doi.org/10.1007/978-3-030-69395-4_16
34. Panda, S.K., Daliyet, S.P., Lokre, S.S., Naman, V.: Distributed ledger technology in the construction industry using corda. The New Advanced Society: Artificial Intelligence and Industrial Internet of Things Paradigm (2022). https://doi.org/10.1002/9781119884392.ch2

35. Panda, S.K., Satapathy, S.C.: An investigation into smart contract deployment on ethereum platform using Web3.js and solidity using blockchain. In: Bhateja, V., Satapathy, S.C., Travieso-González, C.M., Aradhya, V.N.M. (eds.) Data Engineering and Intelligent Computing. Advances in Intelligent Systems and Computing, vol. 1. Springer, Singapore (2021). https://doi.org/10.1007/978-981-16-0171-2_52
36. Panda, S.K., Rao, D.C., Satapathy, S.C.: An investigation into the usability of blockchain technology in internet of things. In: Bhateja, V., Satapathy, S.C., Travieso-González, C.M., Aradhya, V.N.M. (eds.) Data Engineering and Intelligent Computing. Advances in Intelligent Systems and Computing, vol. 1. Springer, Singapore (2021). https://doi.org/10.1007/978-981-16-0171-2_53
37. Panda, S.K., Dash, S.P., Jena, A.K.: Optimization of block query response using evolutionary algorithm. In: Bhateja, V., Satapathy, S.C., Travieso-González, C.M., Aradhya, V.N.M. (eds.) Data Engineering and Intelligent Computing. Advances in Intelligent Systems and Computing, vol. 1. Springer, Singapore (2021). https://doi.org/10.1007/978-981-16-0171-2_54
38. Varaprasada Rao, K., Panda, S.K.: A design model of copyright protection system based on distributed ledger technology. In: Satapathy, S.C., Lin, J.CW., Wee, L.K., Bhateja, V., Rajesh, T.M. (eds.) Computer Communication, Networking and IoT. Lecture Notes in Networks and Systems, vol. 459. Springer, Singapore (2023). https://doi.org/10.1007/978-981-19-1976-3_17
39. Varaprasada Rao, K., Panda, S.K.: Secure electronic voting (E-voting) system based on blockchain on various platforms. In: Satapathy, S.C., Lin, J.CW., Wee, L.K., Bhateja, V., Rajesh, T.M. (eds.) Computer Communication, Networking and IoT. Lecture Notes in Networks and Systems, vol. 459. Springer, Singapore (2023). https://doi.org/10.1007/978-981-19-1976-3_18
40. Nanda, S.K., Panda, S.K., Das, M., Satapathy, S.C.: Automating vehicle insurance process using smart contract and ethereum. In: Chakravarthy, V.V.S.S.S., Flores-Fuentes, W., Bhateja, V., Biswal, B. (eds.) Advances in Micro-Electronics, Embedded Systems and IoT. Lecture Notes in Electrical Engineering, vol. 838. Springer, Singapore (2022). https://doi.org/10.1007/978-981-16-8550-7_23
41. Panda, S.K., Elngar, A.A., Balas, V.E., Kayed, M. (eds.): Bitcoin and Blockchain: History and Current Applications, 1st edn. CRC Press (2020). https://doi.org/10.1201/9781003032588
42. In: Panda, S.K., Jena, A.K., Swain, S.K., Satapathy, S.C. (eds.) Blockchain Technology: Applications and Challenges. Intelligent Systems Reference Library. Springer. https://doi.org/10.1007/978-3-030-69395-4.